Regional Innovation Index of China: 2017

Xielin Liu • Taishan Gao • Xi Wang

Regional Innovation Index of China: 2017

How Frontier Regions Innovate

Springer

Xielin Liu
School of Economics and Management
University of Chinese Academy
of Sciences
Beijing, China

Taishan Gao
Information and Industrial Development
Department
State Information Center
Beijing, China

Xi Wang
Sino-Danish College
University of Chinese Academy of Science
Beijing, China

ISBN 978-981-13-1204-5 ISBN 978-981-13-1205-2 (eBook)
https://doi.org/10.1007/978-981-13-1205-2

Library of Congress Control Number: 2018948010

Printed on acid-free paper

This Springer imprint is published by the registered company Springer Nature Singapore Pte Ltd.
The registered company address is: 152 Beach Road, #21-01/04 Gateway East, Singapore 189721, Singapore

Preface

China is a big country with fast dynasty of innovation model and rich diversity of regional innovation. The regional competition and innovation help China become more and more important S&T and innovation power in the world. Without Huawei, Tencent, and ZTE in Shenzhen, the regional IT giants, Chinese IT industry will not have its astonishing achievement.

The book is intending to analyze the sources of regional innovation that are not much explained by existing literature, books, or papers.

Based on our long time regional innovation index since 2000 in Chinese, this is our attempt to publish our report in English.

Following our long tradition, our book on Chinese regional innovation has three parts: part one is the innovation index of China. We do it based on the regional innovation system framework with rich statistical data in provincial level. The second part is our main thematical report. Each year there would be a main topic: this year, it is analysis of innovation and entrepreneurship model in five leading regions. The last part is the analysis of key innovation for each region.

Though China is mainly a top down country, it is so large that it can make the regional innovation model very diverse. Based on our analysis of innovation in leading regions, we have the following findings:

Finding one: the old rich industrial resource in some regions has played as barrier for their later stage innovation; less resource regions create more market-based innovation. It seems that there is a resource curse against innovation. Less resource stimulates the motivation of special entrepreneurship climate and hunger for innovation. Guangdong is the case, and it has more entrepreneurial spirit than the regions that used to have rich government investment. In there, the leading innovative enterprise such as Huawei, Tencent, and Media are good at market oriented innovation. Zhejiang is similar with less SOE and government investment but has gone through from a rural economy to industrial economy by innovation in industrial cluster and e-business innovation. Vice versa, the regions with lots of government support and investment such as Northeast region, even Tianjin, had taken the opportunity to become industrialized faster, but their innovation still lags behind.

Finding 2: the central government strategy plays an overarching role in leveraging the S&T and innovation. We can see that in last 20 years; the government emphasized the role of S&T and innovation for economic growth through policy, planning, and programming; it creates the milieu for regional government to priority S&T and innovation. The new trend is more and more R&D inputs from regional government, the role of high-tech zone become more and more important in China. Now, talented people are given special role in new industry creation.

Finding 3: in last 30 years, entrepreneurship-based innovation, not the science and technology per se, play a more important role in China regional innovation. Guangdong, Jiansu, and Zhejing are the typical case. This is because the transition process creates a vast space for entrepreneurship in terms of technology, organization, and institution. Jiangsu creates the TVE institutional innovation and Zhejiang creates the cluster-based innovation.

Finding 4: there is a phenomena of development rigidity: the old industrial region, from northeast, to Tianjin, even Shanghai, have shown the rigidity of old hard and soft assets. Those assets are very important in the process of early and middle stage of industrialization. From 1990s to 2000s, those regions carry on the key industrial product, from automobile (First Auto in Changchun) to steel making (in Liaoning and Hebei), chemical engineering (Liaoning and Tianjin), energy supply (Shanxi), and so on. But entering the stage of post industrialization, they become source of the production oversupply. With heavy investment in hardware and software, it is hard for them to transform to a new wave of development based on industrial 4.0 or smart manufacturing.

Finding 5: open innovation and FDI plays the key role in China's regional development and innovation. In Guangdong, Jiangsu, and Beijing, FDI has transferred lots of technology and management expertise to the relevant regions of China. Without FDI, Beijing and Guangdong, Shanghai would not be a good place for IT-based innovation. FDI was especially important in the earlier stages of high tech zone in China. They used to be the key players in high tech zones.

Finding 6: the associate model of government regulation is more better for entrepreneurship and innovation in China. Though China is basically a development state model in which the governments can use plan or policy to support industry in their region, China is so big that we can find various models to co-exist there: Zhejiang and Guangdong are more associate model, while Shanghai and Beijing are more state development model, but even so, those regions are also very open for knowledge and development.

For each year, the report writting is a knowledge draining as well as labor intensive work. We would like to express our thanks to many friends, colleagues, and students who contributed the report in many ways; they are: Yuchen Gao, Xuechen Ding, Caiting Dong, Langmei Zhu, Xiaodan Xu, Tong Wang, Xiaoxi Chang, Cong Zhou, Lingfei Tian, Die Wang, and Yutian Liu.

I am especially indebted to Professor Thomas Schergnell of Austria Institute of Technology for his valued friendship and assistance. His corrections and improvement upgrade our book to a new level.

We have to thank the long-time support of Ministry of Science and Technology, though our research is done in an independent way.

The book is originally a collective work initiated by Chinese S&T Development Strategic Team in 2000s. They are listed as consulting experts as follows.

Beijing, China
April 30, 2018

Xielin Liu

Consulting Experts

Xin Fang, Dean, School of Public and Management, University of Chinese Academy of Sciences

Chunfa Wang, Secretary, China Association for Science and Technology

Zhijian Hu, President, Chinese Academy of Science and Technology for Development

Guangrong You, Director, Research Fellow, Beijing System Engineering Institute

Lan Xue, Dean, Professor, School of Public Policy and Management, Tsinghua University

Rongping Mu, Director, Research Fellow, Institutes of Science and Development, Chinese Academy of Sciences

Changlin Wang, Assistant President, Research Fellow, Academy of Macroeconomic Research, State Development and Reform Commission

Shiji Gao, Director, Research Fellow, Institute of Resource and Environmental Policy, Development Research Center of the State Council

Jianguo Qi, Research Fellow, Institute of Quantitative & Technical Economics, Chinese Academy of Social Sciences

Jun Su, Professor, School of Public Policy and Management, Tsinghua University

Guangling Xiao, Professor, Institute of Science, Technology and Society, Tsinghua University

Taishan Gao, Doctor, Information and Industrial Development De State Information Center, The National Information Center

Contents

About the Authors

Xielin Liu, born in 1957, is a Professor in School of Economics and Management, Director of Research Center for Innovation and Entrepreneurship, University of Chinese Academy of Sciences. He was a research fellow in Chinese Academy of Science and Technology for Development, Ministry of Science and Technology (1996–2006), and Professor in Tsinghua University from 1985–2006.

Born in September 25, 1957; B.Sc., Peking University, 1982; M.Sc, Chinese Academy of Science; Ph.D., 1994, Tsinghua University.

His research areas mainly cover innovation management and policy, globalization of technology, and innovation. He has published a lot of papers in *Research Policy*, T*echnovation*, *Journal of Management Studies,* and *International Journal of Technology Management*. He served as Editor for *Research Policy* and other journals. He coauthored two books in English, *Environmental Innovation in China* (WIT, 2012) and *Innovation, Technology Policy and Regional Development* with Tim Turpin (Edward Elgar, 2003).

Taishan Gao, born in 1987, is an Associate Professor at State Information Center, China. He got his Ph.D. from University of Chinese Academy of Sciences in 2014. His research mainly focuses on regional innovation, sharing economy, and information society. He is the deputy Secretary-General of Centre for Sharing Economy Research. In recent years, he has published several papers in journals like *Research Policy* and *Technological Forecasting and Social Change*.

Xi Wang, born in 1990, is a Ph.D. student in Sino-Danish College, University of Chinese Academy of Sciences. Her main research interests are innovation ecosystem, regional innovation, industrial upgrading, etc. In recent years, she has coauthored several books on enterprises' breakthrough innovation and regional innovation-driven development models of China. Email address: wangxi0916@sina.com

Part I
Analysis of China's Regional Innovation Index in 2017

Chapter 1
The Measure of Regional Innovation Capability

1.1 Introduction

Since the 1990s, scholars have paid more and more attention to the regional innovation system in different countries (Cooke et al. 1997; Metcalfe 1995; Miguélez and Moreno 2015). Theoretically, while enriching the theoretical basis of the innovation system conception, the approach of regional innovation system has its own significance. It provides new dimensions of innovation that national innovation system does not have.

Firstly, the key factors of innovation system have been extended to include the dimension of geography in the system (Furman et al. 2002). In most of the time, following this kind of logic, policy makers take regional polarization and equilibrium of innovative resource allocation as an important research proposition and one of the important policy options (Courvisanos 2009; Dumont 2017). This is of great importance for cases in China.

Secondly, different regions usually have great diversity of climate and geographical properties that matter for knowledge creation and knowledge flow among the regions. China is divided into four major economic regions, including the northern region, the southern region, the northwest region, and the Qinghai-Tibet region, among which the eastern part is most developed. The northern region of China is considered to be the north part alongside the Qinling Mountain-Huaihe River Line, and so does the southern area which is obviously considered to the southern part alongside the Qinling-Huaihe boundary. Different regions share many differences in history, culture, geography, and even climate. For instance, there is a great difference in temperature between north and south in winter. In summer, the temperature is generally high in the country, with more rainfall in the south than that of in the north. There are many plains in the east, with humid climates and convenient transportation. Comparatively speaking, there are plenty of mountainous areas in the west with arid climate and inconvenient transportation. However, the mineral resources are abundant in the west.

X. Liu et al., *Regional Innovation Index of China: 2017*,
https://doi.org/10.1007/978-981-13-1205-2_1

Since the opening door policy released in 1978, the east coastal regions have enjoyed the advantage of shipping transportation for industrial goods. The advantage is especially obvious for manufacturing industry which has made FDI booming in that area. Western regions are comparatively lagging behind the eastern region. At present, the most developed regions are those located in the east and south of China. Therefore, classifications of east, middle, and west regions are often considered to check the different models of development in China. Therefore, in order to complement each other's advantages, China launched the South-to-North Water Diversion Project, the West-to-East Natural Gas Transmission Project, the West-to-East Power Transmission Project, and so on to balance the regional development. Figure 1.1 shows the division of China.

Note: By administration, China is usually considered to be composed of the east area (Beijing, Tianjin, Hebei, Shanghai, Jiangsu, Zhejiang, Fujian, Shandong, Guangdong, and Hainan), central area (Shanxi, Anhui, Jiangxi, Henan, Hubei, Hunan), west area (Inner Mongolia, Guangxi, Chongqing, Sichuan, Guizhou, Yunnan, Xizang,

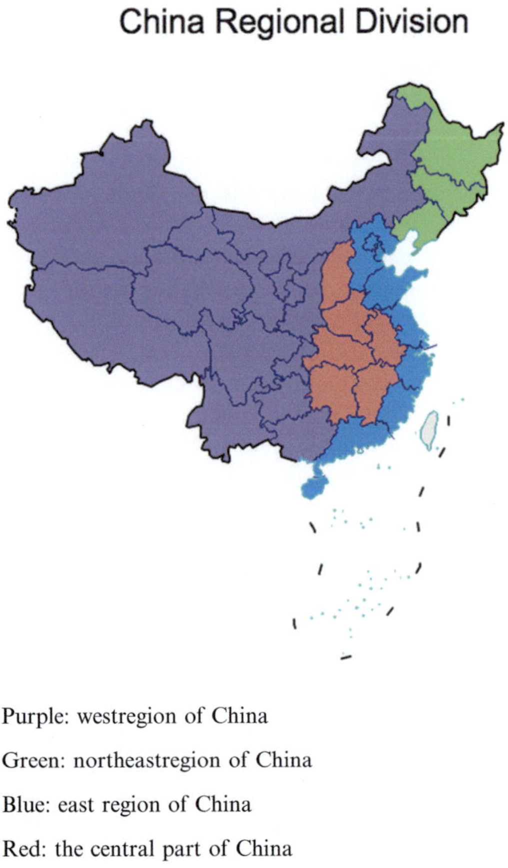

Fig. 1.1 The China regional division
Purple: west region of China
Green: northeast region of China
Blue: east region of China
Red: the central part of China

Shaanxi, Gansu, Qinghai, Ningxia, and Xinjiang), and northeast area (Jilin, Liaoning, Heilongjiang) (National Bureau of Statistics of China(NBS 2011).

Thirdly, regions show significant differences in terms of their institutional development and governmental orientation. China has intensified the transition from a planning economy to a market-oriented economy since the 1980s. The economy of some regions is still dominated or led by state-owned enterprises (SOEs), while some are not. In the period of planning economy, the central government inserted heavy investment in the northeast region, reflected by a number of megaprojects, especially in heavy industries, such as automobile industry, machine tool industry, and shipbuilding industry. Because of these projects, the living standard in the northeast area was higher than the national level in the era of planning economy. Large SOEs used to be the main players there. Most public research institutions and university clusters also locate there. A contrasting region is Zhejiang province; it is known as "the edge of the planning economy," according to "Statistical Yearbook of the Chinese Investment in Fixed Assets (1950–1985)," in 1953–1975 (the "First Five-Year Plan" to the "Fifth Five-Year Plan"); the capital construction investment in there was very low; and the investment per capita ranked the last one in all regions. Figure 1.2 shows the distribution of SOEs in different regions against its innovation capabilities in 2015.

Fourthly, differences of regional innovation and entrepreneurship also root in different cultural characteristics and historical development, which are related to entrepreneurial spirit. This is not sufficiently addressed by mainstream research, such as why some regions obtain more innovative spirits than others though they have similar innovation endorsements from economic theory (Li 2009; Saxenian 1996; Bruton et al. 2017).

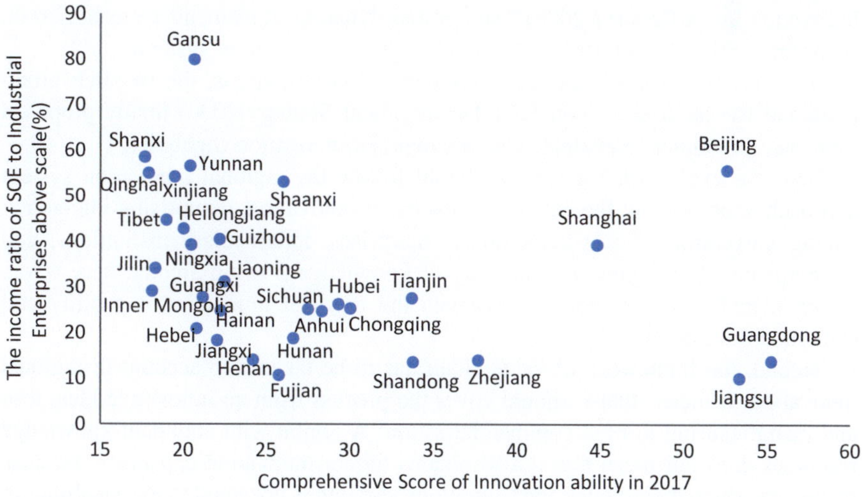

Fig. 1.2 The relation of regional innovation capabilities and revenues from SOEs at the level of provinces in 2017

Due to the vast difference of climate, culture, and demography, southerners may have more entrepreneurial sense. There's little impact of the success or failure of entrepreneurial projects that people are involved on their own lives; on the contrary, to the northerners, entrepreneurship is of less importance, because their lifestyles are different. The north region is mostly plains, and people can be self-sufficient by farming, while the south is mostly mountainous and hilly area, and people need to start businesses to survive.

Lastly, but not the least, there is a competition among areas for catching up the leading regions in China (Jin et al. 2005). The competition is generally considered as conducive for economic growth in China. This also proposes some very interesting research questions: why some regions with less science and technology may have a faster growth rate than the regions with good S&T resources such as Hubei province. Why the market economy is more developed in south part of China than north part of China?

1.2 Evaluation Principles

Evaluating the innovation capability of a region requires a proper set of indicators. The indicator selection, the number of indicator, the selection of weights, and the ratio of subjective and objective indicators all these can affect the final evaluation of innovation capability. Therefore, we were very cautious in the procedures of indicator selection, evaluation methods, and many other aspects. In addition, various references are used in our study, including some well-known domestic and foreign reports such as "The World Competitiveness Yearbook (1996)," "Global Competitiveness Report (2004)," "Global Innovation Index (2011)," and "(China) National Innovation Index Report (2006)," and proper dynamic adjustments were carried on according to the characteristics of China's regional innovation system.

After nearly ten academic meetings with different experts, the research group (National Research Group on S&T Development Strategy 2000) finally proposed four main principles to evaluate China's regional innovation capability:

First, the evaluation framework should follow the regional innovation system approach, emphasizing the networked nature of innovation, comprising interaction among universities, R&D institutions, enterprises, intermediary institutions, and governments. The degree and intensity of knowledge flowing among these actors is considered as a key point to evaluate the regional innovation capability and economic development.

Second, the framework of value chain has to be taken into account in regional innovation. It means that it should cover the process from an innovative idea, test, and manufacturing to final commercialization. A region with abundant knowledge resources does not mean that it also obtains higher innovation capability, because many facts show that science and technology strength is not equal to the capability of technological innovation. Many regions do not have strong technological foundation but still show a great capability of technology innovation. The capability of technological innovation in a region depends on whether the enterprises have enough innovation impetus and innovation capability. Therefore, among the five subsystems of knowledge creation, knowledge transfer, business innovation, innovation environment, and innovation performance, the report gives the enterprise innovation more weight than others such as the knowledge creation and transfer. Third, the framework emphasizes the importance of innovative environmental construction. In the market economy system, the important factor of evaluating the local government's work is not the traditional plan and intervention degree but how to create an environment or ecosystem which is conducive for enterprise innovation. Since the government is far from the market, it cannot directly guide the technological innovation flow of the enterprises. The key of its functional adjustment to promote the enterprise's technological innovation is to achieve the transition from relying on plans to the creation of innovation environment. Fourth, the framework must balance the stock, relative level, and growth rate of a region's development. In "IMD World Competitiveness Yearbook," there is an emphasis on stock and relative levels, but not the growth rate. A feature of our report is the emphasis on growth rates, because we believe that growth rates can be able to reflect the economic potential of a catching-up region. Therefore, since 2007, we have divided the comprehensive value index into strength index, efficiency index, and potential index, which continues until present days. The strength means the real capabilities a region has based on its assets, such as R&D inputs and outputs in absolute size. This will give the large regions some advantage, but it catches what a power a region it is. The efficiency index is the level of strength index in terms of their inputs, such as total R&D inputs/capita, or invention patent/million people. The potential index is the growth rate of strength index, such as the growth rate of total R&D in one region and the growth of granted patent in one region.

Figure 1.3 shows the evaluation framework of China's regional innovation capability.

1.3 Index System

According to the 4 abovementioned principles, a regional innovation capability index system has been proposed (see Table 4.1), including 5 first-level indexes, comprised of 20 second-level indexes, 40 third-level indexes, and 137 fourth-level indicators. The first-level indicators include knowledge creation, knowledge acquisition, enterprise innovation, innovation environment, and innovation performance (Table 1.1). In the system, the knowledge creation measures the capability of regions to constantly create new knowledge; the knowledge acquisition measures the capability of regions to take advantage of all available knowledge in the world; the enterprise innovation measures the regional enterprises' capability of applying new knowledge and launching new products or processes; the innovation environment measures the capability of the region to provide the corresponding environment for

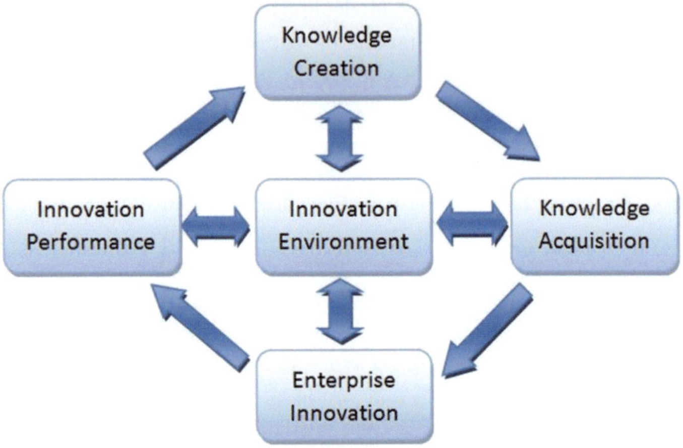

Fig. 1.3 The evaluation framework of China's regional innovation capability

Table 1.1 China's regional innovation capability indicator system

First-level index	Second-level index
1.Knowledge creation	1.1 Research and development input synthetic index
	1.2 Patent synthetic index
	1.3 Research paper synthetic index
2.Knowledge acquisition	2.1 Scientific and technological cooperation synthetic index
	2.2 Technology transfer synthetic index
	2.3 Foreign capital enterprise investment synthetic index
3.Enterprise innovation	3.1 Enterprise research and development investment synthetic index
	3.2 Design capability synthetic index
	3.3 Technical promotion capability synthetic index
	3.4 New product sales revenue synthetic index
4.Innovation environment	4.1 Innovation infrastructure synthetic index
	4.2 Marketing environment synthetic index
	4.3 Quality labor force synthetic index
	4.4 Financial environment synthetic index
	4.5 Entrepreneurship level synthetic index
5.Innovation performance	5.1 Macroeconomy synthetic index
	5.2 Industrial structure synthetic index
	5.3 Industrial international competitiveness synthetic index
	5.4 Employment synthetic index
	5.5 Sustainable development and environmental protection synthetic index

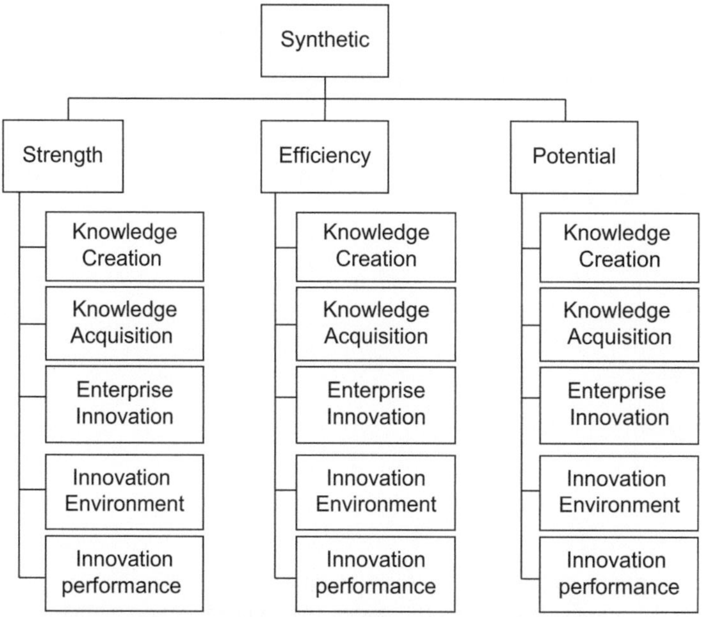

Fig. 1.4 The analytical framework of China regional innovation capability

the generation, flowing, and application of knowledge; and the innovation performance measures the output capacity of regional innovation (Fig. 1.4).

Compared with other index systems, our index system is relatively comprehensive which covers most of the innovation basic indexes (Table 1.2). The most important is that the index system is divided into three parts, strength, efficiency, and potential.

In our index system, not only the change of the total amount can be observed but also the speed and magnitude of changes in each region. Unfortunately, due to the difficulty of the related data acquisition and the lack of direct measures of the system, the dimension of policy and the government efficiency are not evaluated here, although they are also very important for the evaluation of innovation capability. Other reports like GII, IMD, and World Economic Forum use questionnaire system to benchmark the differences in different countries. Also business model innovation is hard to measure. In China, for the leading companies such as Tencent, Alibaba, and Baidu, their core competence falls in the business model innovation (Ghezzi et al. 2013). Since those kinds of innovation cannot be measured by existing indicators, like patent or others, we did not include them either.

In this report, the overall innovation capability of a region is shown by the relative ranking compared to other regions; in this report, we have covered all 31 China provinces, direct-controlled municipality, and autonomous regions. Based on our framework of indicator system, in fact, the result of score is not a direct measure of the innovation capability in the region; it is more of comparative measuring. In

Table 1.2 Innovation capability evaluation index international and domestic well-known reports

Publication	Index levels	Dimensions	First level	Second level	Third level	Fourth level	Analytical approach
China's Regional Innovation Index	4	5	5	20	40	137	Quantitative
China National Innovation Index	3	8	2	8	39	–	Quantitative
China's Urban Innovation Report	2	3	3	21	–	–	Quantitative
World Competitiveness Yearbook	3	4	4	20	327	–	Quantitative + qualitative
Global Competitiveness Report	2	12	12	113	–	–	Quantitative + qualitative
European Innovation Scoreboard	3	3	3	7	24	–	Quantitative + qualitative
Global Innovation Index	4	7	2	7	20	82	Quantitative + qualitative

general, comparing to last year, innovation capability of the provinces (autonomous regions and municipalities directly under the central government) all has been improved, but in terms of comparative measuring, some can go up and some go down.

1.4 Methods for Evaluation

The single indicator is represented by the direct-obtained regional data. The utility value method is taken in the nondimensionalization. The value range is [0,100], which refers to the optimal utility value 100 and the worse 0. The calculation is shown below:

(1) Positive utility indicator

If set i is the indicator i and j is the region j:

x_{ij} represents obtained value of indicator *i* in region *j*.
y_{ij} represents the utility value of indicator *i* in region *j*.
$x_{i\max}$ is the maximum value of this indicator.
$x_{i\min}$ is the minimum value of this indicator.

$$y_{ij} = \frac{x_{ij} - x_{i\min}}{x_{i\max} - x_{i\min}} \times 100$$

The positive utility indicator here means the greater the indicator value, the higher the utility value, such as labor productivity, per capita GDP, number of invention patents, etc.

(2) Negative utility indicator

Negative utility indicator means that the greater the indicator value, the lower the utility value, such as unemployment rate [(unemployed people + laid-off workers)/ local employed people]. Calculate as below:

$$y_{ij} = \frac{x_{i\max} - x_{ij}}{x_{i\max} - x_{i\min}} \times 100$$

(3) Composite indicator

Composite indicator is calculated by two or more single indicators, such as growth rate, average number, etc. The utility value of composite indicator uses the same method as single indicator.

(4) Weight selection

We choose the expert scoring to determine the weight. This method is relatively subjective but is generally used internationally. The experts we choose have all deep attainments in researches of domestic technology policy management, as well as foreign relevant reports.

(5) Weighted comprehensive

The weighted calculation is carried out class by class, illustrated in Fig. 1.5:

a, *b*, *c*, and *d* represent different class.
f(*a*), *f*(*b*)... represent their weights.
x(*a*,*i*) and *x*(*b*,*i*) represent the utility value of indicator in different classes and regions. Calculate from right to left.

To calculate the indicator value of ci (weighted utility value), set *x*(*ci*,*i*) as the comprehensive utility value of region *i* under indicator *ci* and *x*(*di*,*i*) as the utility value of region *i* under indicator *di*. Then,

$$x(c1, i) = x(d1, i)f(d1) + x(d2, i)\,f(d2) + x(d3, i)\,f(d3) + \cdots\cdots$$

Correspondingly, calculate *x*(*c*2,*i*), *x*(*c*3,*i*)...
Then, *x*(*bi*,*i*):

$$x(b1, i) = x(c1, i)f(c1) + x(c2, i)\,f(c2) + x(c3, i)\,f(c3) + \cdots\cdots$$

Then, *x*(*b*2,*i*), *x*(*b*3,*i*)...
Then, *x*(*a*,*i*):

$$x(a, i) = x(b1, i)f(b1) + x(b2, i)\,f(b2) + x(b3, i)\,f(b3) + \cdots\cdots$$

When $i = 1,2,3,\ldots\ldots,31$, calculate the utility values of 31 provinces at class.

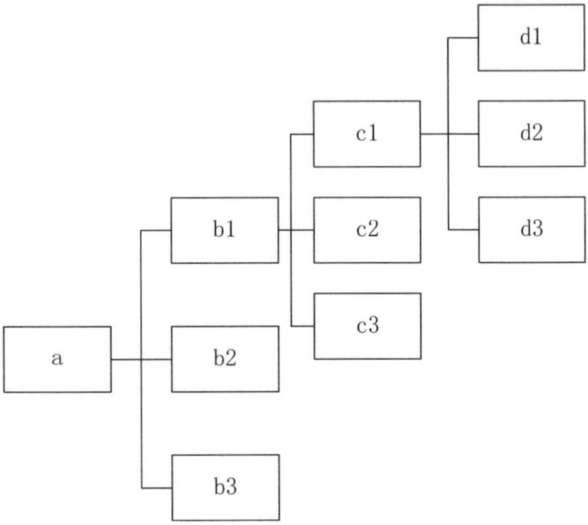

Fig. 1.5 Diagrammatic sketch of index system

1.5 Data Sources

To ensure the research verifiability, all data in this report comes from published statistical yearbooks and government reports, mainly including China Statistical Yearbook, China Statistical Yearbook of Science and Technology, China Statistical Yearbook of High-tech Industry, China Statistical Yearbook of Industrial Economy, statistics and analysis report of Chinese S&T papers, Ministry of Science and Technology, State Intellectual Property Office, State Administration of Industry and Commerce, technology innovation funds of technology-based SMEs, and so on. The data used for the index of 2017 regional innovation is from 2015, so did the index of previous year. Many international reports follow the same principle to select the year of the data. In addition, due to limited information, there is no data from Hong Kong, Macao, and Taiwan.

We have done research on China regional innovation capability for about 17 years since 2000 (National Research Group on S&T Development Strategy 2000). To make the evaluation results truly reflect the economic structural adjustments and the model of economic development, we would make certain changes and adjustments of our index according to the data as well.

References

Bruton, G., Zahra, S.,& Li, C. (2017). Examining entrepreneurship through indigenous lenses. *Entrepreneurship Theory and Practice*, 1–11(editorial).

Chinese Academy of Science and Technology for Development. (2006). *National innovation index report*. Beijing: Scientific and Technical Documentation Press.

Cooke, P., Uranga, M. G., & Etxebarria, G. (1997). Regional innovation systems: Institutional and organizational dimensions. *Research Policy, 26*(4–5), 475–491.

Courvisanos, J. (2009). Political aspects of innovation. *Research Policy, 38*(7), 1117–1124.

Dumont, M. (2017). Assessing the policy mix of public support to business R&D. *Research Policy, 46*(10), 1851–1862.

Furman, J. L., Porter, M. E., & Stern, S. (2002). The determinants of national innovative capacity. *Research Policy, 31*(6), 899–933.

Ghezzi, A., Georgiades, M., Reichl, P., & Mangiaracina, R. (2013). Generating innovative interconnection business models for the future internet. *Journal of Policy, Regulation and Strategy for Telecommunications, Information and Media, 15*(4), 43–68.

IMD World Competitiveness Center. (1996). The IMD world competitiveness yearbook. https://worldcompetitiveness.imd.org

Jin, H., Qian, Y., & Weingast, B. R. (2005). Regional decentralization and fiscal incentives: Federalism, Chinese style. *Journal of Public Economics, 89*(9–10), 1719–1742.

Li, X. (2009). China's regional innovation capacity in transition: An empirical approach. *Research Policy, 38*(2), 338–357.

Metcalfe, J. S. (1995). Technology systems and technology policy in an evolutionary framework. *Cambridge Journal of Economics, 19*(1), 25–46.

Miguélez, E., & Moreno, R. (2015). Knowledge flows and the absorptive capacity of regions. *Research Policy, 44*(4), 833–848.

National Research Group on S&T Development Strategy. (2000). *China regional innovation capability report*. Beijing: Science and Technical Documentation Press.

NBS (2011) Division method of East, Mid, West and Northeast area. China Statistics Press, Beijing.

Saxenian, A. (1996). *Regional advantage: culture and competition in Silicon Valley and route 128*. Cambridge: Harvard University Press.

The Global Innovation Index 2011: Accelerating Growth and Development. Cornell University, INSEAD, and WIPO. (2011). https://www.globalinnovationindex.org/userfiles/file/GII-2011_Report.pdf

World Economic Forum. (2004). *Global competitiveness report*. https://www.weforum.org

Chapter 2
The Overview of Chinese Innovation

2.1 The Basic Profile of Innovation in China

Nowadays, innovation has become key for the improvement of the comprehensive national competitiveness as well as a strong driving force for changes and progress in how the society evolves and how people live. In the latest 19th report of communist party, innovation is regarded as the major force for economic development (Xi, 19th report, 2017).

China is the second largest economy in the world in terms of GDP, only second to the USA (The World Bank 2017). In terms of innovation capability, China is rising rapidly, from No. 29 in 2007 to No. 22 by Global Innovation Index (Global Innovation Index 2017).

The rapid rising of Chinese economy is not occasional. For four decades, from Deng Xiaoping to current Chairman Xi Jingping, China focused on S&T and innovation as the major force for development. The new strategy for the new era of China is innovation-driven development. We can see that China has taken some great steps toward the building of a knowledge-based economy fostering innovation (Scherngell et al. 2014).

The government has set a very clear target for innovation. In 2006, the national medium- and long-term S&T Development Program for 2006–2020 has made it clear that China should invest in R&D to catch up with developed countries, with the goal to raise the R&D expenditures as share of GDP to a 2.5%; the contribution of scientific and technological progress (total factor productivity) to economic growth should improve by more than 60%; foreign technology dependence (technology import/R&D) should be consolidated within 30%. The annual number of invention patents authorized and the number of scientific impact (international scientific paper citations) should be in top 5 around the world (MOST 2006).

From Fig. 2.1, we can see that the R&D quota (R&D expenditures as share of GDP) has been continuously growing even – in contrast to most of other countries – in the period of the world's financial crisis since 2008.

X. Liu et al., *Regional Innovation Index of China: 2017*,
https://doi.org/10.1007/978-981-13-1205-2_2

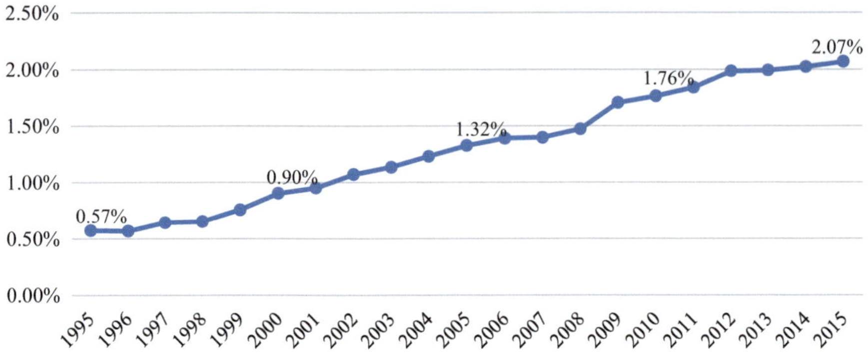

Fig. 2.1 The percentage of R&D in GDP 1995–2015 (%). (*Source*: China Statistical Yearbook 1999–2016. Beijing: China Statistics Press)

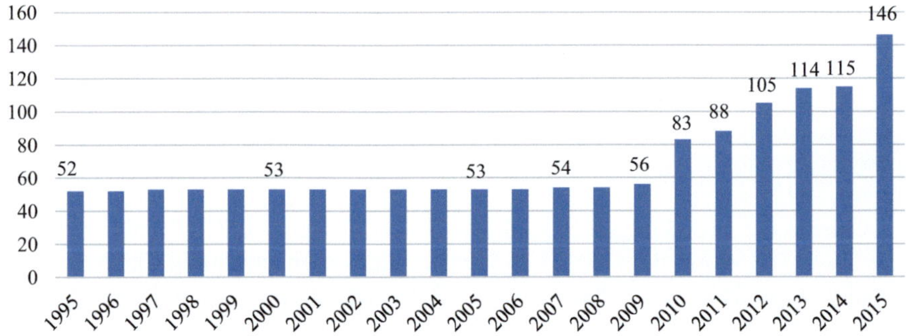

Fig. 2.2 The numbers of high-tech zones 1995–2015. (*Source*: China Statistical Yearbook 1999–2016. Beijing: China Statistics Press)

As an example of this development, in Zhongguancun, a part of Beijing, a high-tech science park has been established since the 1980s, based on the study of the Silicon Valley model. Since then, China has set up a lot of high-tech zones or science parks nationwide. The number and output value of high-tech zones have grown rapidly. In 2013, the percentage of high-tech zones' output value in GDP reached more than 25% (Fig. 2.2, 2.3 and 2.4).

Meanwhile, the strategy for opening up to the outside world in innovation has been intensified in China. Compared with other catching-up countries like Japan and South Korea in the 1970s and 1980s, China is much more open to FDI. The openness benefits China in two important ways. First, it brought China the capital needed in the 1990s (Cheung and Lin 2004). Second, it has promoted the technology transfer from developed countries to China, with direct transfer and spillover effects (Hu and Jefferson 2002; Zhao 2013; Liang 2017). Figures 2.5 and 2.6 shows that China has been continuously attracting FDI.

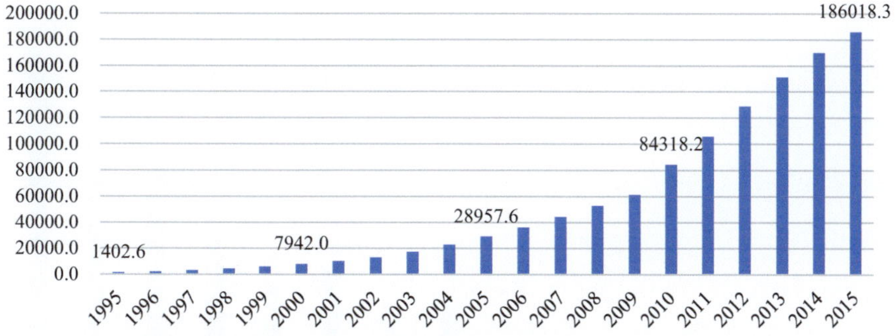

Fig. 2.3 The high-tech zones' output value 1995–2015 (100 million yuan). (*Source*: China Statistical Yearbook 1999–2016. Beijing: China Statistics Press)

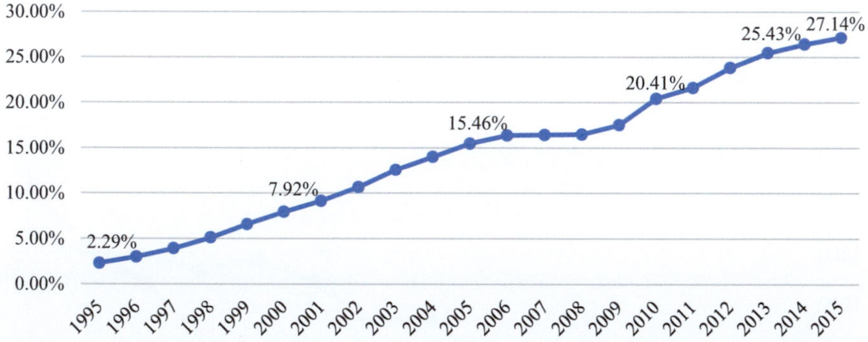

Fig. 2.4 The percentage of high-tech zones' output value in GDP 1995–2015 (%). (*Source*: NBS 1999–2016)

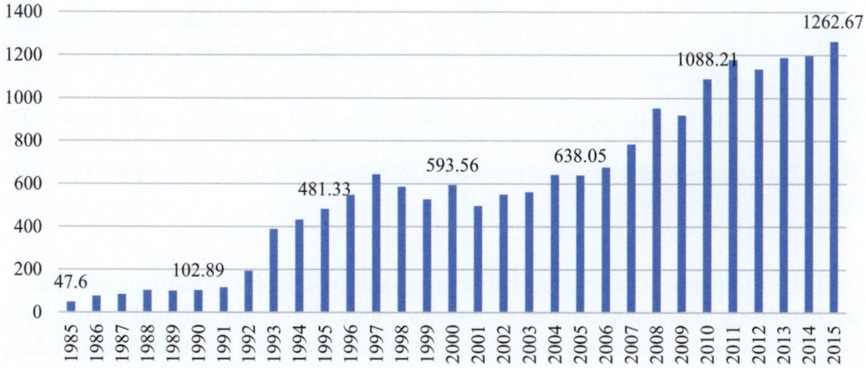

Fig. 2.5 Foreign direct investment actually utilized in China 1985–2015(100 million USD). (*Source*: NBS 1999–2016)

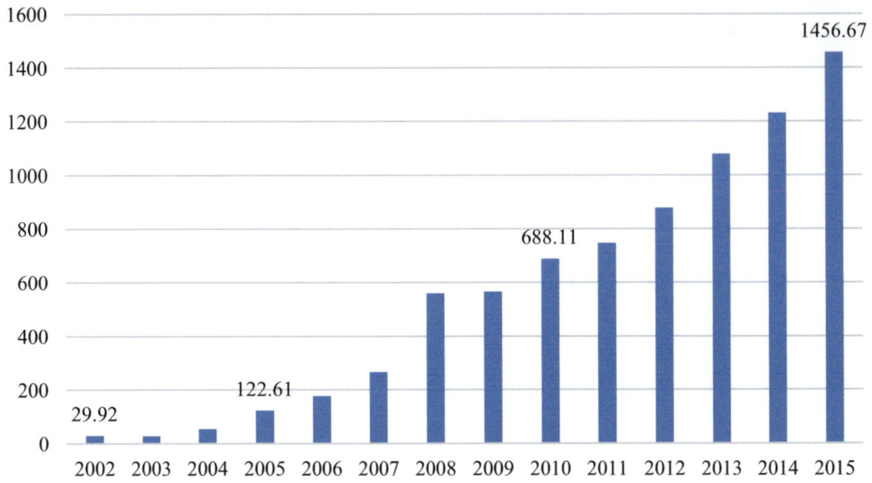

Fig. 2.6 Net overseas direct investment of China 2002–2015(100 million USD). (*Source*: NBS 2003–2016)

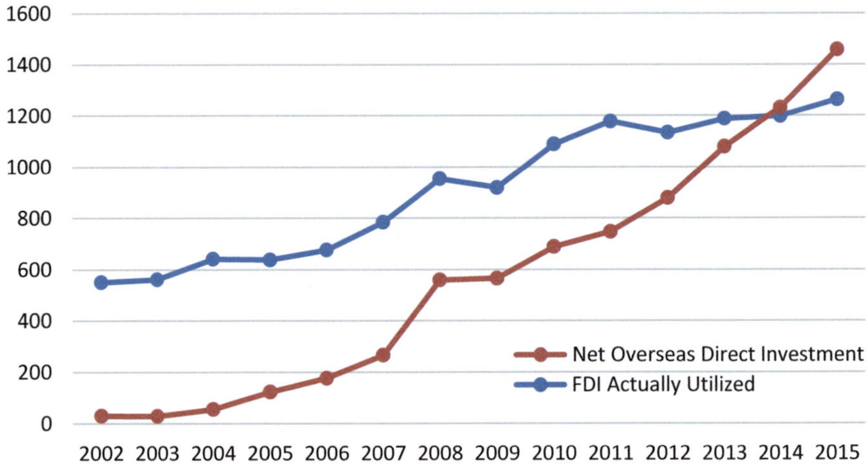

Fig. 2.7 Comparison of FDI actually utilized and net overseas direct investment 2002–2015 (100 million USD). (*Source*: NBS 2003–2016.)

In the same time, Chinese enterprises have improved their technological capability, beginning to invest in overseas countries. Figure 2.7 shows FDI and net overseas direct investment. The competitiveness of Chinese firms seems to be improved by adapting themselves to a more open strategy.

Entering the 2000s, as FDI was so important in China, some industries even were monopolized by multinationals, such as the auto industry, so, around 2003 or so, the government and many researchers came to realize that China has given FDI too

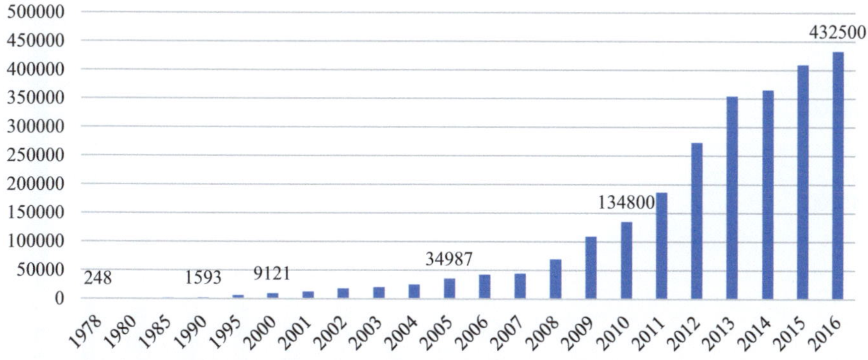

Fig. 2.8 China returnees between 1978–2016 (Nr). (*Source*: NBS 2016; Statistics of Ministry of Education of the People's Republic of China 2016)

much space for industrial development with probably too little spillover effects (Lu 2006); domestic R&D capability and talented people are very important and necessary for the high-tech industry. This was shown in the medium- and long-term plan for S&T for development (MLP) 2006–2020. Since then, China started to strengthen indigenous innovation, by increasing R&D investment and attracting talents from worldwide, in particular calling up Chinese returners to come back and work in home countries. They have achieved higher education in overseas countries, and some of them obtain tacit knowledge in high-tech industry which is exactly the urgent necessity of domestic industry. The returners do not only play a key role in ICT industry, they also bring back innovative business models that are good for Chinese ICT industry. Nowadays, the number of talented people has increased rapidly in China; see Fig. 2.8.

As result of those strategies and policies, we can see that both the S&T output and high-tech industry achieved astonishing accomplishments. There's a sharp increase in the SCI paper publications, PCT patent applications, and high-tech revenues. Figure 2.9 shows the SCI publications from 2000 to 2014. Figure 2.10 shows the PCT patent applications from 2002 to 2016. Figure 2.11 shows the high-tech industry revenue.

Regarding the rise of China's innovation capability, the regions have made great contributions. Note in this context that the economic size of some of the provinces is as much as for some independent European countries. In China, regional competition is highlighted in the competition among local governments, which has stimulated the development of regional economy (Jin et al. 2005). In a knowledge-based economy, regional competition is not only related to the formation of local innovation clusters but also related to the ability of innovation clusters to participate extensively in various related networks (Li 2003, 2004).

Some other experts have pointed out that if the growth of China's economy in the past 10 or 20 years has been driven by the dynamics of regional competition, with the Yangtze River Economic Belt and other new national strategies, competition will

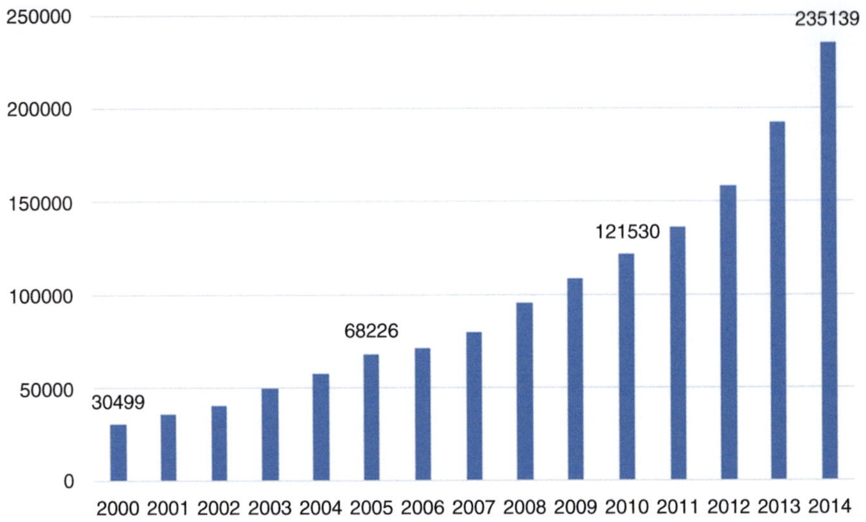

Fig. 2.9 The SCI publications 2000–2014 (piece). (*Source*: NBS 2007–2016. MOST, China S&T Statistics Data Book 2001–2006)

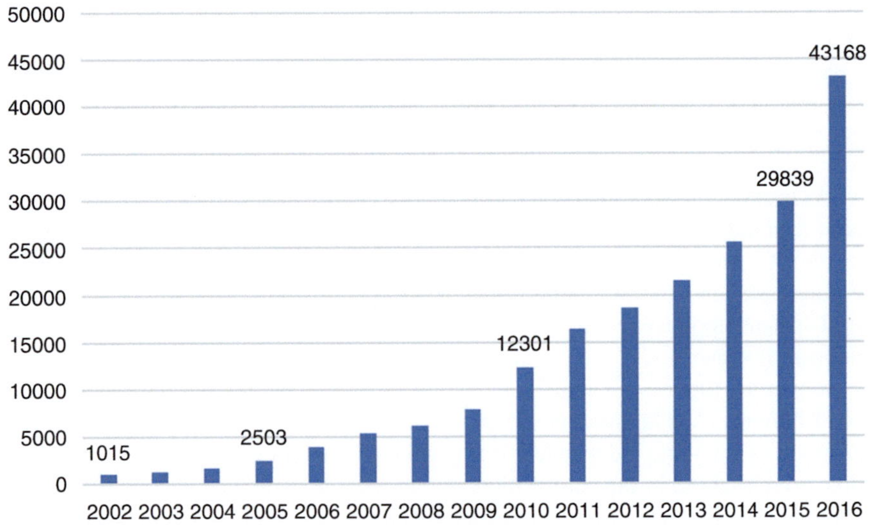

Fig. 2.10 The PCT patent applications 2002–2016 (piece). (*Source*: NBS 2003–2016. WIPO Statistics 2016)

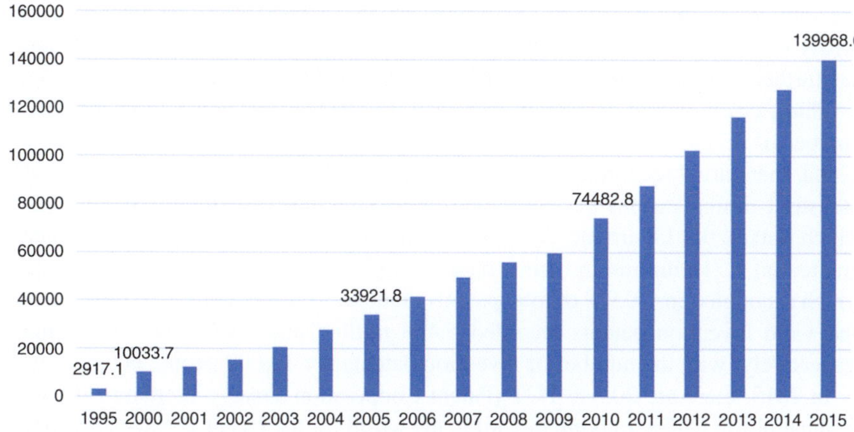

Fig. 2.11 The high-tech industry revenue from principal business (100 million yuan). (*Source*: NBS 2001–2016. MOST, China S&T Statistics Data Book 2001–2016)

become the main theme of regional economic development in the future and also will be the new driving force for China's economic transformation and the upgrade of industry (Tang and Shu 2014; Pi and Wang 2014).

2.2 The Trends of Regional Innovation in China

Based on our analysis in this report and observation of national and regional innovation of China, we can summarize some trends of regional innovation in China in recent years as below.

2.2.1 Regional Innovation Competence Across China Has Been Further Improved

First, the intensity of R&D input continues to be enhanced. In 2015, the R&D quota in China recorded 2.10% (China Statistical Yearbook 2016), an increase of 0.05% over the previous year, and reached the level of a developed country and ranks top among developing countries. This has created favorable conditions for technological innovation. Some regional governments made the level of R&D input as the indicator of local government's performance. The provincial and municipal levels of R&D input have continuously increased. Some regions and local governments even increased 20% on R&D input.

Second, R&D talent pool is continuously expanded. In 2015, nationwide R&D personnel FTE recorded 3,758,800 person-years, with an increase of 1.3% over the previous year, with a growth exceeding 5% in 13 provinces and municipalities across China. The increase in the percentage of highly educated R&D personnel has further facilitated the improvement of overall R&D capacity.

Third, patent output is steadily increased. In 2015, patent applications reached a number of 2.799 million in China. In particular, invention patent applications (in China, there are three kinds of patent; there are invention patent, utility patent, and industrial design patent) exceeded one million for the first time (1.102 million), which ranked the 1st around the world for five consecutive years. In 2015, patents granted hit 1.718 million in China, including 359,000 of invention patents granted, which accounted for 20.9% of total patents granted. At the end of 2015, valid patents and valid invention patents recorded 5.478 million and 1.472 million in China, respectively, with the number of invention patents per 10,000 people is 6.3.

In 2015, there are 30,548 PCT (Patent Cooperation Treaty) international patent applications in China, ranking the 3rd in the world for the recent 2 years since 2013. If talking about regions, it shows that there are 256,400 invention patents granted in 31 provinces and municipalities across China in 2015, with a growth rate of 40% in 28 regions. The rapid growth, structural optimization, and the international ranking of patents show that China's technological capacity has been greatly improved (NBS 2016).

2.2.2 Coordinated Development to Address Regional Imbalance in Innovation Capacity

To have a more balanced and mutual benefit development within China, regional economic coordination has become a strategic option for the government, especially in the east area, which has to enhance the overall competitiveness of the region (Xiao et al. 2015; Scherngell and Hu 2011). Among these coordinated regions, Jingjinji region, Yangtze River Delta, and Guangdong, Hong Kong, and Macao Bay Area are typical and fast-developed ones.

(1) Integration of Beijing-Tianjin-Hebei Regions

In 2014, President Xi Jinping proposed the coordinated development of Beijing-Tianjin-Hebei regions, which is considered as one of China's three major national strategies, aiming to establish a world-class city cluster with Beijing as the center. On March 23, 2015, the central government published the Outline of Collaborative Development of Beijing, Tianjin, and Hebei Province (Ji 2016). The background of integration is that the imbalance in the area of Beijing, Tianjin, and Hebei Province is quite obvious. Beijing is the highest in terms of development and innovation, but Beijing did not produce the spillover effects for the surrounding regions; instead it played a role of syphon age: attract and bring talented people and resources to Beijing (Sun and Li 2016; Mao 2017). There is also a big gap between Hebei to Beijing and Tianjin in terms of public service (Ji 2016).

So, it is of vital importance for the regional integration of Beijing-Tianjin-Hebei to achieve in-depth regional cooperation, free mobility, innovation resource sharing, and collaborative regional innovation system environment. Coordinated development of Beijing-Tianjin-Hebei region has become an important national strategy. There are two important developments following the integration: the first one is that the Zhongguancun science park will open sub-parks in Hubei Province and a new area called Xiongan New District will be built as a place to hold many high-tech firms and research institutes moving from Beijing so as to reduce the pressure of Beijing as a large metropolis (Ji 2016).

(2) Yangtze River Delta

In 2016, the State Council published the development plan of the urban agglomeration in the Yangtze River Delta, aiming to promote the coordinated development and innovation of the urban agglomeration in this area. By 2030, a world-class city cluster with global influence will be fully completed. Shanghai will play the center of a "leading" economic belt around the Yangtze River and its entire basin in a coordinated way (Zhang 2016; Li and Wu 2017). The polarization effect of the core city of Yangtze River Delta urban system has decreased obviously due to this reason; it produced the effect of innovation spillover (Gao, et al. 2016).

(3) Guangdong-Hong Kong-Macao Greater Bay Area

In 2017, Premier Li Keqiang said in the government work report that it is necessary to deepen the Mainland's cooperation with Hong Kong and Macao, to formulate a development plan for the urban agglomerations of Guangdong, Hong Kong, and Macao areas so as to form the unique bay advantages. Actually, the urban agglomeration in the Guangdong-Hong Kong-Macao Greater Bay Area has been an important part of the national development strategy. Its strategic orientation is to be the predominant bay area in world economy. Meanwhile, it is going to become the hub of "The Belt and Road" Initiative.

It will make the best use of the advantage of "One Country, Two Systems" to fortify the position as an international competition power, in order to improve the internationalization level to enhance the developing dynamics and finally to construct the free trade zones along the Pearl River Estuary so as to create a new economy engine for the urban agglomeration in Greater Bay Area (Qin et al. 2017).

2.2.3 Approaching for Global Science Center

Until June 2017, Shanghai Zhangjiang Comprehensive National Science Center, Hefei Comprehensive National Science Center, and Beijing Huairou Comprehensive National Science Center have got the approval. Shanghai Zhangjiang Comprehensive National Science Center will focus on four aspects: (1) building a world-class scientific and technological infrastructure cluster, (2) promoting the deep integration of facilities construction and cross-cutting edge research, (3) building collaborative interdisciplinary innovation network, and (4) exploring and implementing a new management system for major scientific and technological facilities. For now, Shanghai Synchrotron Radiation Facility (SSRF), National Protein Science Research (Shanghai) Facilities, and Shanghai Supercomputer Center have been built completely and are making achievements, such as free electron laser device, super-short laser device, water window free-electron laser platform, SXFEL user station, etc. In the future, a series of large scientific devices will also be built in Shanghai (State Council 2016).

Hefei Science Center will focus on six areas: energy, information, materials, life, environment, and advanced manufacturing, which try to solve major scientific problems, improving original innovation capability and promoting transformational technology. At present, the world's largest wind tunnel laboratory and the world's largest high-speed train model test platform have been completed, which play important roles in the present scientific research. The world's most "bright" high-energy synchrotron radiation source, extreme condition experiment device, earth system numerical simulation device, etc. in a few years, China will build up six to eight major scientific and technological infrastructures.

The development of Huairou Science Center in Beijing is intended (1) to strengthen the top-level design, formulate the development strategy, and plan the development blueprint, (2) to further improve the projects that have been already started and promptly demonstrate a number of new major scientific devices and cross-research platforms, (3) to strengthen reform and innovation in institutional mechanisms and actively seek to build national laboratories, (4) to attract international top talents to build a high-quality scientific research team, and (5) to speed up the construction of supporting facilities and build Huairou Science Center into an innovative center and a livable center (State Council 2016).

2.2.4 *Innovation as an Important Strategy for Corporate Development*

Usually Chinese enterprises spend less money on research and more money on development or reverse engineering. Now, more and more Chinese enterprises understand that innovation is a key factor for enterprises to survive in the fierce market competition and obtain long-term leading advantages. The success or failure of innovation-driven strategy, to a large extent, depends on whether the companies would like to invest in R&D to build up their core competence.

First, corporate R&D expenditure is continuously increased. In 2015, corporate R&D expenditure of China exceeded RMB 1.1 trillion, with an increase of 40.3% over 2012 and an average annual increase of 11.9%, which accounts for 77.4% of

R&D expenditure in the whole society. Particularly, R&D expenditure of large enterprises exceeded RMB 1 trillion for the first time at RMB 1,015.09 billion. The number of corporate R&D personnel reached 4.25 million (NBS 2016).

Second, the construction of innovation platforms at firm level proceeds smoothly. At the end of 2015, accumulatively 1187 national-level corporate (group) technology centers were certified in China, with an increase of 300 over 2012. At the end of 2015, the number of corporate-based national key laboratories was 177, accounting for 36.8% of national key laboratories, and the number of corporate-based national engineering (technology) research centers was 144, occupying 41.6% of national engineering (technology) research centers (NBS 2016).

Besides, in 2015, accumulative capital of VC for emerging industry projects was RMB 55.68 billion in total nationwide. The number of funds for start-ups cumulatively increased to 1233, with a growth of 995 than 2012 (NBS 2016).

Meanwhile, a friendly environment for SME is emerging following mass innovation and entrepreneurship movement encouraged by the central government.

First, the system for business process is greatly simplified. Driven by "internet and governmental services," the government has launched convenience-for-people measures such as "one-counter handling, one-stop services" to cancel or simplify a large batch of unnecessary certifications or red tapes, thus largely reducing errands and troubles for SMEs. Industrial and commercial registration switches from "certification before license" to "license before certification," precondition approval items are reduced by over 87%, "subscription system" is adopted for registered capital instead of "paid-in system," and reform of "multiple certificates in one and one code for one certificate" is deepened. These measures have greatly improved the efficiency of business registration and bring entrepreneurs more convenience.

Second, corporate tax rate has been significantly reduced. By comprehensively implementing VAT in lieu of business tax and the rolling out preferential tax policies for medium-, small-, and micro-sized enterprises, central and provincial governments have canceled, suspended, and exempted over 1100 charge items. Moreover, the corporate-related administrative business charge items established by central government are reduced by 69% and the government-managed funds reduced by 30%. Corporate burden is reduced by over RMB 2 trillion from 2013 to 2016 accumulatively (Ministry of Finance 2017).

Third, a favorable innovation and start-up atmosphere has been established. According to statistics, 12,000 new enterprises were set up on average on a daily basis over the previous 3 years, and more than 15,000 enterprises were registered on average on a daily basis from January to April 2017, with corporate activity maintained around 70%. There were more than 4000 co-working spaces, with over 3000 technology enterprise incubators and more than 400 accelerators, constituting a chain of start-up incubation services (Keqiang Li 2017).

2.2.5 *Innovation Performance Improves Continuously*

First, new product development has achieved new breakthroughs. In 2015, large enterprises' achieved sales revenue for new products is RMB15.1 trillion, 5.6% more than the previous year. In particular, 21 regions met an increase in sales revenue from new products. High-tech manufacturing industry with value added got an increase of 10.2% over the previous year, which is 4.1% higher than the value-added growth of large enterprises in the same period, accounting for 11.8% of the total value added of industrial enterprises above designated size. Revenue from main businesses has achieved RMB 13.7 trillion throughout the year. In 2012–2015, high-tech manufacturing industry has made a contribution of 19.9% to the growth of main business revenue of industrial enterprises. The steady development of high-tech industry laid a solid foundation for a structural optimization of the industry.

Second, export of high-tech products has increased. In 2015, exports of China's high-tech products recorded USD 655.29 billion. In particular, Guizhou, Qinghai, Gansu, Hunan, etc. met significant increases in trade, reflecting the improvement of international competitiveness of relevant industries driven by innovation.

Third, innovation facilitates the more efficient utilization of energy to some extent. In 2014, energy consumption of China was 4.26 billion with the consideration of TCE, with an increase of 2.2% over the previous year. Energy consumption per GDP was 0.76 TCE/RMB 10,000, which was a decrease of 4.8% than the previous year. The emission of major pollutants from waste gas was significantly reduced based on the observation in 30 regions across China (NBS 2014).

References

Cheung, K. Y., & Lin, P. (2004). Spillover effects of FDI on innovation in China: Evidence from the provincial data. *China Economic Review, 15*(1), 25–44.

Gao, L., Zhu, S., & Li, J. (2016). Innovation ability, spatial dependence and the evolution of growth poles of Yangtze River Delta urban system. *Science & Technology Progress and Policy, 33*(5), 40–44.

Global Innovation Index. (2017). *Innovation feeding the world*. Derived from https://www.globalinnovationindex.org/gii-2017-report. Cornell University, INSEAD, and WIPO (2017).

Hu, A., & Jefferson, G. H. (2002). FDI impact and spillover: Evidence from China's electronic and textile industries. *The World Economy, 25*(8), 1063–1076.

Ji, F., (2016). *Outline of collaborative development of Beijing, Tianjin and Hebei Province. Connect to Jingjin—to understand integration of Jing-Jin-Ji and promote the coordinated development of regional economy* (13th workshop of connect to Jing-jin and Bohai area along capital). Langfang applied economic association, Langfang.

Jin, H., Qian, Y., & Weingast, B. R. (2005). Regional decentralization and fiscal incentives: Federalism, Chinese style. *Journal of Public Economics, 89*(9–10), 1719–1742.

Li, L. (2004). Innovation clusters, cooperation networks and regional competitiveness. *Journal of Yunnan University of Finance and Economics (Chinese), 2004*(5), 99–103.

Li, K. (2017). Be "perfect partner" in innovation cooperation—Speech on "Sino-German forum build up innovation together". Derived from: http://www.sohu.com/a/145372303_115402

Li, Y., & Wu, F. (2017). Understanding city-regionalism in China: Regional cooperation in the Yangtze River Delta. *Journal Regional Studies, 4*, 1–12.

Liang, F. H. (2017). Does foreign direct investment improve the productivity of domestic firms? Technology spillovers, industry linkages, and firm capabilities. *Research Policy, 46*(1), 138–159.

Lu, F. (2006). *Towards Indigenous Innovation*. Nanning: Guangxi Normal University Press.

Mao, H. (2017). Innovation of mechanism and regional policy for promoting coordinated development of the Beijing-Tianjin-Hebei. *Progress in Geography, 36*(1), 2–14.

Ministry of Finance. (2017). The responsible person of the Ministry of Finance answers questions about the list of national government funds and administrative fees directory. Derived from http://www.gov.cn/xinwen/2017-06/29/content_5206809.htm

Ministry of Science and Technology of the People's Republic of China (MOST). (2001). China Science &Technology Statistics Data Book. Derived from http://www.sts.org.cn/sjkl/kjtjdt/index.htm

National Bureau of Statistics (NBS). (1999). *China statistical yearbook 2016*. Beijing: China Statistics Press.

NBS. (2016). *The innovation driving force of science and technology innovation is remarkable – the state of science and technology innovation in China since the 18th National Congress*. Derived from. http://www.gov.cn/xinwen/2016-03/09/content_5051482.htm

Pi, S., & Wang, X. (2014). *China's regional economic development has entered a new stage of competition and cooperation*. Derived from http://money.163.com/14/0925/18/A70RIR4900254TI5.html

Qin, C., Liu, L., & Qin, W. (2017). On the development strategy of the urban agglomeration in the Guangdong-Hong Kong-Macao greater bay area. *Regional Economic Review, 5*, 113–118.

Scherngell, T., & Hu, Y. (2011). Collaborative knowledge production in China. Regional evidence from a gravity model approach. *Regional Studies, 45*(6), 755–772.

Scherngell, T., Borowiecki, M., & Hu, Y. (2014). Effects of knowledge capital on total factor productivity in China: A spatial econometric perspective. *China Economic Review, 29*, 82–94.

State Council. (2016). The decision on building Beijing as National Science, Technology and Innovation Center in Beijing. Document no.52.

Sun, L., & Li, Y. (2016). Coordinated innovation based on regional integration of Beijing-Tianjin-Hebei. *Journal of Hebei University of Technology (Chinese), 8*(1), 1–6.

Tang, W., & Shu, P. (2014). Yangtze River layout and Chinese economy strategy. *Business View (Chinese), 1*(12), 14–15.

The State Council. (2016). *To lead the development of urban agglomeration in Yangtze River economic belt with scientific planning*. Derived from http://www.gov.cn/xinwen/2016-05/12/content_5072453.htm

The World Bank. (2017). *World development indicators*.

WIPO. (2016). *World intellectual property organization database of PCT*. Derived from http://www.wipo.int/pct/en/

Xi, J. (2017). Secure a decisive victory in building a moderately prosperous society in all respects and strive for the great success of socialism with Chinese characteristics for a New Era. 19th National Congress of the Communist Party of China, Beijing.

Xiao, J., Zhong, L., & National Development and Reform Commission Institute of Land Development and Regional Economic Research Group. (2015). Research on Beijing-Tianjin-Hebei region development and corporation. *Economic Research Reference (Chinese), 49*, 3–29+43.

Zhang, B. (2016). Research on the coordinated development model of the urban agglomeration in the Yangtze River Delta. *Zhejiang University of Finance and Economics (Chinese), 12*(1), 45–49.

Zhao, S. (2013). Privatization, FDI inflow and economic growth: Evidence from China's provinces, 1978–2008. *Applied Economics, 45*(15), 2127–2139.

Chapter 3
Rankings of Innovation Capability by Region

3.1 Rankings by Comprehensive Index

Regarding comprehensive rankings of regional innovation capability in 2017, we can see Fig. 3.1. Guangdong, Jiangsu, and Beijing rank top 3, and top 10 regions are Guangdong, Jiangsu, Beijing, Shanghai, Zhejiang, Shandong, Tianjin, Chongqing, Hubei, and Anhui in order. In particular, this year, Hubei makes progress, listed rank 9th.

With regard to ranking changes, nine regions see a rise in ranking in 2017, including Guangdong, Hubei, Hunan, Liaoning, Jiangxi, Ningxia, Yunnan, Tibet, and Qinghai, with the greatest progress observed in Hubei and Tibet with three positions up. Nine regions see a decline in ranking, including Jiangsu, Anhui, Shaanxi, Guizhou, Guangxi, Gansu, Heilongjiang, Jilin, Inner Mongolia, and Shanxi. Among all, Heilongjiang had got the greatest drop of four places (Table 3.1).

In terms of regional differences, eastern provinces remain to be the strongest in terms of innovation capability, while central and western regions undergo great changes in rankings. In particular, Shanxi has seen a decline in ranking for consecutively 3 years, down two places in 2017, ranking the 30th in China. Inner Mongolia slips down one place, ranking the 28th in China. Jilin and Heilongjiang continue to see a decline in rankings. In general, regions with strong innovation capability have diverse drivers for innovation and maintain relative stability. Under increasing economic downward pressure and in the context of urgent need for transformation of development mode, it becomes more and more difficult to sustain traditional investment-driven development modes, and the resource-based provinces are facing with great pressure during the process of transformation, resulting in increasing divides among regions.

Among the five provinces and municipalities in Northern China, Beijing and Tianjin take the lead, while Hebei, Shanxi, and Inner Mongolia are in lower place. Especially, Shanxi and Inner Mongolia have seen a decline in rankings for the recent 3 years, facing increasing pressure for transformation.

X. Liu et al., *Regional Innovation Index of China: 2017*,
https://doi.org/10.1007/978-981-13-1205-2_3

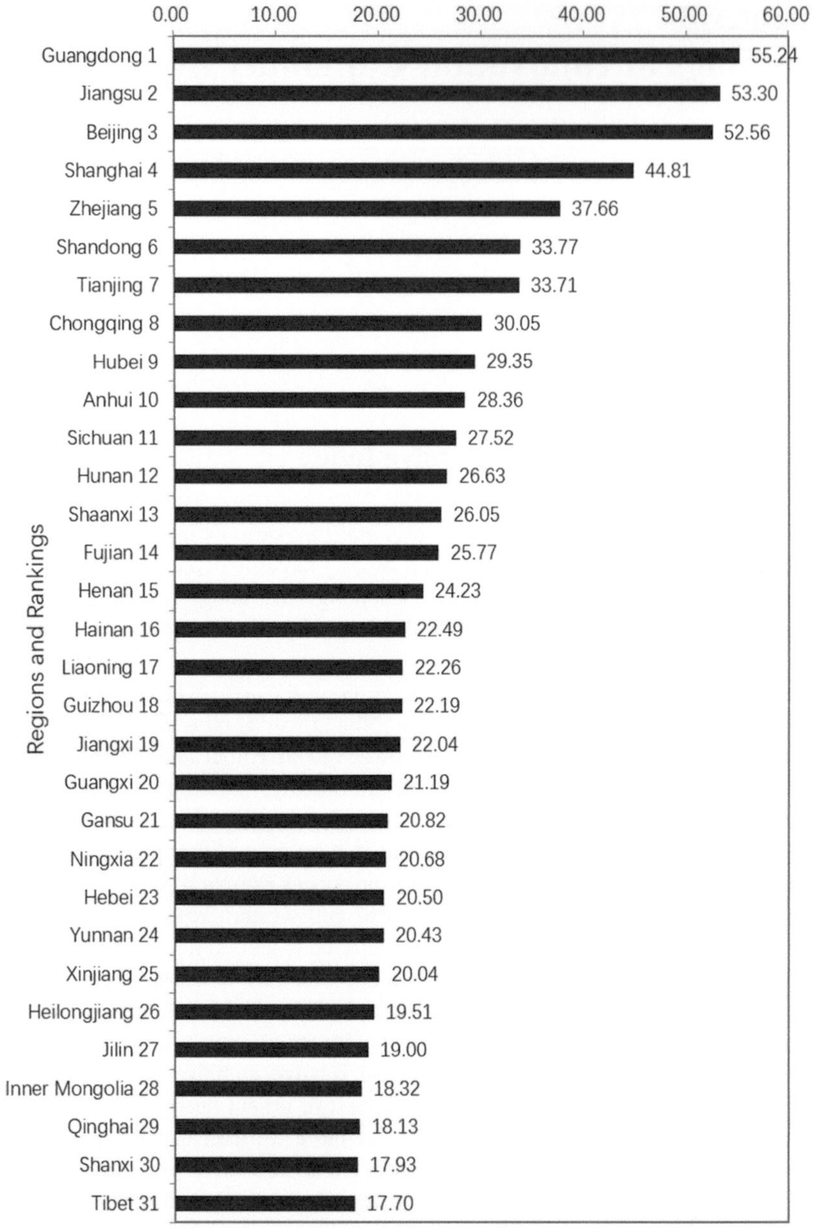

Fig. 3.1 Comprehensive index of regional innovation capabilities in 2017

The seven provinces and municipalities in Eastern China rank top in terms of innovation capability in China. Except Fujian and Jiangxi, the other four places rank among top 10 in China and maintain stable rankings.

Table 3.1 Changes in rankings of innovation capabilities by region in 2017

Region	Ranking in 2017	Ranking in 2016	Change	Region	Ranking in 2017	Ranking in 2016	Change
Guangdong	1	2	1	Liaoning	17	18	1
Jiangsu	2	1	−1	Guizhou	18	17	−1
Beijing	3	3	0	Jiangxi	19	21	2
Shanghai	4	4	0	Guangxi	20	19	−1
Zhejiang	5	5	0	Gansu	21	20	−1
Shandong	6	6	0	Ningxia	22	24	2
Tianjin	7	7	0	Hebei	23	23	0
Chongqing	8	8	0	Yunnan	24	26	2
Hubei	9	12	3	Xinjiang	25	25	0
Anhui	10	9	−1	Heilongjiang	26	22	−4
Sichuan	11	11	0	Jilin	27	27	0
Hunan	12	13	1	Inner Mongolia	28	28	0
Shaanxi	13	10	−3	Qinghai	29	31	2
Fujian	14	14	0	Shanxi	30	29	−1
Henan	15	15	0	Tibet	31	30	−1
Hainan	16	16	0				

Note: With regard to change in rankings in the table, rise is expressed by a positive number, while decline is expressed by a negative number

Great differences are observed among the three provinces in Southern China in terms of innovation capability. Except Guangdong, Hainan and Guangxi Zhuang Autonomous regions have seen a decline in rankings to varying degrees. In the future, it should use the innovation advantage of the Pearl River Delta to drive the coordinated innovation and integrated development of Guangdong and Guangxi.

The three provinces in Central China have seen an improvement in rankings, with Hubei listed among top 10 in China. As one of the target regions of key national support, the three provinces in Northeastern China still underperform in terms of construction of innovation capability, with a decline trend observed in Jilin and Heilongjiang and a moderate improvement in Liaoning.

As the bellwether in Northwestern China, Shaanxi ranks at middle position in China. Ningxia Hui Autonomous Region has seen a higher ranking for consecutively 3 years with outstanding performance. Xinjiang Uygur Autonomous Region and Qinghai still need to continue to enhance innovation capability.

The five provinces and municipalities in Southwestern China have seen steady rises in rankings. To be specific, Chongqing and Sichuan keep rankings unchanged; Guizhou sees a slight decline, while Yunnan move up two places (Table 3.2).

In terms of first-level index (Table 3.3), Beijing ranks 1st in China in terms of the knowledge creation with a comprehensive score of 74.1 in 2017, which is far higher than other provinces and municipalities. Jiangsu and Shanghai rank 2nd and 3rd, respectively, closely followed by Guangdong, Zhejiang, and Anhui. Shanghai ranks 1st in terms of score of the knowledge acquisition comprehensive indicator with

Table 3.2 Region-specific comparison of provinces in 2017

Region	Province/municipality	2017	2016	2015	Region	Province/municipality	2017	2016	2015
Northern China	Beijing	3	3	3	Central China	Hubei	9	12	12
	Tianjin	7	7	7		Hunan	12	13	11
	Hebei	23	23	23		Henan	15	15	17
	Inner Mongolia	28	28	21	Northeastern China	Liaoning	17	18	15
	Shanxi	30	29	25		Heilongjiang	26	22	24
Eastern China	Jiangsu	2	1	1		Jilin	27	27	27
	Shanghai	4	4	4	Northwestern China	Shaanxi	13	10	14
	Zhejiang	5	5	5		Gansu	21	20	20
	Shandong	6	6	6		Ningxia	22	24	28
	Anhui	10	9	9		Xinjiang	25	25	29
	Fujian	14	14	10		Qinghai	29	31	30
	Jiangxi	19	21	19	Southwestern China	Chongqing	8	8	8
Southern China	Guangdong	1	2	2		Sichuan	11	11	16
	Hainan	16	16	13		Guizhou	18	17	22
	Guangxi	20	19	18		Yunnan	24	26	26
		Tibet	31	30	31				

Table 3.3 Comprehensive and first class index of regional innovation capability in 2017

	Comprehensive value	Knowledge creation	Knowledge acquisition	Corporate innovation	Innovation environment	Innovation performance						
Region	Comprehensive value	Ranking	Comprehensive value	Ranking	Comprehensive value	Ranking	Comprehensive value	Ranking	Comprehensive value	Ranking	Comprehensive value	Ranking
Weight		0.15	0.15	0.25	0.25	0.20						
Guangdong	55.24	1	39.07	4	38.81	4	66.22	1	54.98	1	66.28	1
Jiangsu	53.30	2	47.28	2	42.99	3	62.43	2	46.88	3	62.14	2
Beijing	52.56	3	74.10	1	47.42	2	46.30	4	51.05	2	49.97	3
Shanghai	44.81	4	42.21	3	54.06	1	42.83	5	38.73	4	49.90	4
Zhejiang	37.66	5	34.63	5	22.86	7	50.07	3	34.97	6	38.90	9
Shandong	33.77	6	28.93	9	18.37	12	39.04	6	37.03	5	38.26	10
Tianjin	33.71	7	27.28	11	26.04	6	36.56	8	28.52	9	47.22	5
Chongqing	30.05	8	26.02	12	20.15	10	34.27	10	23.92	15	42.86	6
Hubei	29.35	9	23.79	13	18.75	11	29.78	11	29.85	7	40.30	8
Anhui	28.36	10	31.99	6	11.60	24	38.82	7	22.48	19	32.46	16
Sichuan	27.52	11	30.21	8	15.30	15	24.98	12	28.99	8	36.02	12
Hunan	26.63	12	20.44	16	14.04	18	35.30	9	23.71	17	33.51	14
Shaanxi	26.05	13	30.41	7	16.19	13	21.41	19	25.83	12	36.25	11
Fujian	25.77	14	19.57	19	21.16	8	24.76	13	25.64	13	35.31	13
Hebei	24.23	15	15.96	24	11.24	26	21.38	20	26.15	11	41.34	7
Hainan	22.49	16	15.93	26	9.54	30	24.14	14	28.15	10	27.97	20
Liaoning	22.26	17	22.22	14	20.44	9	18.90	24	22.47	20	27.58	21
Guizhou	22.19	18	19.64	18	15.89	14	20.88	21	20.38	26	32.70	15
Jiangxi	22.04	19	13.77	29	14.41	17	20.75	22	25.48	14	31.26	18
Guangxi	21.19	20	28.59	10	9.17	31	16.82	27	19.57	28	32.12	17

(continued)

Table 3.3 (continued)

	Comprehensive value	Knowledge creation	Knowledge acquisition	Corporate innovation	Innovation environment	Innovation performance						
Region	Comprehensive value	Ranking	Comprehensive value	Ranking	Comprehensive value	Ranking	Comprehensive value	Ranking	Comprehensive value	Ranking	Comprehensive value	Ranking
Gansu	20.82	21	19.47	20	10.19	27	22.06	16	21.73	24	27.09	22
Ningxia	20.68	22	15.96	24	26.10	5	22.68	15	18.40	30	20.52	31
Henan	20.50	23	18.47	21	9.84	29	21.64	17	23.79	16	24.49	28
Yunnan	20.43	24	20.06	17	10.06	28	21.46	18	22.33	21	24.81	27
Xinjiang	20.04	25	17.91	22	13.09	21	18.93	23	22.09	23	25.66	25
Heilongjiang	19.51	26	21.89	15	15.08	16	13.84	31	23.41	18	23.28	29
Jilin	19.00	27	17.24	23	13.66	20	17.46	25	16.99	31	28.75	19
Inner Mongolia	18.32	28	11.91	31	11.63	23	16.32	28	22.22	22	25.75	24
Qinghai	18.13	29	13.95	28	11.29	25	15.85	29	20.77	25	25.92	23
Shanxi	17.93	30	15.43	27	13.91	19	15.57	30	20.12	27	23.03	30
Tibet	17.70	31	13.02	30	11.84	22	16.84	26	19.13	29	24.89	26

comprehensive value of 54.06. Beijing and Jiangsu rank 2nd and 3rd, respectively, closely followed by Guangdong. Guangdong ranks the 1st in China in terms of corporate innovation capability with comprehensive value of 66.22, surpassing Jiangsu for the first time in recent years and closely followed by Zhejiang, Beijing, and Shanghai. Guangdong ranks the 1st in China over Beijing with innovation environment comprehensive value of 54.98, followed by Jiangsu, Shanghai, and Shandong. Guangdong shows the best performance in innovation performance with comprehensive value of 66.28, with Jiangsu and Beijing ranking the 2nd and 3rd, respectively, closely followed by Shanghai, Tianjin, and Chongqing. As a whole, the gaps among provinces and municipalities in terms of knowledge acquisition and innovation environment seem smaller than that in terms of knowledge creation, corporate innovation, and innovation performance.

3.2 Rankings by Strength

In this report, we divide regional innovation capability into three parts: the strength, the efficiency, and the potential of innovation. By breaking down the overall indicator system into three structurally different components, the difference and dynamics of innovation can be seen more clearly (see Table 3.4 for specific rankings).

Here, strength refers to innovative resources that a region owns, such as the absolute science and technology input level, availability of human resources in research, and the output level of innovation, such as the number of patents and the number of new products.

Efficiency refers to the benefits generated by unit input in a region, such as the number of papers or patents produced by unit scientific and technical personnel and R&D funding. This is a common way for measuring innovation capability in many international reports such as World Economic Forum and IMD Global Competitiveness Report.

Potential refers to the growth rate than the previous year. Usually, it is difficult to achieve faster growth in regions with stronger innovation strength. Therefore, regions with strong strength are not always regions with great potential; in contrast, some latecomer regions perform better in relation to this aspect.

From Table 3.4, it can be seen that in 2017 innovation ranking by strength indicator, Guangdong ranks 1, Jiangsu 2, and Beijing, Shandong, and Zhejiang rank 3, 4, and 5, respectively. Among the top 10 provinces in comprehensive ranking, there are eight provinces listed in top 10 provinces ranking by strength as well, while Shanghai, Tianjin, Chongqing, Hubei, and Anhui (ranking 4th, 7th, 8th, 9th, and 10th in comprehensive ranking) rank 6th, 12th, 17th, 7th, and 9th by strength indicator.

Among provinces whose rankings by strength are higher than comprehensive rankings, the provinces with ranking gap (no less than three places) from largest to

Table 3.4 2017 regional innovation capacity ranking by strength, efficiency, and potential

Region	Comprehensive	Knowledge creation	Knowledge acquisition	Corporate innovation	Innovation environment	Innovation performance												
	Strength	Efficiency	Potential	Strength	Efficiency	Potential	Strength	Efficiency	Potential	Strength	Efficiency	Potential	Strength	Efficiency	Potential	Strength	Efficiency	Potential
Guangdong	1	4	18	3	19	13	2	12	16	1	3	15	1	3	20	1	5	19
Jiangsu	2	5	27	2	7	19	1	6	26	2	6	20	2	10	31	2	1	14
Beijing	3	1	26	1	1	23	4	2	4	10	1	26	4	1	27	4	4	28
Shandong	4	16	24	6	24	21	6	27	13	4	14	24	3	16	21	3	12	24
Zhejiang	5	7	25	4	12	16	5	11	24	3	5	21	5	7	26	6	10	18
Shanghai	6	2	28	5	2	18	3	1	10	5	2	27	6	2	29	5	2	21
Hubei	7	10	9	10	16	20	10	15	11	8	11	10	9	9	10	8	13	4
Sichuan	8	12	16	7	9	11	11	26	15	13	15	19	8	13	13	11	9	17
Anhui	9	13	8	8	6	2	14	31	19	6	7	9	13	30	17	14	18	10
Henan	10	26	17	13	28	28	16	28	20	9	30	17	7	29	15	7	8	6
Fujian	11	14	20	16	26	9	8	9	14	11	17	25	10	18	24	10	11	16
Tianjin	12	3	12	12	13	3	9	3	6	12	4	18	15	5	22	9	3	15
Liaoning	13	15	31	11	11	30	7	7	31	15	19	31	11	19	30	16	15	31
Hunan	14	11	15	14	21	17	21	13	27	7	9	13	14	23	18	13	14	9
Shaanxi	15	8	23	9	4	27	13	20	8	18	16	23	17	8	19	18	7	20
Hebei	16	31	13	17	27	7	20	30	21	14	25	12	12	27	11	23	26	25

Chongqing	17	6	5	15	8	5	12	5	22	16	8	1	19	12	8	12	6	5
Jiangxi	18	22	7	23	29	14	17	18	18	17	26	3	18	11	6	15	16	13
Heilongjiang	19	17	30	18	5	31	18	8	30	24	24	30	16	21	9	26	25	29
Guangxi	20	18	21	19	3	15	24	29	12	19	21	28	20	28	14	17	20	8
Jilin	21	28	22	20	15	29	19	25	9	21	31	2	24	26	28	20	17	27
Yunnan	22	25	6	21	23	1	26	24	23	25	20	5	22	14	12	25	28	11
Shanxi	23	21	29	22	22	25	23	19	17	20	22	29	23	20	25	28	21	30
Inner Mongolia	24	29	19	26	31	10	22	17	28	22	27	16	21	17	16	27	19	23
Gansu	25	19	11	24	10	22	25	21	29	26	13	14	27	24	4	21	27	12
Guizhou	26	20	3	25	18	4	27	14	3	23	18	11	25	22	7	24	22	1
Xinjiang	27	27	4	26	17	12	28	16	7	27	23	6	26	15	5	30	30	2
Hainan	28	9	14	29	14	24	29	23	25	29	10	22	28	4	3	19	23	22
Ningxia	29	23	2	28	25	8	15	4	2	28	12	8	29	31	1	31	31	7
Tibet	30	30	1	31	30	6	30	10	1	31	28	4	31	25	2	22	24	26
Qinghai	31	24	10	30	20	26	31	22	5	30	29	7	30	6	22	29	29	3

smallest in order are Shanxi (23/31, comprehensive strength raking/comprehensive ranking), Hebei (16/23), Heilongjiang (19/26), Jilin (21/28), Henan (10/15), Liaoning (13/17), Sichuan (8/11), and Fujian (11/14). This means that those regions do have the S&T and business capabilities, but they cannot be transformed into the innovation capabilities.

Among provinces whose rankings by strength are lower than comprehensive rankings, the provinces with ranking gaps (no less than three places) from largest to smallest in order are Hainan Province (28/16), Guizhou Province (26/18), Chongqing (17/8), Ningxia Hui Autonomous Region (29/22), Tianjin (12/7), Gansu (25/21), and Tibet Autonomous Region (30/27).

3.3 Rankings by Efficiency

In the efficiency ranking, usually the metropolitan regions will usually have an advantage. The research and education level per capita will matter. Beijing and Shanghai rank first and second. Tianjin and Guangdong rank third and fourth, respectively. Jiangsu ranks fifth. As seen from Table 2.4, some provinces whose efficiency rankings are higher than comprehensive rankings are Shanxi (21/31), Heilongjiang (17/26), Hainan (9/16), Shaanxi (8/13), and Tianjin (3/7); this means that their good R&D and education level do not sufficiently contribute to innovation. Some provinces whose efficiency rankings are lower than comprehensive rankings with ranking gap from largest to smallest in order are Henan (26/15), Shandong (16/6), Hebei (31/23), Guangdong (4/1), Jiangsu (5/2), Anhui (13/10), Jiangxi (22/19), and the Tibet Autonomous Region (30/27). It means that those regions have a good entrepreneurship spirit though they do not have good R&D and education resources comparatively.

3.4 Rankings by Potential

In rankings by potential, some latecomer regions show a good performance. Here, Tibet Autonomous Region ranks the first; Ningxia Hui Autonomous Region and Guizhou rank the second and the third, respectively. This shows that the latecomer regions upgrade their capabilities and catch up. There are 15 provinces whose potential rankings are higher than comprehensive rankings. Among them, Tibet Autonomous Region's ranking gap between potential ranking and comprehensive rankings is 26 places. The ranking gaps between potential ranking and comprehensive rankings of Xinjiang Uygur Autonomous Region, Ningxia Hui Autonomous Region, Qinghai, Yunnan, Guizhou, and Jiangxi are 21, 20, 20, 18, 15, and 12 places, respectively; there are 15 provinces whose potential rankings are lower than comprehensive rankings. Among them, Jiangsu, Shanghai, Beijing, Zhejiang, Shandong, and Guangdong have the largest ranking gaps and the strongest comprehensive innovation capability as well. Because of their large economic size, the indicator growth rates in such regions are relatively slow.

3.5 Regions with Ranking Changes

The overall landscape of comprehensive ranking of regional innovation capability saw a slight change in 2017 but with some remarkable changes in certain provinces. This chapter makes a selective analysis on provinces of Hubei, Shaanxi, and Heilongjiang,

3.5.1 Hubei (12 → 9)

In 2017, Hubei ranked ninth in terms of the innovation capability and went up three places compared with the previous year. Seen from first-level index, Hubei made improvements in the ranking of four second-level indicators, i.e., knowledge acquisition, corporate innovation, innovation environment, and innovation performance, including knowledge acquisition and corporate innovation up one place, innovation environment up two places, and innovation performance up four places (Table 3.5). It could tell that the Hubei's innovation capability was enhanced on the whole.

In terms of basic indicators, Hubei's growth rate of papers published domestically and internationally moved up 9 and 10 places, respectively, and the growth rate of papers co-authored with authors from foreign countries and others provinces moved up 7 and 17 places; the external R&D spending of industrial enterprises above the designated size on average climbed 11 places or 47.27% over the previous year; the ranking of external R&D funding spending of industrial enterprises above the designated size climbed 6 places or 51.51% over the previous year, and the average external corporate R&D spending increased by 47.27% over the previous year, up 11 places in terms of the ranking; practitioners in the sci-tech service industry moved up two places, an increase of 7.32% over the previous year. A series of data shows that Hubei has made significant progress in promoting the scientific and technological cooperation.

3.5.2 Shaanxi (10 → 13)

In 2017, Shaanxi ranked 13th in terms of the innovation capability, down three places compared with the previous year. Seen from first-level indicators, except for the innovation environment, Shaanxi recorded a decline in the ranking of four indicators, i.e., knowledge creation, knowledge acquisition, corporate innovation, and innovation performance (Table 3.6).

Table 3.5 2016–2017 ranking changes in innovation capability of Hubei

Year	Comprehensive ranking	Knowledge creation	Knowledge acquisition	Corporate innovation	Innovation environment	Innovation performance
2016	12	13	13	12	9	12
2017	9	13	12	11	7	8

Table 3.6 2016–2017 ranking changes in innovation capability of Shaanxi

Year	Comprehensive ranking	Knowledge creation	Knowledge acquisition	Corporate innovation	Innovation environment	Innovation performance
2016	10	6	10	15	12	9
2017	13	7	14	19	12	11

In terms of basic indicators in the knowledge creation field, Shaanxi saw a decline in the rankings by ten indicators, including "invention patent applications accepted (excluding enterprises)" and "invention patents granted from per RMB100 million internal R&D spending" down seven places, "invention patent applications accepted per 10,000 R&D personnel" down nine places, and "invention patent applications from per RMB100 million internal R&D spending" down 16 places; in the knowledge acquisition field, "amount paid by industrial enterprises above the designated size for introduction of overseas technologies" slipped down five places, and "number of sci-tech papers written by authors from the same province but in different units per 100,000 R&D personnel" slipped down eight places; in the corporate innovation field, "the share of sales revenue from new products in sales revenue of large industrial enterprises slipped down four places; in addition, the ranking positions of Shaanxi's industrial sewage and industrial waste gas emission indicators also slipped down.

3.5.3 Heilongjiang (22 → 26)

In 2017, Heilongjiang ranked 26th in regional innovation index, down four places compared with the previous year. Seen from first-level indicators, Heilongjiang saw an overall decline in the ranking of all indicators, including the knowledge creation indicator up one place; corporate innovation indicator down four places, ranking the last in China; and innovation performance down one place, ranking 29th (Table 3.7).

In terms of basic indicators, Heilongjiang recorded a decline in the rankings of 19 indicators in respect of the corporate innovation evaluation indicators, including the technical revamp expenditure of large industrial enterprises down 31.61% over the previous year, the average technical revamp expenditure of large industrial enterprises down 29.25%, the external R&D spending of large industrial enterprises down 18.18%, the number of R&D personnel of large industrial enterprises down 14.62%, and the total internal R&D spending of large industrial enterprises down 7.89%. In respect of the innovation performance evaluation indicators, 15 indicators ranked down; from the absolute value, Heilongjiang saw a decrease of 34.8% in the exports of high-tech products, down two places, and decrease of 1.6% in the main operating revenue of high-tech industry, down one place in 2015.

Meanwhile, we also see that Heilongjiang has been increasing government R&D input, and the invention patents applications and invention patents granted have also been enhanced, but the innovation spillover effect has not yet been fully revealed.

Table 3.7 2016–2017 ranking changes in innovation capability of Heilongjiang

Year	Comprehensive ranking	Knowledge creation	Knowledge acquisition	Corporate innovation	Innovation environment	Innovation performance
2016	22	16	17	27	18	28
2017	26	15	17	31	18	29

3.6 Analysis of Basic Indicators Relevant for Regional Innovation

3.6.1 *Rankings of R&D Input by Local Government*

The top 10 regions in terms of governmental R&D input in 2015, in turn, are Beijing, Shanghai, Sichuan, Shaanxi, Jiangsu, Guangdong, Shandong, Tianjin, Liaoning, and Hubei (Table 3.8).

In respect of ranking position, Sichuan and Shaanxi swap their places compared to the previous year, Tianjin moves up three places, while Hubei slips down two places. Chongqing, seeing the greatest rise in position, ranks 19th from the previous 23rd; both Yunnan and Hebei move up three places; Hunan, Fujian, Gansu, and Jiangxi slip down two places, while the rest see small changes. In terms of input intensity, the inputs of Jiangxi, Hainan, and Qinghai decline slightly, while the remaining provinces, municipalities, and autonomous regions see increases to varying degrees.

3.6.2 *Ranking of Invention Patents Granted by Region*

Invention patents granted are one of key indicators of innovation capability. The top 10 regions in terms of invention patents granted in 2015, in turn, are Jiangsu, Beijing, Guangdong, Zhejiang, Shanghai, Shandong, Anhui, Sichuan, Hubei, and Shaanxi, among which Jiangsu, surpassing Beijing, ranks the first nationwide (Table 3.9).

In terms of ranking change, the overall ranking in 2015 is subject to small changes; specifically, compared to the previous year, Jiangsu and Guangxi Zhuang Autonomous Region move up two places; Anhui, Hubei, Fujian, Jiangxi, and Ningxia Hui Autonomous Region move up one place; Beijing, Guangdong, Sichuan, Shaanxi, Henan, Chongqing, Hebei, Guizhou, and Hainan slip down one place; and other provinces and municipalities remain unchanged.

In terms of invention patents granted, all provinces and municipalities see significant rises except Tibet Autonomous Region.

3.6.3 *Ranking of the Number of Domestic Paper Publication by Region*

The number of papers published represents the scientific output capability of one region. The top 10 regions in terms of number of domestic papers of each region in 2015, in turn, are Beijing, Jiangsu, Guangdong, Shanghai, Shaanxi, Hubei, Zhejiang, Sichuan, Shandong, and Liaoning (Table 3.10).

Table 3.8 R&D input by local government, 2014–2015 (in RMB100 million)

	Governmental R&D input	Ranking	Ranking change	Region	Governmental R&D input	Ranking	Ranking change				
Region	2015	2014	2015	2014			2015	2014	2015	2014	
Beijing	791.64	700.07	1	1	0	Henan	48.33	45.5	17	16	−1
Shanghai	340.80	292.36	2	2	0	Yunnan	37.83	24.91	18	21	3
Sichuan	230.22	193.23	3	4	1	Chongqing	36.45	23.25	19	23	4
Shaanxi	220.22	199.98	4	3	−1	Fujian	33.99	29.8	20	18	−2
Jiangsu	153.34	133.31	5	5	0	Gansu	29.76	26.77	21	19	−2
Guangdong	145.85	116.66	6	6	0	Jiangxi	25.99	26.01	22	20	−2
Shandong	111.02	101.38	7	7	0	Guangxi	24.97	23.48	23	22	−1
Tianjin	104.76	74.58	8	11	3	Shanxi	24.29	20.33	24	24	0
Liaoning	103.98	92.7	9	9	0	Inner Mongolia	16.10	14.16	25	25	0
Hubei	102.39	97.84	10	8	−2	Guizhou	16.03	13.27	26	26	0
Anhui	86.42	85.42	11	10	−1	Xinjiang	13.82	12.3	27	27	0
Zhejiang	75.29	70.65	12	12	0	Ningxia	5.47	5.39	28	29	1
Heilongjiang	66.55	62.52	13	13	0	Hainan	5.39	5.69	29	28	−1
Hebei	53.61	42.71	14	17	3	Qinghai	3.93	4.2	30	30	0
Jilin	51.20	45.53	15	15	0	Tibet	2.67	1.98	31	31	0
Hunan	50.89	50.09	16	14	−2						

Table 3.9 Invention patents granted by region, 2014–2015 (Nr)

	Invention patents granted	Ranking	Ranking change	Region	Invention patents granted	Ranking	Ranking change				
Region	2015	2014	2015	2014			2015	2014	2015	2014	
Jiangsu	36,015	19,671	1	3	2	Guangxi	4017	1933	17	19	2
Beijing	35,308	23,237	2	1	−1	Chongqing	3964	2321	18	17	−1
Guangdong	33,477	22,276	3	2	−1	Hebei	3840	2286	19	18	−1
Zhejiang	23,345	13,372	4	4	0	Shanxi	2432	1559	20	20	0
Shanghai	17,601	11,614	5	5	0	Jilin	2240	1434	21	21	0
Shandong	16,881	10,538	6	6	0	Yunnan	2079	1423	22	22	0
Anhui	11,180	5184	7	8	1	Jiangxi	1639	1033	23	24	1
Sichuan	9105	5682	8	7	−1	Guizhou	1501	1047	24	23	−1
Hubei	7766	4855	9	10	1	Gansu	1238	812	25	25	0
Shaanxi	6812	4885	10	9	−1	Xinjiang	950	605	26	26	0
Hunan	6776	4160	11	11	0	Inner Mongolia	797	458	27	27	0
Liaoning	6569	3975	12	12	0	Ningxia	442	243	28	29	1
Fujian	5730	3426	13	14	1	Hainan	417	380	29	28	−1
Henan	5384	3493	14	13	−1	Qinghai	207	110	30	30	0
Tianjin	4624	3279	15	15	0	Tibet	40	50	31	31	0
Heilongjiang	4024	2454	16	16	0						

Table 3.10 Number of domestic papers published by region, 2014–2015 (pcs)

	Number of domestic papers	Ranking	Ranking change	Region	Number of domestic papers	Ranking	Ranking change				
Region	2015	2014	2015	2014			2015	2014	2015	2014	
Beijing	66,999	67,557	1	1	0	Heilongjiang	11,345	12,579	17	17	0
Jiangsu	46,555	48,616	2	2	0	Guangxi	9296	10,232	18	18	0
Guangdong	31,481	33,787	3	3	0	Fujian	8949	9000	19	19	0
Shanghai	29,708	31,210	4	4	0	Gansu	8738	8792	20	20	0
Shaanxi	26,452	28,257	5	5	0	Jilin	8704	8498	21	21	0
Hubei	23,855	24,623	6	6	0	Xinjiang	8077	7424	22	23	1
Zhejiang	23,063	24,494	7	7	0	Shanxi	7910	7149	23	24	1
Sichuan	22,897	23,214	8	9	1	Yunnan	7559	7860	24	22	−2
Shandong	22,367	23,360	9	8	−1	Jiangxi	6621	6648	25	25	0
Liaoning	19,291	19,514	10	10	0	Guizhou	5355	5479	26	26	0
Henan	18,148	18,843	11	11	0	Inner Mongolia	4287	3829	27	27	0
Hebei	17,671	17,341	12	12	0	Hainan	3037	2951	28	28	0
Hunan	14,447	15,536	13	13	0	Ningxia	2041	1998	29	29	0
Tianjin	13,546	13,775	14	14	0	Qinghai	1249	1259	30	30	0
Anhui	12,863	13,493	15	16	1	Tibet	230	263	31	31	0
Chongqing	12,640	13,537	16	15	−1						

In terms of number of papers published, only seven regions including Ningxia Hui Autonomous Region, Hainan, Xinjiang Uygur Autonomous Region, Inner Mongolia Autonomous Region, and Shanxi see a rise in the number of domestic papers in 2015, and the remaining 24 provinces and municipalities see declines in the number of domestic papers published due to the emphasized talent evaluation orientation of publishing international papers by colleges, universities, and scientific research institutions in recent years. Studies suggest, however, that one region's spillover effect of paper output is closely related to the absorbing ability of local innovative actors, that is, not all regions are suitable for international cooperation; particularly some Central and Western China regions shall further strengthen local knowledge cooperation, so as to create more notable spillover effect on the improvement of innovation capability of local enterprises (Qiu and Liu, 2017).

3.6.4 Ranking of the Number of International Paper Publication by Region

The number of international papers published is an important indicator which could reflect the scientific level of one region and its ability to cooperate at an international level. The top 10 regions in terms of number of international papers in 2015, in turn, are Beijing, Jiangsu, Shanghai, Shaanxi, Guangdong, Hubei, Zhejiang, Sichuan, Shandong, and Liaoning (Table 3.11).

In terms of absolute number, the number of international papers published nationwide in 2015 totaled 447,162, up 13.5% compared to the previous year; except for Ningxia Hui Autonomous Region, the remaining 30 provinces and municipalities see rises. The significant increase in the number of international papers shows that the domestic academic research is increasingly recognized by the international community and embraces more international presence.

3.6.5 Ranking of Total Internal Corporate R&D Spending by Large Enterprises by Region[1]

The intensity of corporate R&D spending represents the degree of emphasis placed on innovation by enterprises. The top 10 regions in terms of total internal corporate R&D spending by large enterprises in 2015, in turn, are Guangdong, Jiangsu, Shandong, Zhejiang, Shanghai, Hubei, Henan, Tianjin, Hunan, and Fujian (Table 3.12).

[1]The large enterprise hereafter means the enterprise with more than 20 million sales revenues in a year by National Bureau of Statistics or called above certain scale enterprise

Table 3.11 Number of international papers by region, 2014–2015 (pcs)

	Number of international papers	Ranking	Ranking change	Region	Number of international papers	Ranking	Ranking change				
Region	2015	2014	2015	2014			2015	2014	2015	2014	
Beijing	81,896	72,672	1	1	0	Henan	9572	8214	17	16	-1
Jiangsu	43,846	38,340	2	2	0	Fujian	8230	7003	18	18	0
Shanghai	37,134	34,191	3	3	0	Hebei	7911	6028	19	19	0
Shaanxi	24,683	21,948	4	4	0	Gansu	6582	5697	20	20	0
Guangdong	24,313	19,338	5	7	2	Jiangxi	5106	3901	21	22	1
Hubei	23,470	19,979	6	6	0	Yunnan	4538	3946	22	21	-1
Zhejiang	22,408	20,051	7	5	−2	Shanxi	4133	3701	23	23	0
Sichuan	19,421	16,711	8	10	2	Guangxi	3076	2805	24	24	0
Shandong	18,647	17,617	9	8	−1	Xinjiang	1848	1471	25	25	0
Liaoning	18,595	17,256	10	9	−1	Inner Mongolia	1644	1358	26	26	0
Hunan	16,027	14,922	11	11	0	Guizhou	1244	1129	27	27	0
Heilongjiang	13,701	13,162	12	12	0	Hainan	673	570	28	28	0
Tianjin	13,103	11,893	13	13	0	Ningxia	345	350	29	29	0
Jilin	12,663	10,917	14	14	0	Qinghai	250	214	30	30	0
Anhui	12,349	10,635	15	15	0	Tibet	30	13	31	31	0
Chongqing	9724	8089	16	17	1						

Table 3.12 Total internal corporate R&D spending by large enterprises, 2014–2015 (in RMB100 million)

Region	Corporate R&D spending	Ranking	Ranking change	Region	Corporate R&D spending	Ranking	Ranking change				
	2015	2014	2015	2014			2015	2014	2015	2014	
Guangdong	1520.55	1375.29	1	2	1	Shaanxi	172.58	160.69	17	17	0
Jiangsu	1506.51	1376.54	2	1	−1	Jiangxi	147.5	128.46	18	18	0
Shandong	1291.77	1175.55	3	3	0	Inner Mongolia	118.63	108.03	19	20	1
Zhejiang	853.57	768.15	4	4	0	Shanxi	100.89	124.7	20	19	-1
Shanghai	474.24	449.22	5	5	0	Heilongjiang	88.04	95.58	21	21	0
Hubei	407.27	362.95	6	6	0	Jilin	86.15	78.94	22	23	1
Henan	368.83	337.23	7	7	0	Guangxi	76.92	84.88	23	22	-1
Tianjin	352.67	322.81	8	9	1	Yunnan	61.96	51.66	24	24	0
Hunan	352.55	310.04	9	11	2	Gansu	48.61	46.44	25	25	0
Fujian	346.98	315.38	10	10	0	Guizhou	45.73	41.01	26	26	0
Anhui	322.14	284.73	11	12	1	Xinjiang	36.62	35.78	27	27	0
Hebei	285.81	260.67	12	13	1	Ningxia	20.05	18.65	28	28	0
Beijing	244.09	233.5	13	14	1	Hainan	11.18	11.1	29	29	0
Liaoning	241.88	324.23	14	8	−6	Qinghai	6.5	9.25	30	30	0
Sichuan	223.81	196.01	15	15	0	Tibet	0.26	0.29	31	31	0
Chongqing	199.66	166.47	16	16	0						

In terms of absolute number, total internal corporate R&D spending by large enterprises of all regions nationwide in 2015 recorded RMB 1001.395 billion, up 8.21% compared to the previous year; by region, 25 provinces and municipalities in total nationwide see increases in input, among which Yunnan and Chongqing see the greatest rise, both rose up 19.9%; six regions including Qinghai, Liaoning, and Shanxi see decreases in corporate R&D input, among which Liaoning and Shanxi decreased by 25.4% and 19.1%, respectively.

3.6.6 Ranking of New Product Sales Revenue from Large Enterprises by Region

Sales revenue from new products of an enterprise represents the enterprise's innovation performance. The top 10 regions in terms of sales revenue from new products of large enterprises in 2015, in turn, are Jiangsu, Guangdong, Zhejiang, Shandong, Shanghai, Hunan, Anhui, Henan, Tianjin, and Hubei (Table 3.13).

Total sales revenue from new products of large enterprises in 2015 recorded 15085.65 billion RMB, up 5.57% compared to the previous year, among which 10 regions see declines in sales revenue from new products and 21 regions see increases. Qinghai, Ningxia Hui Autonomous Region, and Chongqing see rapid increases, up 166.3%, 47.8%, and 25.6% respectively; 10 regions including Gansu, Liaoning, Beijing, Shanxi, and Heilongjiang see declines in sales revenue from new products, with Liaoning down 17.3% and Shanxi down 9.9%.

3.6.7 Ranking of Educational Expenditure by Region

Educational expenditure is a basic indicator reflecting the importance attached by local government for education as a whole in the first place but indirectly also to R&D and innovation. The top 10 regions in terms of educational expenditure in 2015, in turn, are Guangdong, Jiangsu, Shandong, Henan, Zhejiang, Sichuan, Beijing, Hebei, Hunan, and Anhui (Table 3.14).

Thirteen regions see rises in 2015 ranking, among which Xinjiang and Beijing having significant increases, up four places and three places, respectively; eight regions see declines in their position, among which Liaoning slips down seven places and Ningxia, Heilongjiang, Chongqing, Shanxi, Anhui, and Hunan slip down two places.

Table 3.13 Sales revenue from new products of large enterprises by region, 2014–2015 (in RMB100 million)

Region	Sales revenue from new products of enterprise	Ranking	Ranking change	Region	Sales revenue from new products of enterprise	Ranking	Ranking change				
	2015	2014	2015	2014			2015	2014	2015	2014	
Jiangsu	24463.27	23540.93	1	1	0	Jiangxi	2058.6	1756.38	17	17	0
Guangdong	22642.5	20313.32	2	2	0	Jilin	1822.75	1659.99	18	18	0
Zhejiang	18839.14	16507.86	3	3	0	Guangxi	1633.37	1348.42	19	19	0
Shandong	14698.43	14555.82	4	4	0	Shaanxi	1040.99	1126.76	20	20	0
Shanghai	7470.93	8446.96	5	5	0	Shanxi	833.34	924.68	21	21	0
Hunan	7349.8	6310.37	6	6	0	Inner Mongolia	664.84	557.32	22	23	1
Anhui	5882.23	5280.88	7	8	1	Gansu	574.1	719.35	23	22	-1
Henan	5789.42	5168.95	8	10	2	Yunnan	513.2	518.26	24	25	1
Tianjin	5727.77	5665.11	9	7	−2	Heilongjiang	511.05	527.28	25	24	−1
Hubei	5676.92	5274.59	10	9	−1	Xinjiang	494.39	483.79	26	26	0
Chongqing	4535.12	3610.78	11	13	2	Guizhou	394.48	408.37	27	27	0
Beijing	3564.04	4247	12	11	−1	Ningxia	282.69	191.28	28	28	0
Fujian	3525.55	3511.71	13	14	1	Hainan	133.09	148.26	29	29	0
Hebei	3476.24	3334.03	14	15	1	Qinghai	22.82	8.57	30	30	0
Liaoning	3337.35	4036.96	15	12	−3	Tibet	5.64	0	31	31	0
Sichuan	2892.48	2711.3	16	16	0						

Table 3.14 Educational expenditure by local government, 2014–2015 (in RMB 100 million)

Region	Educational expenditure	Ranking	Ranking change	Region	Educational expenditure	Ranking	Ranking change				
	2015	2014	2015	2014			2015	2014	2015	2014	
Guangdong	2022	2477.55	1	1	0	Fujian	715.18	822.8	17	17	0
Jiangsu	1671.68	1986.28	2	2	0	Liaoning	705.34	930.21	18	11	−7
Shandong	1581.12	1779.62	3	3	0	Guizhou	676.11	679.98	19	20	1
Henan	1318.18	1557.71	4	4	0	Xinjiang	585.81	598.99	20	24	4
Zhejiang	1208.14	1449.04	5	5	0	Shanxi	576.23	691.82	21	19	−2
Sichuan	1177.93	1380.55	6	6	0	Inner Mongolia	573.57	612.16	22	22	0
Beijing	968.36	999.84	7	10	3	Chongqing	553.89	656.56	23	21	−2
Hebei	892.65	1029.81	8	9	1	Tianjin	553.28	569.96	24	25	1
Hunan	880.93	1078.46	9	7	−2	Heilongjiang	527.68	600.63	25	23	−2
Anhui	862.98	1041.3	10	8	−2	Gansu	463.31	481.1	26	27	1
Shanghai	796.53	906.97	11	12	1	Jilin	446.03	548.03	27	26	−1
Yunnan	793.07	900.69	12	13	1	Hainan	199.21	222.29	28	28	0
Hubei	763.4	897.23	13	14	1	Qinghai	184.37	156.94	29	30	1
Jiangxi	739.5	828.5	14	16	2	Tibet	150.76	120.67	30	31	1
Shaanxi	730.52	892.69	15	15	0	Ningxia	150.26	157.89	31	29	−2
Guangxi	716.31	779.42	16	18	2						

In terms of absolute input value, except for Qinghai and Tibet, the remaining 29 regions see declines in educational expenditure to varying degrees. This is a not good signal for innovation.

3.6.8 Ranking of the Number of High-Tech Enterprises by Region

The number of high-tech enterprises is an important indicator for a region's innovation capability, as the high-tech sector faces the highest innovation pressure and is the most knowledge-intensive one. The top 10 regions in terms of number of high-tech enterprises in 2015, in turn, are Guangdong, Jiangsu, Zhejiang, Shandong, Anhui, Henan, Hubei, Shanghai, Sichuan, and Hunan (Table 3.15).

The number of high-tech enterprises recognized nationwide as at 2015 totaled 29,631, up 6.1% compared to 1692 in the previous year; by region, 27 provinces and municipalities nationwide see increases in the number of high-tech enterprises, Beijing remains unchanged, while Liaoning, Heilongjiang, and Tibet Autonomous Region see declines, down 43, 3, and 1, respectively, compared to the previous year.

3.6.9 Ranking of Share of Value Added of Tertiary Industry in GDP by Region

The share of value added of tertiary industry in GDP reflects the industrial structure of one region, and the change in share represents the level of industrial structure upgrading of the region. The top 10 regions in terms of share of value added of tertiary industry in GDP in 2015, in turn, are Beijing, Shanghai, Tibet Autonomous Region, Hainan, Shanxi, Tianjin, Guangdong, Zhejiang, Jiangsu, Chongqing, and Heilongjiang (Table 3.16).

In terms of ranking positions, the structure of output value in 2015 shows significant changes; specifically 11 regions see rises in ranking positions, among which Shanxi had a significant rise for successive 2 years, up 7 places, Liaoning moves up 5 places, while Qinghai and Shaanxi move up 4 places; 13 regions see declines in ranking positions, among which Guangxi Zhuang Autonomous Region slips down 7 places, Guizhou down 5 places, Inner Mongolia Autonomous Region, Ningxia Hui Autonomous Region, Chongqing, and Jiangxi down 3 places.

The shares of value added of tertiary industry of 31 regions nationwide in 2015 all see stable increases, among which Shanxi sees an increase of 8.68 percentage points. This shows that China has entered the stage of post industrialization.

Table 3.15 Number of high-tech enterprises by region, 2014–2015

	Number of high-tech enterprises	Ranking	Ranking change	Region	Number of high-tech enterprises	Ranking	Ranking change				
Region	2015	2014	2015	2014			2015	2014	2015	2014	
Guangdong	6194	5874	1	1	0	Chongqing	561	460	17	17	0
Jiangsu	4903	4852	2	2	0	Shaanxi	475	435	18	18	0
Zhejiang	2603	2437	3	3	0	Jilin	406	393	19	19	0
Shandong	2268	2114	4	4	0	Guangxi	313	296	20	20	0
Anhui	1198	1036	5	6	1	Guizhou	226	193	21	21	0
Henan	1176	1068	6	5	−1	Heilongjiang	179	182	22	22	0
Hubei	1037	920	7	8	1	Yunnan	177	148	23	23	0
Shanghai	1020	1003	8	7	−1	Shanxi	139	134	24	24	0
Sichuan	999	911	9	9	0	Gansu	124	117	25	25	0
Hunan	953	900	10	10	0	Inner Mongolia	107	95	26	26	0
Jiangxi	923	792	11	13	2	Hainan	51	49	27	27	0
Fujian	844	796	12	12	0	Xinjiang	42	34	28	29	1
Beijing	805	805	13	11	−2	Qinghai	41	36	29	28	−1
Hebei	633	556	14	16	2	Ningxia	31	24	30	30	0
Liaoning	604	687	15	14	−1	Tibet	8	9	31	31	0
Tianjin	591	583	16	15	−1						

Table 3.16 Share of value added of tertiary industry in GDP by region, 2014–2015 (%)

	Share of tertiary industry	Ranking	Ranking change	Region	Share of tertiary industry	Ranking	Ranking change				
Region	2015	2014	2015	2014			2015	2014	2015	2014	
Beijing	79.65	77.95	1	1	0	Xinjiang	44.71	40.83	17	20	3
Shanghai	67.76	64.82	2	2	0	Ningxia	44.45	43.38	18	15	−3
Tibet	53.8	53.47	3	3	0	Hunan	44.15	42.19	19	17	−2
Hainan	53.26	51.85	4	4	0	Sichuan	43.68	38.7	20	23	3
Shanxi	53.18	44.5	5	12	7	Hubei	43.1	41.45	21	19	−2
Tianjin	52.15	49.57	6	5	−1	Fujian	41.56	39.6	22	21	−1
Heilongjiang	50.73	45.77	7	10	3	Qinghai	41.41	37.04	23	27	4
Guangdong	50.61	48.99	8	6	−2	Shaanxi	40.74	37.01	24	28	4
Zhejiang	49.76	47.85	9	7	−2	Inner Mongolia	40.45	39.52	25	22	−3
Gansu	49.21	44.02	10	13	3	Henan	40.2	37.1	26	26	0
Jiangsu	48.61	47.01	11	8	−3	Hebei	40.19	37.25	27	25	−2
Chongqing	47.7	46.78	12	9	−3	Jiangxi	39.1	36.8	28	29	1
Liaoning	46.19	41.77	13	18	5	Anhui	39.09	35.39	29	31	2
Shandong	45.3	43.48	14	14	0	Jilin	38.83	36.17	30	30	0
Yunnan	45.14	43.25	15	16	1	Guangxi	38.8	37.86	31	24	−7
Guizhou	44.89	44.55	16	11	−5						

Chapter 4
Analysis of Regional Innovation Capability

4.1 Relation Between Innovation Capability and Economic Development

The innovation capability of a region has a close relationship with its economic development, household consumption, and education background. It can be seen from Table 4.1 and Fig. 4.1 that, according to both per capita GDP and household consumption data which may reflect the economic development level and the population education background which reflect the education level, regions with leading innovation capability are generally better than lagging ones. These factors sometimes even play a greater role in improving regional innovation capability than science and technology input.

4.2 R&D Input and Its Intensity

4.2.1 R&D Input and Its Sources and Structure

In 2015, China's government R&D spending grew by 14.28% compared with the previous year at an obviously faster pace. By different regions, Beijing has the highest government R&D input reaching RMB 79,164 million and far outpaces other regions, which is deeply related to the fact that there are a many universities and scientific research institutions in Beijing. It is a fact that R&D input in China mainly rests in a limited number of regions. The top five regions together (Beijing, Shanghai, Sichuan, Shaanxi, and Jiangsu) account for 57.62% of nationwide total R&D input.

In terms of growth rate of R&D input, Chongqing has the highest growth rate (56.78%), followed by Yunnan (51.85%) and Tianjin (40.46%). Moreover, the other 16 regions like Tibet Autonomous Region, Hebei, and Guangdong realized a growth

X. Liu et al., *Regional Innovation Index of China: 2017*,
https://doi.org/10.1007/978-981-13-1205-2_4

Table 4.1 Economic development, household consumption, and education background by region in 2015

Region	Per capital GDP/(RMB/person)	Household consumption (in RMB)	Percentage of college or above degree holders in population aged 6 or above
Tianjin	107960	32594.72	23.33
Beijing	106497	39200.39	42.34
Shanghai	103796	45815.73	28.7
Jiangsu	87995	31682.44	16.42
Zhejiang	77644	28711.54	14.66
Inner Mongolia	71101	20834.87	16.14
Fujian	67966	20828.04	13
Guangdong	67503	26364.97	11.99
Liaoning	65354	23693.11	16.93
Shandong	64168	20684.2	12.63
Chongqing	52321	18859.75	12.56
Jilin	51086	14629.98	13.17
Hubei	50654	17429.39	14.99
Shaanxi	47626	15363.35	17.74
Ningxia	43805	17209.57	15.24
Hunan	42754	16288.85	11.89
Qinghai	41252	15167.34	10.45
Hainan	40818	17019.13	10.85
Hebei	40255	12829.11	10.21
Xinjiang	40036	13683.81	14.55
Heilongjiang	39462	16443.3	13.28
Henan	39123	14507.26	8.74
Sichuan	36775	14773.97	10.99
Jiangxi	36724	14488.92	10.57
Anhui	35997	13940.96	11.73
Guangxi	35190	13856.67	9.21
Shanxi	34919	14363.68	13.74
Tibet	31999	8755.75	7.11
Guizhou	29847	12876.28	8.45
Yunnan	28806	13400.5	9.48
Gansu	26165	11867.67	12.52

rate above 10%. Compared to the previous year, Qinghai, Hainan, and Jiangxi see declines in government R&D input, among which Qinghai has cut its input by 6.37%, which is the biggest fall among all (Table 4.2). This pattern shows that some lagging regions are catching up in terms of R&D input.

In 2015, nationwide corporate R&D input reached 1,001,395 million RMB, up 8.21% compared to 2014. The corporate R&D input of Guangdong reached 152,055

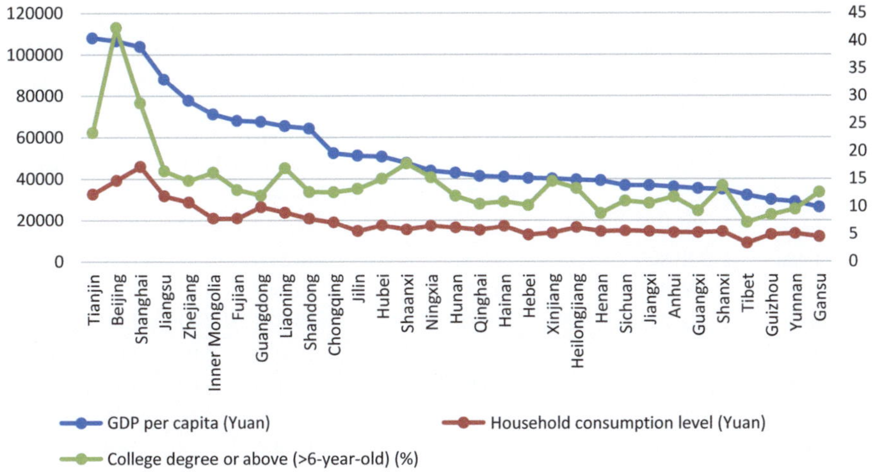

Fig. 4.1 Economic development, household consumption, and education background by region in 2015. (*Source*: China Statistical Yearbook 2016. Beijing, China Statistics Press)

million RMB and accounted for 15.18% of nationwide input, this enabling Guangdong to surpass Jiangsu and become No.1 by total corporate R&D input. With more than 560 billion RMB of corporate R&D input, the sum of the top five regions (Guangdong, Jiangsu, Shandong, Zhejiang, and Shanghai) combined accounted for over 56% of nationwide corporate R&D input.

Except Tibet, in 2015, a total of 25 regions showed a trend of significant increase in corporate R&D spending and 11 regions witnessed a growth rate over 10%. Yunnan, Chongqing, and Jiangsu made the top three in the list with growth rates at 19.94%, 19.94%, and 14.82%, respectively (Table 4.3).

The R&D input structure of a region is closely associated with the distribution of its innovating actors. In 2015, except for Beijing, Shaanxi, Sichuan, and Tibet, the other 27 provinces and municipalities see more corporate R&D input than government R&D input (Fig. 4.2). Beijing has distinct characteristics and takes the lead in terms of government R&D input, which is far beyond corporate R&D input; this is strongly linked to densely distributed scientific research institutions and universities in Beijing; Shaanxi also attracts a batch of "Project 985/211" universities and research institutes; Jiangsu, Guangdong, Shandong, and Zhejiang also have distinct characteristics and see their corporate R&D input considerably higher than government R&D input.

Different R&D input structures reflect diversity of innovation-driven sources. Beijing and Shaanxi are more associated with science-driven innovation or state-driven innovation while Jiangsu and Guangdong are dominated by technology-driven innovation or market-driven innovation.

Table 4.2 2014–2015 government R&D input by region

	Government R&D input/in RMB100 million	Share in nationwide government R&D input/%				
Region	2015	2014	Growth rate	2015	2014	Change
China total	3013.2	2636.7	14.28%			
Beijing	791.64	700.07	13.08%	26.27	26.55	−0.28
Shanghai	340.80	292.36	16.57%	11.31	11.09	0.22
Sichuan	230.22	193.23	19.14%	7.64	7.33	0.31
Shaanxi	220.22	199.98	10.12%	7.31	7.58	−0.27
Jiangsu	153.34	133.31	15.02%	5.09	5.06	0.03
Guangdong	145.85	116.66	25.02%	4.84	4.42	0.42
Shandong	111.02	101.38	9.50%	3.68	3.84	−0.16
Tianjin	104.76	74.58	40.46%	3.48	2.83	0.65
Liaoning	103.98	92.7	12.17%	3.45	3.52	−0.07
Hubei	102.39	97.84	4.65%	3.4	3.71	−0.31
Anhui	86.42	85.42	1.18%	2.87	3.24	−0.37
Zhejiang	75.29	70.65	6.57%	2.5	2.68	−0.18
Heilongjiang	66.55	62.52	6.44%	2.21	2.37	−0.16
Hebei	53.61	42.71	25.52%	1.78	1.62	0.16
Jilin	51.20	45.53	12.45%	1.7	1.73	−0.03
Hunan	50.89	50.09	1.60%	1.69	1.9	−0.21
Henan	48.33	45.5	6.21%	1.6	1.73	−0.13
Yunnan	37.83	24.91	51.85%	1.26	0.94	0.32
Chongqing	36.45	23.25	56.78%	1.21	0.88	0.33
Fujian	33.99	29.8	14.05%	1.13	1.13	0
Gansu	29.76	26.77	11.16%	0.99	1.02	−0.03
Jiangxi	25.99	26.01	−0.09%	0.86	0.99	−0.13
Guangxi	24.97	23.48	6.34%	0.83	0.89	−0.06
Shanxi	24.29	20.33	19.49%	0.81	0.77	0.04
Inner Mongolia	16.10	14.16	13.71%	0.53	0.54	−0.01
Guizhou	16.03	13.27	20.83%	0.53	0.5	0.03
Xinjiang	13.82	12.3	12.39%	0.46	0.47	−0.01
Ningxia	5.47	5.39	1.49%	0.18	0.2	−0.02
Hainan	5.39	5.69	−5.27%	0.18	0.22	−0.04
Qinghai	3.93	4.2	−6.37%	0.13	0.16	−0.03
Tibet	2.67	1.98	34.63%	0.09	0.08	0.01

Source: China Statistical Yearbook on Science and Technology 2016. China Statistical Yearbook 2016. Beijing: China Statistics Press

Table 4.3 2014–2015 R&D expending of large enterprises by region

	Enterprise R&D input/in RMB100 million	Share in nationwide corporate R&D input/%				
Region	2015	2014	Growth rate/%	2015	2014	Change
China	10013.95	9254.23	8.21%			
Guangdong	1520.55	1375.29	10.56%	15.18	14.86	0.32
Jiangsu	1506.51	1376.54	9.44%	15.04	14.87	0.17
Shandong	1291.77	1175.55	9.89%	12.9	12.7	0.2
Zhejiang	853.57	768.15	11.12%	8.52	8.3	0.22
Shanghai	474.24	449.22	5.57%	4.74	4.85	−0.11
Hubei	407.27	362.95	12.21%	4.07	3.92	0.15
Henan	368.83	337.23	9.37%	3.68	3.64	0.04
Tianjin	352.67	322.81	9.25%	3.52	3.49	0.03
Hunan	352.55	310.04	13.71%	3.52	3.35	0.17
Fujian	346.98	315.38	10.02%	3.46	3.41	0.05
Anhui	322.14	284.73	13.14%	3.22	3.08	0.14
Hebei	285.81	260.67	9.64%	2.85	2.82	0.03
Beijing	244.09	233.5	4.54%	2.44	2.52	−0.08
Liaoning	241.88	324.23	−25.40%	2.42	3.5	−1.08
Sichuan	223.81	196.01	14.18%	2.23	2.12	0.11
Chongqing	199.66	166.47	19.94%	1.99	1.8	0.19
Shaanxi	172.58	160.69	7.40%	1.72	1.74	−0.02
Jiangxi	147.5	128.46	14.82%	1.47	1.39	0.08
Inner Mongolia	118.63	108.03	9.81%	1.18	1.17	0.01
Shanxi	100.89	124.7	−19.09%	1.01	1.35	−0.34
Heilongjiang	88.04	95.58	−7.89%	0.88	1.03	−0.15
Jilin	86.15	78.94	9.13%	0.86	0.85	0.01
Guangxi	76.92	84.88	−9.38%	0.77	0.92	−0.15
Yunnan	61.96	51.66	19.94%	0.62	0.56	0.06
Gansu	48.61	46.44	4.67%	0.49	0.5	−0.01
Guizhou	45.73	41.01	11.51%	0.46	0.44	0.02
Xinjiang	36.62	35.78	2.35%	0.37	0.39	−0.02
Ningxia	20.05	18.65	7.51%	0.2	0.2	0
Hainan	11.18	11.1	0.72%	0.11	0.12	−0.01
Qinghai	6.5	9.25	−29.73%	0.06	0.1	−0.04
Tibet	0.26	0.29	−10.34%	0.003	0.003	0

Source: China Statistical Yearbook on Science and Technology 2016. Beijing: China Statistics Press.

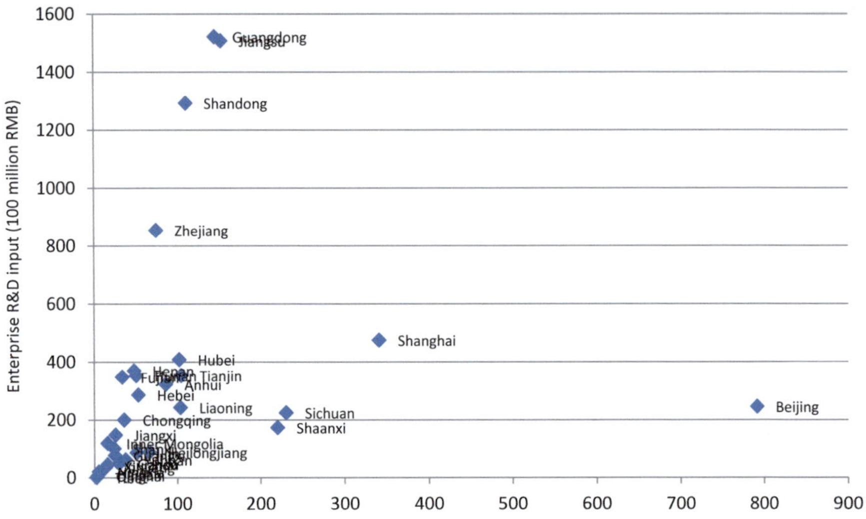

Fig. 4.2 R&D input and its sources and structure by region in 2015

4.2.2 R&D Input Intensity

Both the share of government R&D input in GDP and the share of R&D input of large enterprises in their sales revenue in most of regions have been enhanced to some extent compared to the previous year, but with a rather small overall change (Table 4.4).

In terms of the share of government R&D input in GDP, in 2015, there were 21 provinces and municipalities seeing their government R&D input increased, seven reduced and three unchanged. Beijing and Tianjin had the biggest rise and both improved such share by 0.16 percentage point while Anhui and Qinghai saw the most significant decline and both reduced such share by 0.02 percentage point.

In terms of business R&D input intensity, in 2015, there were 25 provinces and municipalities seeing their corporate R&D input increased, 5 reduced, and 1 unchanged. The biggest rise shows Zhejiang improving its share by 0.16 percentage points, while Qinghai saw the most significant decline by 0.11.

4.3 The Allocation of Government R&D Funding

The allocation of government R&D funding by region in 2015 shows the following main characteristics (Table 4.5 and Figure 4.3): Government R&D funding in most regions follows the order of research institutes, universities, and enterprises. The vast majority of R&D resources were allocated to universities and scientific research

Table 4.4 2014–2015 government and corporate R&D inputs by region

	Share of government R&D input in GDP/%	Share of total internal R&D spending of enterprises above designated size in sales revenue/%				
Region	2015	2014	Change	2015	2014	Change
Anhui	0.39	0.41	−0.02	0.82	0.77	0.05
Beijing	3.44	3.28	0.16	1.29	1.18	0.11
Fujian	0.13	0.12	0.01	0.88	0.85	0.03
Gansu	0.44	0.39	0.05	0.56	0.5	0.06
Guangdong	0.2	0.17	0.03	1.28	1.19	0.09
Guangxi	0.15	0.15	0	0.38	0.45	−0.07
Guizhou	0.15	0.14	0.01	0.46	0.47	−0.01
Hainan	0.15	0.16	−0.01	0.67	0.63	0.04
Hebei	0.18	0.15	0.03	0.63	0.55	0.08
Henan	0.13	0.13	0	0.5	0.5	0
Heilongjiang	0.44	0.42	0.02	0.75	0.71	0.04
Hubei	0.35	0.36	−0.01	0.94	0.88	0.06
Hunan	0.18	0.19	−0.01	1	0.93	0.07
Jilin	0.36	0.33	0.03	0.39	0.34	0.05
Jiangsu	0.22	0.2	0.02	1.02	0.97	0.05
Jiangxi	0.16	0.17	−0.01	0.45	0.41	0.04
Liaoning	0.36	0.32	0.04	0.73	0.66	0.07
Inner Mongolia	0.09	0.08	0.01	0.63	0.55	0.08
Ningxia	0.19	0.2	−0.01	0.58	0.53	0.05
Qinghai	0.16	0.18	−0.02	0.3	0.41	−0.11
Shandong	0.18	0.17	0.01	0.89	0.82	0.07
Shanxi	0.19	0.16	0.03	0.69	0.7	−0.01
Shaanxi	1.22	1.13	0.09	0.88	0.82	0.06
Shanghai	1.36	1.24	0.12	1.39	1.27	0.12
Sichuan	0.77	0.68	0.09	0.58	0.51	0.07
Tianjin	0.63	0.47	0.16	1.26	1.14	0.12
Tibet	0.26	0.22	0.04	0.19	0.25	−0.06
Xinjiang	0.15	0.13	0.02	0.45	0.38	0.07
Yunnan	0.28	0.19	0.09	0.63	0.5	0.13
Zhejiang	0.18	0.18	0	1.35	1.19	0.16
Chongqing	0.23	0.16	0.07	0.96	0.89	0.07

Source: China Statistical Yearbook 2016. Beijing: China Statistics Press

institutes; especially scientific research institutes have a leading position in obtaining government R&D resources, in spite of the corporate transformation for application-type institutes. This is rooted in the tradition of planning economy, that pubic institute, not the university, plays a more important role in S&T.

Table 4.5 Allocation of government R&D funding by region in 2015

Region	Government R&D input/ in RMB100 million	Share (%)				Universities and scientific research institutes in total
		Large industrial enterprises	Universities	Scientific research institutes	Others	
Anhui	86.42	23.46	45.35	22.93	8.27	68.28
Beijing	791.64	3.09	78.28	14.3	4.33	92.58
Fujian	33.99	25.24	34.62	31.79	8.34	66.41
Gansu	29.76	5.51	76.43	14.11	3.95	90.54
Guangdong	145.85	24.25	29.23	31.92	14.6	61.15
Guangxi	24.97	12.82	42.4	34.76	10.02	77.16
Guizhou	16.03	32.35	32.13	30.51	5.01	62.64
Hainan	5.39	10.51	57.03	26.63	5.83	83.66
Hebei	53.61	17.12	64.69	14.8	3.38	79.49
Henan	48.33	24.02	47.57	24.59	3.82	72.16
Heilongjiang	66.55	29.83	33.63	34.82	1.72	68.45
Hubei	102.39	13.9	48.09	34.96	3.05	83.05
Hunan	50.89	36.66	26	33.11	4.22	59.11
Jilin	51.2	6.4	53.9	36.28	3.43	90.18
Jiangsu	153.34	16.28	44.69	33.77	5.26	78.46
Jiangxi	25.99	22.92	42.71	27.37	7.01	70.08
Liaoning	103.98	23.01	53.05	22.22	1.72	75.27
Inner Mongolia	16.1	20.97	51.07	15.05	12.91	66.12
Ningxia	5.47	22.12	28.49	28.7	20.69	57.19
Qinghai	3.93	11.81	51.25	23.38	13.56	74.63
Shandong	111.02	29.15	37.52	22.06	11.27	59.58
Shanxi	24.29	13.85	53.01	29.99	3.15	83
Shaanxi	220.22	18.58	69.69	10.88	0.85	80.57
Shanghai	340.8	8.62	69.95	16.24	5.19	86.19
Sichuan	230.22	9.48	79.55	9.73	1.25	89.28
Tianjin	104.76	23.35	35.14	34.6	6.91	69.74
Tibet	2.67	0.08	52.33	36.56	11.03	88.89
Xinjiang	13.82	8.09	56.66	25.02	10.24	81.68
Yunnan	37.83	16.87	50.81	20.78	11.54	71.59
Zhejiang	75.29	20.45	26.68	45.26	7.61	71.94
Chongqing	36.45	21.87	44.66	28.58	4.88	73.24

Source: China Statistical Yearbook on Science and Technology 2016. Beijing: China Statistics Press

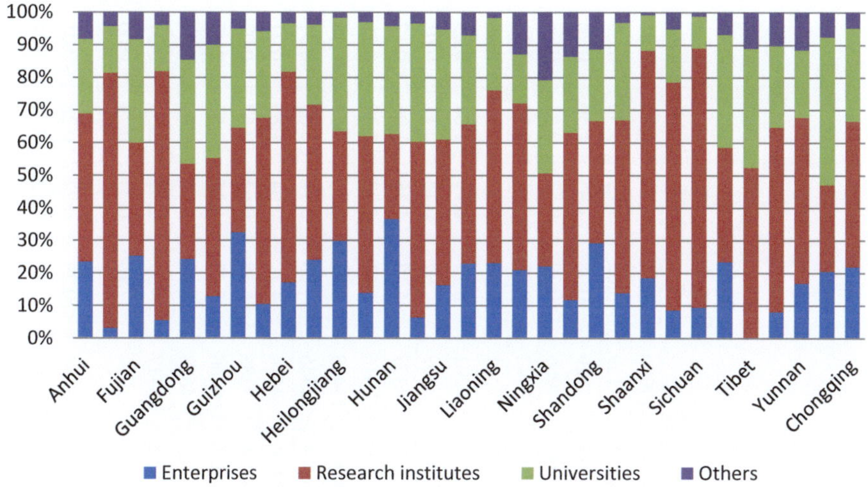

Fig. 4.3 The allocation of government R&D funding by region in 2015 (%). (*Source*: China Statistical Yearbook on Science and Technology 2016. Beijing: China Statistics Press)

Research institutes and universities in 11 regions received more than 80% of government R&D funding input, especially Beijing, Gansu, and Jilin with over 90% of government R&D funding input, standing at 92.58%, 90.54%, and 90.18%, respectively. Sichuan spends their 79.55% of government R&D funding just in research institutes.

In general, the structure of government R&D funding is in line with international development trends, the government funds mainly used for supporting basic R&D activities with greater externality, rather than directly applied to enterprises' development of competitive products. However, in some regions such as Beijing and Sichuan, enterprises are really short of government R&D funding. Currently, China is limited in its means to support corporate innovation capability. But the policy orientation is to establish an innovation system with enterprises as main actors. Therefore, future R&D input may be tilted toward enterprises, and the institutional reform of institutes and the changes in utilization of government R&D funding are also the important policy issues for achieving the innovation-driven development. Certainly, the structure of government R&D input should not only adapt to the level of economic development, but also the characteristics of industrial structure and advantages of resource endowments of the region, so as to maximize the efficiency of capital input.

Table 4.6 2014–2015 patent applications by region

	Three types of patents	Invention patents										
	Patents (nr.)	Share of the national total/%	Patents	Share of the national total/%								
Region	2015	2014	Growth rate%	2015	2014	Change	2015	2014	Growth rate%	2015	2014	Change
Anhui	127709	99160	28.79	4.88	4.54	0.34	68314	49960	36.74	7.14	6.33	0.81
Beijing	156312	138111	13.18	5.97	6.32	−0.35	88930	78129	13.82	9.29	9.89	−0.60
Fujian	83146	58075	43.17	3.18	2.66	0.52	17663	12529	40.98	1.85	1.59	0.26
Gansu	14584	12020	21.33	0.56	0.55	0.01	5504	4986	10.39	0.57	0.63	−0.06
Guangdong	355939	278358	27.87	13.60	12.73	0.87	103941	75147	38.32	10.86	9.52	1.34
Guangxi	43696	32298	35.29	1.67	1.48	0.19	30815	22237	38.58	3.22	2.82	0.40
Guizhou	18295	22467	−18.57	0.70	1.03	−0.33	7538	8203	−8.11	0.79	1.04	−0.25
Hainan	3127	2416	29.43	0.12	0.11	0.01	1211	969	24.97	0.13	0.12	0.01
Hebei	44060	30000	46.87	1.68	1.37	0.31	11259	8332	35.13	1.18	1.06	0.12
Henan	74373	62434	19.12	2.84	2.86	−0.02	21338	19646	8.61	2.23	2.49	−0.26
Heilongjiang	34611	31856	8.65	1.32	1.46	−0.14	14663	13468	8.87	1.53	1.71	−0.18
Hubei	74240	59050	25.72	2.84	2.7	0.14	30204	22536	34.03	3.16	2.85	0.31
Hunan	54501	44194	23.32	2.08	2.02	0.06	19499	14474	34.72	2.04	1.83	0.21
Jilin	14800	11933	24.03	0.57	0.55	0.02	6154	5288	16.38	0.64	0.67	−0.03
Jiangsu	428337	421907	1.52	16.37	19.3	−2.93	154608	146660	5.42	16.15	18.57	−2.42
Jiangxi	36936	25594	44.32	1.41	1.17	0.24	5722	4688	22.06	0.60	0.59	0.01
Liaoning	42153	37860	11.34	1.61	1.73	−0.12	19332	18417	4.97	2.02	2.33	−0.31
Inner Mongolia	8876	6359	39.58	0.34	0.29	0.05	2254	1924	17.15	0.24	0.24	0.00
Ningxia	4394	3532	24.41	0.17	0.16	0.01	2626	2183	20.29	0.27	0.28	−0.01
Qinghai	2590	1534	68.84	0.10	0.07	0.03	1103	660	67.12	0.12	0.08	0.04

Shandong	193220	158619	21.81	7.38	7.25	0.13	93475	77298	20.93	9.76	9.79	−0.03
Shanxi	14948	15687	−4.71	0.57	0.72	−0.15	5680	6107	−6.99	0.59	0.77	−0.18
Shaanxi	74904	56235	33.20	2.86	2.57	0.29	17322	24399	−29.01	1.81	3.09	−1.28
Shanghai	100006	81664	22.46	3.82	3.73	0.09	46976	39133	20.04	4.91	4.96	−0.05
Sichuan	110746	91167	21.48	4.23	4.17	0.06	40437	29926	35.12	4.22	3.79	0.43
Tianjin	79963	63422	26.08	3.06	2.9	0.16	28510	23391	21.88	2.98	2.96	0.02
Tibet	309	248	24.60	0.01	0.01	0.00	128	92	39.13	0.01	0.01	0.00
Xinjiang	12250	10210	19.98	0.47	0.47	0.00	3024	2360	28.14	0.32	0.3	0.02
Yunnan	17603	13343	31.93	0.67	0.61	0.06	6301	4732	33.16	0.66	0.6	0.06
Zhejiang	307264	261435	17.53	11.74	11.96	−0.22	67674	52406	29.13	7.07	6.64	0.43
Chongqing	82791	55298	49.72	3.16	2.53	0.63	35086	19418	80.69	3.67	2.46	1.21

Source: China Statistical Yearbook on Science and Technology 2016. China Statistical Yearbook 2016. Beijing: China Statistics Press

4.4 Innovation Capability Based on Patent Application

Generally speaking, regions with leading innovation capability also have a leading advantage in the number of patent applications. As shown in Table 4.6, patent applications in Jiangsu reached 428,337 in 2015, including 154,608 invention patents, or 16.37% and 16.15% of the patent applications nationwide, respectively, far more than other provinces. The total patent applications of Jiangsu, Guangdong, Zhejiang, Shandong, and Beijing accounted for 55.07% of the national total, including the invention patent applications for 53.13% of the national total.

In terms of the growth rate in patent applications, Qinghai showed the highest one of 68.84%; while Guizhou and Shanxi experienced a significant decline of 18.57% and 4.71%, respectively, compared with 2014. In contrast, Guangdong showed the largest increase in the proportion of total patent applications in the national total in 2015 (0.87 percentage points), ranking the top for 2 consecutive years in terms of growth, while Jiangsu recorded the largest decline (2.93 percentage points), dropping for 3 consecutive years. Guangdong also recorded the largest increase in the proportion of invention patent applications in the national total, up 1.34 percentage points, while Jiangsu saw the largest decline, down 2.42 percentage points.

The figures presented in Table 4.7 illustrate another important feature of China's patent activities, that is, in most regions, patent applications by large industrial enterprises (whether in terms of invention patents or the sum of three kinds of patents) occupy a lower share in the total patent applications of all regions. In 2015, China's applications of three kinds of patents of industrial enterprises above the designated size accounted for 24.4% of the national total, down 4.44 percentage points over the previous year; the invention patent applications by the enterprises accounted for 25.66% of the national invention patent applications, a decline of 4.72 percentage points over the previous year.

Regions with the highest proportion of enterprise applications in all three kinds of patents were Anhui (35.7%), Hunan (33.35%), Ningxia (32.52%), Fujian (29.97%), and Guangdong (29.79%), respectively; provinces with the highest proportion of invention patents of large industrial enterprises in its invention patents were Guangdong (49.71%), the Inner Mongolia Autonomous Region (45.74%), Jiangxi (44.08%), Fujian (38.95%), and Hunan (38.93%), respectively. In Guangdong, the proportion of enterprise invention patents is significantly higher than other regions, and the position of enterprise in innovation is prominent.

Table 4.7 Patent applications by large industrial enterprises by region in 2015

	All	Industrial enterprises above the designated size				
	Quantity/nr.	Quantity/nr.	Share in the provincial total/%			
Region	Three types of patents	Invention	Three types of patents	Invention	Three types of patents	Invention
Guangdong	355939	103941	106038	51672	29.79%	49.71%
Inner Mongolia	8876	2254	2585	1031	29.12%	45.74%
Jiangxi	36936	5722	8561	2522	23.18%	44.08%
Fujian	83146	17663	24916	6880	29.97%	38.95%
Hunan	54501	19499	18175	7591	33.35%	38.93%
Hebei	44060	11259	10396	3393	23.60%	30.14%
Anhui	127709	68314	45598	19967	35.70%	29.23%
Ningxia	4394	2626	1429	761	32.52%	28.98%
Jiangsu	428337	154608	119927	41744	28.00%	27.00%
Xinjiang	12250	3024	2340	788	19.10%	26.06%
Guizhou	18295	7538	3782	1953	20.67%	25.91%
Hainan	3127	1211	441	312	14.10%	25.76%
Zhejiang	307264	67674	80512	17242	26.20%	25.48%
Henan	74373	21338	16518	5250	22.21%	24.60%
Hubei	74240	30204	17315	7227	23.32%	23.93%
Yunnan	17603	6301	3751	1493	21.31%	23.69%
Shanxi	14948	5680	3569	1303	23.88%	22.94%
Shanghai	100006	46976	21725	10740	21.72%	22.86%
Tianjin	79963	28510	16721	6507	20.91%	22.82%
Liaoning	42153	19332	9190	4131	21.80%	21.37%
Shandong	193220	93475	42289	19621	21.89%	20.99%
Sichuan	110746	40437	21912	8085	19.79%	19.99%
Chongqing	82791	35086	20239	6758	24.45%	19.26%
Shaanxi	74904	17322	7521	3036	10.04%	17.53%
Qinghai	2590	1103	305	144	11.78%	13.06%
Jilin	14800	6154	1972	787	13.32%	12.79%
Gansu	14584	5504	2230	698	15.29%	12.68%
Heilongjiang	34611	14663	3902	1752	11.27%	11.95%
Beijing	156312	88930	20024	10281	12.81%	11.56%
Tibet	309	128	17	14	5.50%	10.94%
Guangxi	43696	30815	4613	2005	10.56%	6.51%

Source: China Statistical Yearbook on Science and Technology 2016. China Statistical Yearbook 2016. Beijing: China Statistics Press

Part II
The Pattern of Regional Innovation in China

Chapter 5
Guangdong: From OEM to ODM and More

5.1 Ranking and Profile of Innovation in Guangdong

In 2017, the innovation capacity of Guangdong province ranked 1st in China and has surpassed Jiangsu Province for the first time in nearly 10 years. Given the rapid development of Guangdong along different aspects in recent years, in particular economically, this comes not that surprising. For instance, Guangdong has intensified its efforts on R&D input, technology transfer, and innovation platform cultivation which resulted in double growth of some indicators. It is now ranked 1st in China by various indicators such as "value added of tertiary industry," "revenue of high-tech industry," and "exports of high-tech products." Its outstanding innovation capability is demonstrated by its higher level of open innovation, high intensity of foreign trade, better market dynamics, active entrepreneurial and start-up activities, and better innovation ecosystem. In particular, innovation-driven enterprises make cities like Shenzhen, located in the south of Guangdong, prosperous, and they are becoming a significant driving force for economic transformation and upgrading of industry in Guangdong, but even in the wider China.

In terms of sub-index of regional innovation, knowledge creation and knowledge acquisition both rank 4th; enterprise innovation, innovation environment, and innovation performance rank 1st. Among them, the enterprise innovation and innovation environment have moved one and two places ahead, respectively, as compared with the previous year (Table 5.1, Fig. 5.1).

In Guangdong, a number of meanwhile globally recognized and well established firms are located, mainly belonging to the ICT sector, such as Huawei or Tencent. According to "2016 China Top 500 Enterprises list," issued by China Enterprise Confederation, there are 48 enterprises in Guangdong on the list, which are mainly private enterprises, such as Huawei, Media, TCL, Tencent, BYD, etc. (Table 5.2).

Among the top 20 enterprises, there are 15 private and 5 state-owned enterprises, and according to the "List of Top 100 Guangdong innovation enterprises," 86 of 100 enterprises are private ones and 14 are state-owned enterprises. Private

X. Liu et al., *Regional Innovation Index of China: 2017*,
https://doi.org/10.1007/978-981-13-1205-2_5

Table 5.1 Guangdong innovation capability comprehensive value

	2017 comprehensive value	2017 sub-index rank			
Name of indexes	Index value	Rank	Strength	Efficiency	Potential
Comprehensive value	55.24	1	1	4	18
1 Knowledge creation	39.07	4	3	19	13
1.1 R&D input	33.45	5	3	7	19
1.2 Patent	53.06	4	3	15	3
1.3 Research paper	22.35	15	4	31	19
2 Knowledge acquisition	38.81	4	2	12	16
2.1 Science and technology cooperation	25.78	14	8	23	10
2.2 Technology transfer	43.76	2	1	13	15
2.3 Foreign-funded enterprise investment	44.87	4	3	5	19
3 Enterprise innovation	66.22	1	1	3	15
3.1 Enterprise R&D Input	72.29	2	2	6	20
3.2 Design capability	69.99	1	1	2	14
3.3 Technology improvement	50.37	3	2	16	12
3.4 New product sales income	67.32	2	2	6	9
4 Innovation environment	54.98	1	1	3	20
4.1 Innovation infrastructure	51.27	1	2	3	23
4.2 Market environment	65.39	3	1	3	3
4.3 Employee quality	47.66	2	1	28	19
4.4 Financial environment	46.16	3	2	7	20
4.5 Entrepreneurship	64.44	1	1	3	17
5 Innovation performance	66.28	1	1	5	19
5.1 Macroeconomic	72.69	2	1	8	18
5.2 Industry structure	73.46	1	1	1	22
5.3 Industrial international competitiveness	62.09	1	1	10	13
5.4 Employment	68.99	1	1	1	22
5.5 Sustainable development and environment	54.14	25	30	19	14

enterprises have become the main driver of patenting in Guangdong; the state-owned enterprises should not be ignored. Four of which are in the top ten, including ZTE, GREE, China Southern Power Grid, and China Guangdong Nuclear Power Group (Table 5.3).

ICT industry is the most developed one in Guangdong. The top 100 patenting enterprises are mainly associated with the industries of computer, communication,

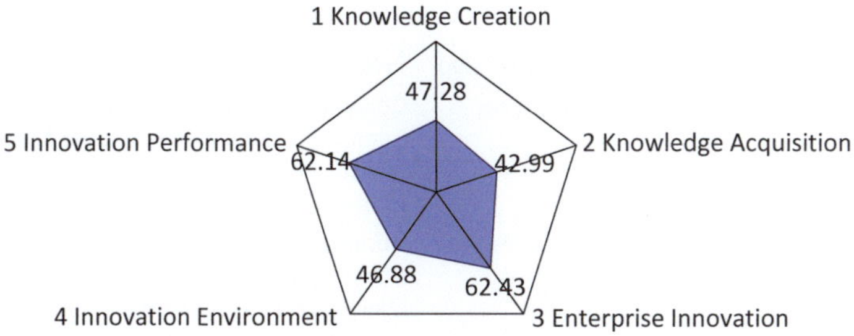

Fig. 5.1 Guangdong innovation capability spider chart

Table 5.2 Guangdong enterprises in the "China Top 500 Enterprises list 2016" (Top 20)

	Enterprise	Operation revenue (billion yuan)	Rank
1	Huawei Technologies	395.01	27
2	Guangzhou Automobile Group	216.44	64
3	Media Group	138.44	110
4	Hengda Real Estate Group	133.13	113
5	TCL Group	104.58	136
6	Tencent Technology	102.86	140
7	ZTE Corporation	100.19	150
8	GREE Electric Appliances	97.75	154
9	Guangsha Holding Group	89.71	162
10	BYD	80.01	175
11	Guangzhou Pharmaceutical Group	74.86	187
12	Teng Bang Group	65.48	206
13	Guangxin Holdings Group	64.25	210
14	Guangwu Holdings Group	63.59	214
15	OCT Group	50.69	260
16	China Guangdong Nuclear Group	50.60	261
17	Shenzhen Ai Shi De	49.57	270
18	Shenzhen Dasheng Agricultural Group	49.11	272
19	Guangdong Wens Foodstuffs Group	48.24	278
20	Shenzhen Feima International Supply Chain	47.64	281

Source: China Enterprise Confederation 2017

and other electronic equipment manufacturing, electrical machinery and equipment manufacturing, and software and information technology services.

The leading companies in Guangdong all put a strong emphasis on innovation activities, as shown by their great number of patents. In the ranking of China's top ten companies on patent application in 2016, five high-tech companies located in Guangdong, Huawei (4906), ZTE (3941), Oppo (3778), GREE (3299), and Nubia (2921), are listed, which hold 50% of total patents (see Fig. 5.2).

Table 5.3 List of Top 100 Guangdong patent innovation enterprises (top 20)

Rank	Enterprise	Ownership	Industry
1	Huawei Technologies	Private	Computer, communications, and other electronic equipment manufacturing/Internet and related services/software and information technology services
2	ZTE Corporation	State-owned	Computer, communications, and other electronic equipment manufacturing/Internet and related services/software and information technology services
3	Midea Group	Private	Electrical machinery and equipment manufacturing
4	Gree Electric Appliances	State-owned	Electrical machinery and equipment manufacturing
5	TCL Group	Private	Electrical machinery and equipment manufacturing/computer, communications and other electronic equipment manufacturing
6	Tencent Technology	Private	Internet and related services/software and information technology services
7	BYD	Private	Automobile manufacturing/electrical machinery and equipment manufacturing
8	Hongfujin Precision Industry	Private	Special equipment manufacturing/computer, communications and other electronic equipment manufacturing
9	China Southern Power Grid	State-owned	Electricity, heat production and supply
10	China Guangdong Nuclear Group	State-owned	Electricity, heat production and supply
11	Guangdong Oppo Mobile Communications	Private	Computer, communications and other electronic equipment manufacturing
12	Shenzhen Mindray Bio-Medical Electronics	Private	Special equipment manufacturing
13	Yulong Computer Telecommunication Technology	Private	Computer, communications, and other electronic equipment manufacturing
14	Comba Telecom Systems	Private	Special equipment manufacturing/computer, communications, and other electronic equipment manufacturing
15	Kingfa Science & Technology	Private	Rubber and plastic products
16	Ocean's King Lighting Science & Technology	Private	Electrical machinery and equipment manufacturing
17	Skyworth Group	Private	Electrical machinery and equipment manufacturing/computer, communications, and other electronic equipment manufacturing
18	Nationz Technologies	Private	Internet and related services/software and information technology services

(continued)

Table 5.3 (continued)

Rank	Enterprise	Ownership	Industry
19	GRG Banking Equipment	State-owned	Special equipment manufacturing/software and information technology services
20	Futaihua Industrial (Shenzhen)	Private	Special equipment manufacturing/computer, communications, and other electronic equipment manufacturing

Source: Guangdong Intellectual Property Office 2017

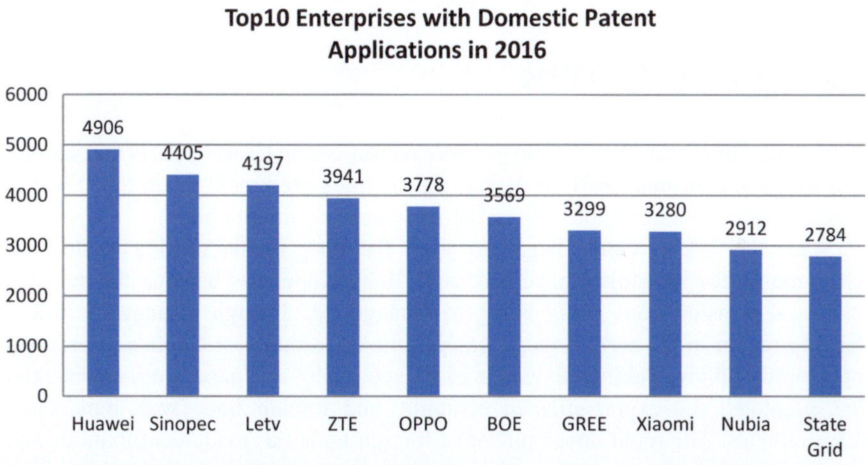

Fig. 5.2 Top 10 domestic invention patent applications in 2016. (*Source*: SIPO 2017)

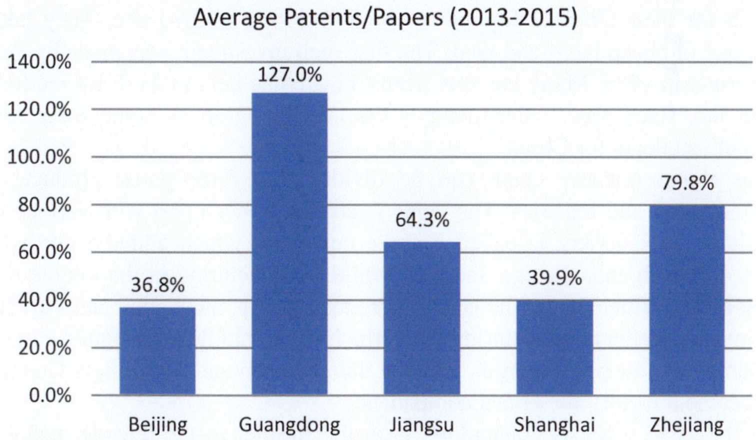

Fig. 5.3 Average patents/papers 2013–2015. (*Source*: NBS 2014–2016)

Figure 5.3 shows the average quantity of patents/papers of 3 years, and Guangdong is the only one region that owns more patents than scientific papers, pointing to the specific strength in terms of innovation, rather than pure knowledge creation.

Guangdong is the region in China that has most developed toward the principle of open innovation, probably related to the traditional openness of the region, e.g., in terms of international trade. Since the 1980s, its manufacturing industry was mostly OEM to assist foreign companies in the simple production sector as the result of FDI.

Guangdong is also a market-oriented innovation mode with massive Internet-based and customer-oriented innovation companies. The rapid development of network media and smartphone industries led to the establishment of a lot of well-known companies, such as Huawei, Oppo, and Vivo.

5.2 From OEM to ODM

In the past, Guangzhou got limited government support in terms of R&D investment. But as the region next to Hong Kong, the central government positioned Guangdong, especially Guangzhou city, as the main hub of foreign trade.

Since the 1980s, China began to open the door to the outside world, and Guangdong, for its geographical link with Hong Kong, has become the base for industrial transfer from Hong Kong to Guangdong. Though Guangzhou was a lagging region in the early stage, it grasped the opportunity of China's reform and opening-up policy. Taking advantage of its geography and popularity, Guangzhou has promoted the export-oriented economy and foreign trade with remarkable achievements. The rapid development of foreign trade has made contributions and provided significant support to the provincial economy and modernization of Guangdong.

In the 1980s, many multinational firms settled in Guangdong, using the region as the hub for their Original Equipment Manufacturing (OEM) site. They take the advantage of cheap labor and land. The first such investment was made by P&G in Guangzhou in 1988. Many laborers from China come here to work for the foreign-related job. Even now, Guangdong is the largest region as home for graduated university students for China.

The global industry chain can be divided into three parts: product R&D, manufacturing, and logistics. The value is created in each part with various factor intensities. This variety is called a smile curve, on which added value is more reflected in both ends, design and sale, while manufacturing in the middle has the lowest added value. Due to the lack of core technology, most companies in Guangdong worked in the manufacturing part, which is the middle of the smile curve with comparatively low profit margin. Without indigenous brand and design, Guangzhou has become a factory for global companies.

At the same time, as Guangdong is in the frontier to the outside, many local business people would know earlier what products the markets prefer. They started to imitate those foreign brands; this also made Guangdong as the Shanzhai headquarter of China.

The learning-by-exporting effect means that export trade can improve the productivity level of companies. After entering the international market, various demands and fierce competitions force the companies have more opportunities to learn advanced foreign technology and management, which promoted company's learning and innovation capacity (Aw et al. 2000); Meanwhile, companies would face more intense competition in the international market than in the domestic market. Under these circumstances, they would improve the productivity faster than those without export business. Many empirical cases of developing economies support this learning-by-exporting hypothesis, which means developing countries can improve their productivity by participating in international competitions (Aw et al. 2000; Alvarez and López 2005; De Loecker 2007). Therefore, learning-by-exporting effects are often used as one of key ways for developing countries to adopt advanced technology from developed countries, as well as to improve the indigenous innovation capability. During this process, it gained a lot of experiences. This is quite helpful for the companies in the early stage of China's reform to understand international market, management, and product design. Compared with those companies without international opportunity, the companies with foreign cooperation experiences have irreplaceable advantages.

However, with the development of society and economy, Guangzhou is now not limited by its foreign trade. It has turned its attention to technology, innovation, and entrepreneurship and become an important regional innovation center in China through successful high-tech transformation. To improve its market competitiveness, especially in international markets, Guangdong has to establish and promote its own brands, strengthen the brand building, and improve the economic power.

FDI in Guangdong also helps a lot for the transformation of OEM to ODM. Guangdong owns the largest foreign capital in China. In 2015, it absorbed 26.875 billion US dollars of foreign investments. Given the intensive FDI, it gives Guangdong the great opportunity to absorb technology in advanced manufacturing and high-end service industry, as well as form a new pattern of international cooperation; it will further improve the productivity of Guangdong.

After the catching-up process by relying on OEM and absorbing technology, Guangdong began to ground its development more on talented people. The talent is the possible creator of new knowledge, builder of new theory, inventor of new technology, and explorer of new industry. It is the decisive factor in the competition among companies, cities, and even countries. So, Guangdong needs to train and introduce innovative talents using industry-university-research cooperation. For example, Guangdong can encourage the company with indigenous brand to take the lead and form the industry-university-research mode with provincial university, such as South China University of Technology and Sun Yat-sen University. By playing their respective advantages, the industrialization of research outcomes could benefit S&T research in return.

At the same time, to improve the international competitiveness of its indigenous brands, Guangzhou also needs to promote the transition of the low-end brands to the middle- or high-end brands besides the R&D. At July 2, 2015, Ant Financial Services Group released the very first report "Employment Flow of University Students" based on Internet information. The report points out that the number of students flowing into Guangdong ranks the first in China. According to "2015

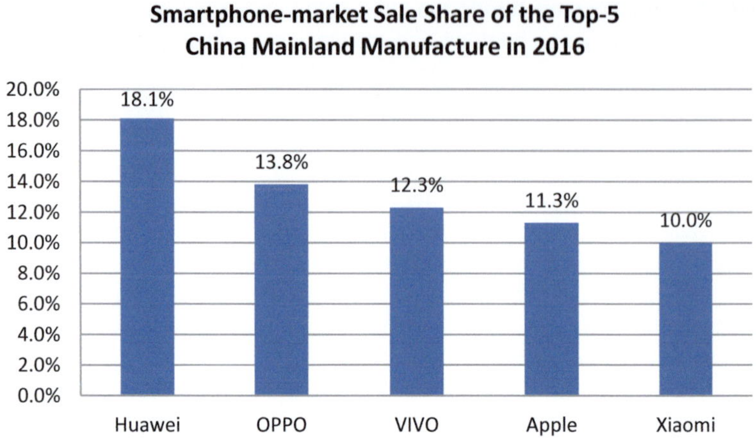

Fig. 5.4 Sales share of top five smartphone markets in Mainland China in 2016. (*Source*: GfK 2017)

national statistic of foreign students," Guangdong attracts 23,015 foreign students, which is 8.57% increase compared with 21,198 in 2014. High-quality and fair talent development environment could help Guangdong to introduce and gather domestic and international talents and refine talent team and structure, which could provide solid support of human resource for S&T development.

To break the brand and core-tech monopoly of foreign capital, as well as to achieve the added value in the high-end industrial chain, lots of domestic companies have taken a new track: to invest R&D and build its own brand. There are two alternatives for indigenous brands to develop in the future. One is Japan/Germany's route, which is refined high-quality manufacturing. However, this route may not fit Chinese culture, which emphasizes the speed more than excellent manufacturing. The other one is intelligent manufacturing route, which is more suitable for China. This is also a leapfrogging strategy for manufacturing enterprises: R&D for new design and new manufacturing.

We can see the transformation of from OEM to ODM in some products. The typical example industry is smartphone. Shenzhen used to be the center for copy product of new and smartphone, but now, just take Dongguan city as an example. Dongguan holds an important position in the global manufacturing industry. Unlike before, Dongguan now produces smartphones with indigenous brands. Every six smartphones produced worldwide has been produced in Dongguan. Figure 5.4 shows the market share of the top five smartphone companies in mainland China in 2016.

As an indicator of ODM, in 2016, 9 companies in Guangdong have been among the world's top 500 companies, 3 more than the previous year. These nine companies are all in the core area of the Pearl River Delta, of which five are located in Shenzhen, three in Guangzhou, and one in Foshan. Large local companies of Guangdong have consolidated their position in the world's top 500. China Ping, an insurance group, is ranked 55 ahead than before, which is its first time to be listed as world top 500. Furthermore, there are 48 companies in Guangzhou among China's top 500. It is 9.6% of the total number. Up to 2016, nearly one quarter of Guangzhou's top 500 companies reached size of above 10 billion yuan; 16 companies are with more than size of 100 billion yuan. From the industry distribution, 52% of these large-scale companies are from manufacturing industry, 40% are from service industry, 7% are from circulation, and 1% are integrated.

So far, Guangdong has already started its innovation development transition: from low-end manufacturing to indigenous brand, from low cost to high tech, and from labor intensive to talent intensive.

Galanz Group: From OEM to ODM

Galanz Group has been founded at September 28, 1978. From the textile industry to microwave oven industry, and then to current leading white goods company, Galanz has always been a famous brand in the international market representing the China made.

At the beginning of its establishment, Galanz was a small factory in Shunde, Guangdong, to produce eiderdown products. In 1992, Galanz dropped its long-term profit-making eiderdown business and entered the microwave oven manufacturing industry. At that time, the microwave oven technology in China was in the mature period. Taking advantage of the technical export by multinational corporations, Galanz formed its initial technology base. The company even achieved the advanced level of Europe and America through assimilation, learning, improvement, integration, and self-development.

By the end of 2016, Galanz has applied 2208 domestic and international patents in total and established its business institutions in over 10 countries and regions. By the end of May 2017, Galanz has registered its trademark in 143 countries and regions globally. Its products and services have extended to nearly 200 countries and regions.

The indigenous innovation and internationalization of the Galanz Group can be summarized into three phases (Wang et al. 2008*):*

(1) *Introduce technology, and rapidly expand the domestic market.*

In 1992, according to the investigation, Galanz decided to drop the profitable eiderdown business and entered the microwave oven manufacturing industry. In the same year, the company invested 4 million USD to introduce the most advanced microwave oven production line and related technology from Japan, as well as senior technical managers. Based on low labor costs, it was quickly turned into high-efficient production capacity. In 1995, Galanz ranked No.1 nationally, with 25% share of domestic microwave oven market. In this phase, the priority for Galanz is the domestic market. It used OBM mode since the beginning to expand and rapidly occupied the strong position nationally. However, in this phase, its technology relied on the introduction, so the technical base was weak.

(2) *Assimilation and internationalization based on OEM.*

In 1997, Galanz already owned 47.6% of the domestic market. Validated by the ISO9001, German GS, the EU's CE, the USA's UL, Denmark's DEMKI, Norway's NEMKO, and many other international quality certifications, Galanz proposed the "world factory" strategy and started its internationalization based on the OEM. In 1997, the company worked with a French company Fillony and got the big order of 100,000 microwave ovens in 1998. Galanz also adopted a lot of technology from its French customer (Ng and Li 2010*). Since then, Galanz began to expand in the international market. Taking advantage of the antidumping sanctions against LG and other Korean brands in that period, Galanz quickly entered the European market, broke the monopoly of Korea and Japan, and set up its European branch. In 1999, the company became the largest microwave oven production base in the world. Based on the British, French, and German markets, Galanz gradually expanded to the African, Latin American, and North American markets.*

(3) *Technical breakthrough and internationalization by OBM and ODM.*

By the end of 2000, Galanz already invested 200 million yuan in R&D. Two hundred proprietary technologies were developed. In 2000, it proposed the strategy to change from "world factory" to "global famous home appliance manufacturing center." This strategy aimed at weakening the "low added value" impression brought by the "factory" and strengthening the new image by emphasizing "Galanz made." In this phase, all the exported high-end microwave oven design of Galanz used its own proprietary technologies and finished the transformation from OEM to ODM. The company's value chain was effectively enhanced. In July 2001, The Galanz American Home Appliance R&D center successfully developed the world's first digital oven combining light wave and microwave. This was the first proprietary technology of the Galanz Group with independent intellectual property rights and brought the entire microwave oven industry into the era of digital light wave.

5.3 From Immigration to an Entrepreneurial Region

Shenzhen city is the hub of innovation of Guangdong. But Shenzhen is a city of immigration. Many young people from different places come here to realize their dreams. Like Silicon Valley and other areas, the immigration culture is good for entrepreneurship: people here do now know each other, and there is no hierarchy, no bureaucracy, and no regulation.

So, Shenzhen is a city of entrepreneurship.

The booming of entrepreneurship spirit in Guangdong can be seen from the GEM (Growth Enterprise Market), also known as the second board market or the second stock market, which is a kind of security market differing from the main board. GEM is designed to provide financing methods and growth space for those entrepreneurial companies, small-medium firms, and high-tech companies that cannot be listed on the main board temporarily. The number of companies on GEM could reflect the entrepreneurial atmosphere of a certain region. According to the statistics of GEM in

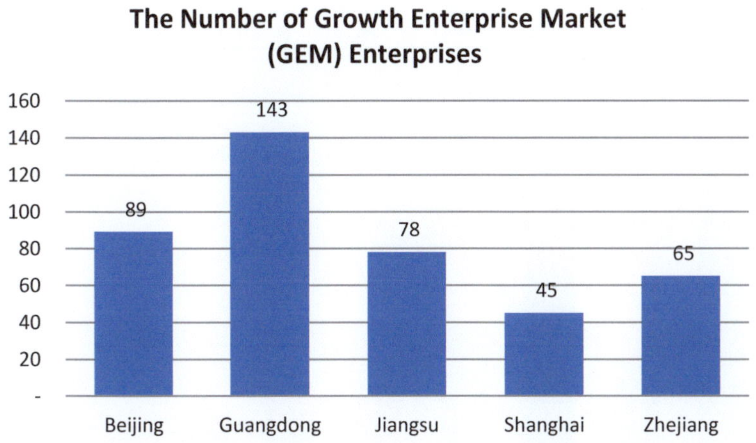

Fig. 5.5 Company quantity on GEM of six provinces in 2016. (*Source*: WIND)

2016 by WIND, a big databank of China, Guangdong is ranked 1st nationwide with 143 companies listed in GEM. It demonstrates the strong entrepreneurial environment of Guangdong as shown in Fig. 5.5.

SMEs are an important element of Guangdong's economy in many areas, such as GDP and employment. The output value of private SMEs created in 2005 is 39.6% of the total provincial GDP, about 860.2 billion yuan. So far, Guangdong has 760,000 SMEs in total with the largest quantity and output value in the whole country. Guangdong should keep its practice to attach importance to the various innovative SMEs. In 2016, 365 group innovation spaces are accounted for in Guangdong, among which 74 are brought into the national incubator management system. It is ranked 1st countrywide. Figure 5.6 shows the distribution of technology incubators in the Pearl River Delta.

The entrepreneurial environment is quite comprehensive in Guangdong, and the gene of entrepreneurship is continuous. Outstanding entrepreneurs of different ages and industries continue to emerge, who also keep up with the rapid change of market, and realize companies' transformation. Take Yongping Duan, the CEO of OPPO, as an example. In March 1989, Mr. Duan began his director career in a small factory under Yihua Group in Zhongshan city. Before that, this factory had 2 million losses. Since Mr. Duan took over, the factory started to produce home video game machine. Mr. Duan decided to create the own brand "SUBOR." After only 3 years, this small factory reached the output value of 100 million yuan and was officially named Zhongshan Subor Digital Sound. Among various game machine brands, SUBOR was always superior. Unexpectedly, when SUBOR was well developed, Mr. Duan chose to leave at September 18, 1995; Mr. Duan established BBK Electronics in Dongguan. His arrival brought BBK to higher and higher level. Now, BBK has become one of the most famous brands in China's cordless phone, VCD, and educational electronic product industries. Mr. Duan is not limited to the development of BBK brand only. He then actively enters the mobile

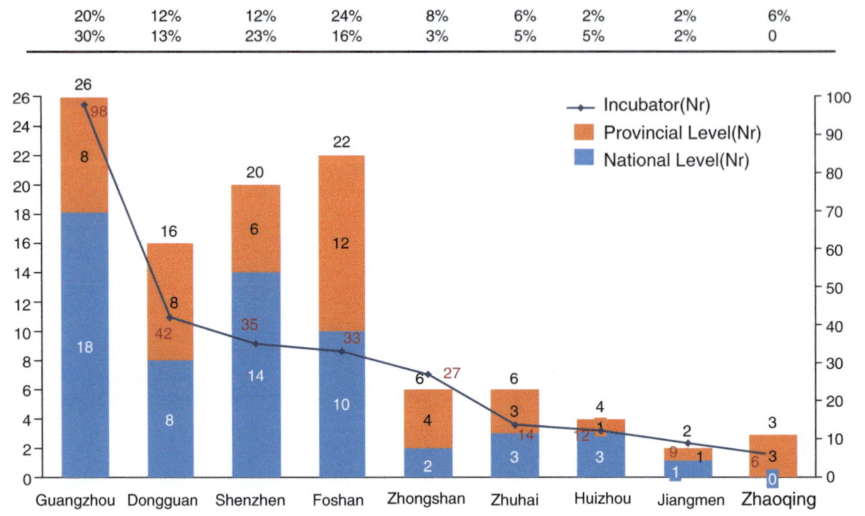

Fig. 5.6 Distribution of tech incubators in the Pearl River Delta. (*Source*: South Daily et al. 2016)

phone industry with Oppo and Vivo, two mobile phone brands. Based on the "Global Smart Phone Report" released by the IDC, the total global shipment of Oppo and Vivo in the third quarter of this year is more than Apple. They have become the giants of the global mobile phone industry.

Huawei

Huawei is the world's leading information and communication technology (ICT) solution provider. It was established in 1987, and the headquarters is located in Shenzhen, Guangdong. In 2013, for the first time, Huawei surpassed Ericsson, the world's largest telecom equipment dealer. By the end of 2016, Huawei owned 180,000 employees, provided products and solutions for over 170 countries, and served 45 global operators of the top 50 as well as one third of the world's population. In July 2017, for the first time, Huawei was named no. 83 in the Fortune Global 500 with its 78.5108 billion USD operation revenue, 46 places higher than the no. 129 in the previous year. The international sales revenue has become the main support of the company.

Huawei has experienced three transformations: from producer to solution provider, from domestic market to international market, and from operators to end consumers.

(1) *The first transformation: from producer to solution provider. In the 1990s, giant companies such as Nokia, Ericsson, Motorola, and Siemens almost monopolized the whole communication market. The only way for Huawei to succeed was by starting from the rural area and then expandeding to cities. However, what the customer really wants is not only a product but a customer solution. IBM awakened Huawei in one sentence: the solution is to use 10% of the product*

to meet 100% of the customer's needs. By implementing it, the industry-level market of Huawei was greatly expanded.

(2) *The second transformation: internationalization and globalized developing strategy. After its successful expansion from rural to urban area, the mature products of Huawei (such as PBX) already dominated the domestic market. Facing the saturated domestic market and improvement of talent pool and management system, Huawei began its second transformation in 1998: differential global competition strategy. By integrating the global resources and participating in the competition, Huawei built 26 capacity centers in the world, such as the wireless communication technology center in Sweden, the engineering technology center in Japan, and the software engineering center in India. The costs were optimized.*

(3) *The third transformation: from operators (big B) to industrial customers (small B) and then to end consumers (C), the split of three BG services. Previously, the costumers of Huawei were only at the operator level, including China Telecom, China Mobile, and so on. After the transformation, Huawei does not only run business with operators but also with many industrial customers, corporate customers, and end consumers. From the view of the ecological chain, company with both channels' and end consumers' experiences can lead to success (Qin* 2017*).*

Huawei's strategy is unusual in a developing country. Since 1992, Huawei has insisted on investing 10% of the annual sales into R&D. In 2016, its R&D personnel were around 80,000, accounting for 45% of the total employees; the R&D expenditure is 76.391 billion yuan, 14.6% of the total income; in the past 10 years, the total investment on R&D is over 313 billion yuan. By the end of December 31, 2016, Huawei obtained 62,519 patent licenses, 57,632 Chinese patents, and 39,613 foreign patents; 90% of the patents are invention patents. By the end of December 31, 2016, Huawei joined more than 360 standard organizations/industrial alliances/open-source communities and served as more than 300 important positions. In the world's top 10 corporations with the most R&D investment in 2016, Huawei ranked No. 9 with 9.2 billion USD (about 63 billion RMB), higher than many giant corporations, such as Apple and Cisco. Huawei has continued to increase investment on the basic research and innovation for the future and has made many research achievements in the hot frontiers of ICT, hoping to drive the industrial development and the success of business mode with technological breakthroughs.

5.4 Using New S&T and "Internet Plus" for Leapfrogging

Recently, Guangdong government released a series of policies and regulations in 2015, such as "Incentive Scheme of Subsidy for R&D of Companies" and "Scheme of Post Coupon Subsidy of S&T Innovation," to encourage governments at all levels to support the companies to build innovation mechanisms in a better and faster way.

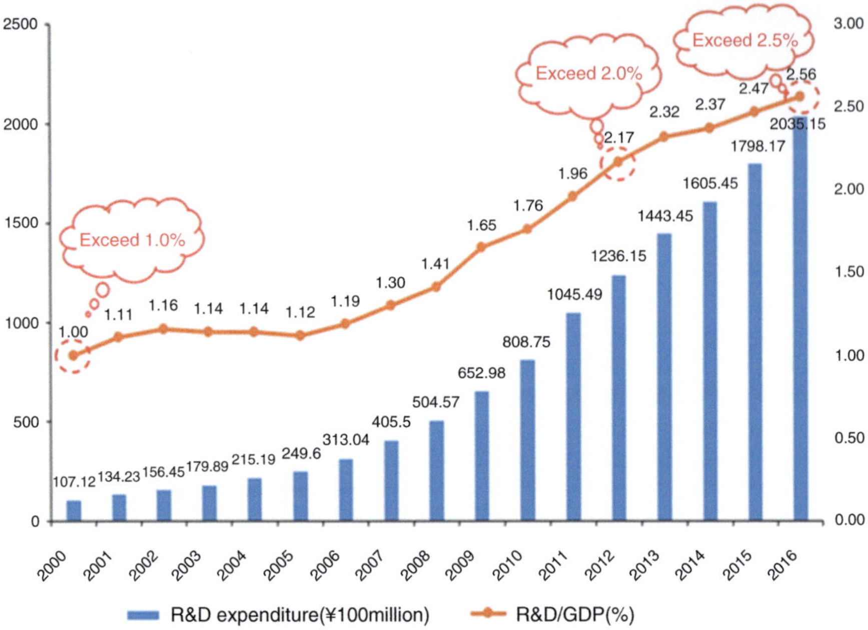

Fig. 5.7 R&D/GDP in Guangdong (2000–2016). (*Source*: Statistics Bureau of Guangdong 2000–2016)

In addition to the implementation of tax preference, indirect instruments, such as R&D expense deduction and other preferential policies for high-tech companies, have been implemented. Guangdong has promoted many inclusive policies to incent innovation. One is to establish R&D subsidy funds for companies and guide them to build a R&D reserve fund system. This year's subsidy funds' budget of total 1.4 billion yuan has been allocated already; another is to establish the training subsidy funds for high-tech companies. The budget of this year has achieved 2 billion yuan; meanwhile, special funds have been set up this year for the risk compensation of incubators. Forty million yuan have been allocated.

Firstly, the budget for R&D inputs has been increased. This is especially so after the global financial crisis, it hurts Guangdong greatly. So, Guangdong has to find a new way for its sustainable development. They began to look for technology as the way of leapfrogging, especially in ICT industry. From 2000 to 2005, R&D investment had been increased stably in Guangdong. Moreover, since 2006, the total R&D from both government and business has been increased quickly (see Fig. 5.7).

A positive signal in Guangdong is that the business sector has also increased its R&D expenditures, even basic research in some sectors. We can see from the Table 5.4 that even some emerging companies began to support basic research, such as Haige and Midea.

Table 5.4 Guangdong innovation and entrepreneurship fund – established fund

Name of fund	Fund size	Guangdong government budget	Leading social capital	Fund partners	Investment orientation
Guangdong Yuke Haige integrated circuit parent fund	4.5 billion	1.5 billion	Triple	Guangzhou Haige Communications Group, Bank of China	Supporting: integrated circuit equipment, materials, packaging projects. Acquire overseas acquisitions if conditions permit, and improve the level of development of Guangdong IC industry
Guangdong New Media Industry Fund (limited partner)	3.2 billion	1.0 billion	Triple	Shanghai Pudong Development Bank, the government of Yuexiu District	Focus on supporting: new media development projects of media companies, media convergence and development of key foundational projects, traditional media industry transformation and upgrading key projects, restructuring and reform of state-owned cultural enterprises

Source: Zheng (2017)

Secondly, they began to set up new industrial R&D institutes for industrial technology. They set up a new Guangdong Academy of Sciences, in their spending, with nine major technology projects targeted industrial development.

Moreover, Guangdong also deepens industry-university research collaboration and sets up more technology business incubators and other transformation platforms to breed new products and companies (Hu 2017).

Thirdly, industry there is quickly upgrading by Internet technology. This is not a huge revolution, but a moderate change. Guangdong takes advantage of the Internet thinking to redefine manufacturing. It is not a reshuffle, but with the Internet as combination. For example, smartphone breaks down the functional phone, but not the whole mobile phone industry. With the rapid development of Internet technology, the Internet is not an indigenous industry outside the real economy. From the information Internet to social media to industrial business Internet, the trend of integration is more and more obvious. The Internet has been integrated into various industries and organizations to assist their upgrading and efficiency. The Internet thinking directly impacts the reconstruction of a company's value chain. It promotes the information level of R&D, manufacturing, equipment production, sales, and service. It also accelerates the development of manufacturing to the digital, network,

Fig. 5.8 Representatives of transformed companies in Guangdong. (*Source*: South Daily et al. 2016)

intelligent, and service-oriented direction and incents the development of e-commerce, modern logistics, information services, etc. At the same time, with the development of information technology and emerged Internet, many new business modes have been created, which provide an excellent opportunity for Guangdong's innovation development. Figure 5.8 lists part of transformed companies.

5.4.1 Midea

The annual report of Midea Group in the year of 2015 shows that its operating income is 138.4 billion yuan, only 4.3 billion more than that of 2011. However, its annual revenue is 13.625 billion yuan, which is twice as 6.641 billion revenues in 2011. In nowadays declined household appliance industry, Midea Group has realized its inverse growth as "output value stays but profit doubled" with the help of the Internet. The principle of Midea's transformation and upgrading is "product leading, efficiency driving, and global running." At present, Midea has established 11 R&D centers around the world, including the one in the USA, Japan, Germany, Italy, Singapore, and other regions. It has also set up a new R&D center in Silicon Valley, San Francisco, specializing in R&D of artificial intelligence and smart home technology. By the end of 2016, Midea owns the most patents among all household appliance companies in the world. Midea combines Internet technology to develop automation and information. The traditional manufacturing has been transformed to intelligent manufacturing. The digitization is the essence of intelligent manufacturing. That means all companies are connected by digitization from R&D of the product, to supply chain, manufacturing, logistics, etc. The data created in every part is analyzed and modeled. Using algorithm, Midea builds a digital world with the digitization of physical-form machines, equipment, production lines, people, vehicles, channels, and logistics. In this digital world, algorithm and model are used to processing analysis and creating value. Then the new value produced will return to the physical-form value chain. This Internet-combined transformation and upgrading way of Midea is worthy for other companies from manufacturing industry

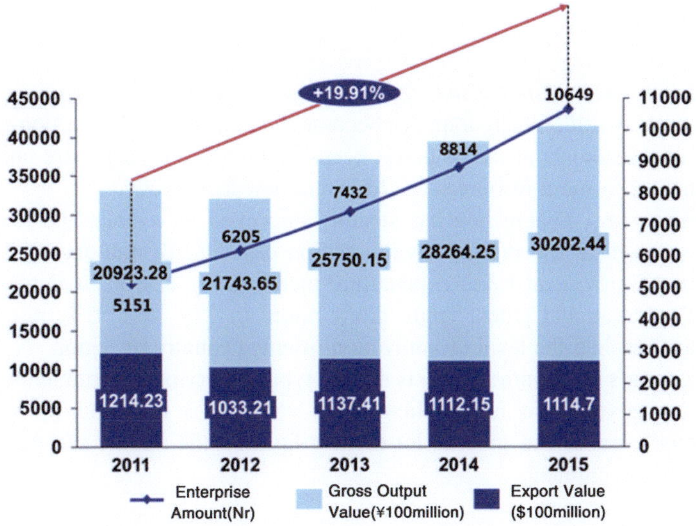

Fig. 5.9 Quantity, total value, and export of Guangdong's high-tech companies in the Pearl River Delta. (*Source*: South Daily et al. 2016)

and even companies from other industries to learn. Connecting products and data by digital world could greatly improve the production efficiency and bring the strong driving force to company's innovation development.

Fourth, the government targets the high-end equipment manufacturing and digital manufacturing as new sectors for the future.

Guangdong has formed the technology innovation cluster with "breakthrough innovation with large companies and massive innovations with small companies." On one side, many large companies, such as Huawei, ZTE, Tencent, GREE, and Midea, are already or being close to the frontier of international technology, which takes the lead of innovation. One the other side, many small- and medium-sized companies have collaborated with large companies to cluster innovation. Technology innovation therefore becomes the new engine for the development of the province.

In the year of 2016, the number of high-tech companies in Guangdong is 19,857, which ranks 1st nationwide. In 2015, there are 10,560 high-tech companies in the Pearl River Delta. This number increases to 18,880 in 2016. Within only 1 year, the high-tech companies in the Pearl River Delta had increase of more than 8000, with 22 additions per day. Guangdong has not only a huge amount of high-tech companies but also good market performances due to its demand-oriented and quick update products. For example, the robot industry of Guangzhou is weak in the market because of the hardware and material shortage. But it can produce robot that meets the most market needs. Figure 5.9 illustrates the development of Guangdong's high-tech industry.

5.5 Conclusion

A region with less investment from the central government, taking the opportunity of open up to the world, Guangdong has become the most innovative region in China. But the more astonishing fact is that it keeps going and entering more innovation-driven stage with more resources on R&D and new technology.

In recent years, Guangzhou has steadily improved its science and technology strength, as well as the indigenous innovation capacity. By the end of 2016, the proportion of provincial R&D expenditure increased by 2.58%; invention patents and PCT international patent applications raised to 21% and 55%, respectively, which have reached the level of innovation-oriented country or region.

Guangdong's development also is not the typical story of government top down, with heavy government intervention. Vice versa, it is a bottom-up story. So, Guangdong's model can give us lot of new light on the impact of innovation to the regional government.

References

Alvarez, R., & López, R. A. (2005). Exporting and performance: Evidence from Chilean plants. *Canadian Journal of Economics, 38*(4), 1384–1400.

Aw, B. Y., Chung, S., & Roberts, M. J. (2000). Productivity and turnover in the export market: Micro-level evidence from the Republic of Korea and Taiwan (China). *World Bank Economic Review, 14*(1), 65–90.

China Enterprise Confederation. (2017). *2016 China Top 500 Enterprises List.*

De Loecker, J. (2007). Do exports generate higher productivity? Evidence from Slovenia. *Journal of International Economics, 73*(1), 69–98.

GfK. (2017). Derived from http://tech.sina.com.cn/zl/post/detail/t/2017-02-07/pid_8509782.htm

Guangdong Intellectual Property Office. (2017). List of Top 100 Guangdong patent innovation enterprises 2016

Hu, C. (2017). Stick on the Implementation of Innovation-driven Development Strategy. *People's Daily*. Aug 30. National Bureau of Statistics of China (2014–2016) China's statics yearbook. Beijing: China Statistics Press.

Ng, S., & Li, B. (2010). *Operation strategy Galanz case*. The University of Western Ontario Aug05:9B10D005.

Qin, J. (2017). *Case analysis of Huawei*. Derived from https://www.toutiao.com/a6475920639373885965/

SIPO. (2017). *State intellectual property office press conference*. Jan 19. Derived from https://www.toutiao.com/i6377280595298353665/

South Daily., et al. (2016). *Report of Pearl River Delta Enterprise Innovation*, Gaungzhou. http://www.360doc.com/content/17/0506/18/29955225_651621054.shtml

Statistics Bureau of Guangdong (SBG). (2000–2016). *Guangdong statistical yearbook 2000*. Beijing: China Statistics Press.

Wang, J., Mao, Y., & Qiu, N. (2008). From OEM to ODM to OBM, the indigenous innovation and internationalization of Galanz – A case study of technical capacity construction and company upgrading. *Management World, 6*, 148–155.

Zheng, H. (2017). Presentation of Guangdong innovation trend at conference of China regional innovation, 2017. Beijing.

Chapter 6
Jiangsu: From Traditional Industry to High-Tech Industry – The Role of TVE and FDI

6.1 The Ranking and Profile of Innovation

In 2017, Jiangsu Province ranks 2nd in our regional innovation of index of China. It used to be the No. 1 for 8 years and has only been surpassed by Guangdong province in this year. In the sub-index ranking system, knowledge creation, enterprise innovation, and innovation performance, Jiangsu ranks 2nd; in knowledge acquisition and innovation environment, it ranks 3rd. The indexes of knowledge acquisition, enterprise innovation, and innovation environments drop one position from the last year (Table 6.1, Fig. 6.1).

There are 44 enterprises from Jiangsu Province in "China Top 500 Enterprises List 2016." The top 20 enterprises include not only state-owned enterprises in traditional construction (Pacific Construction Group) and steel industry bus also private companies, such as Suning and Hengli. These private companies were listed top even nationwide. It shows that private economy is a significant supporting part of economic development in Jiangsu Province. In 2016, the added value of private economy in Jiangsu accounted for 55% of the province's GDP, contributing 57% to the GDP growth (Table 6.2).

In "China Top 1000 Enterprises in Innovation Capability List 2016" (Table 6.3), there are 125 enterprises from Jiangsu, which shows Jiangsu's strong innovation capacity. The top 20 enterprises are mainly from engineering-based industries and efficiency-driven industries, including electronic apparatus and equipment manufacturing, electric power, and heating power.

Owning its deep industrial foundation and comparative advantages, Jiangsu Province has formed a distinctive development model. In terms of GDP and GDP/PC, Jiangsu ranks, respectively, No 1 and 2 in the country. Jiangsu has well-developed industry quality and structure as well as strong talent power. In terms of innovation level, Jiangsu possesses certain global leading advantages in the fields such as information technology, new energy and new energy vehicles, energy conservation and environmental protection, biotechnology, and new medicine and

X. Liu et al., *Regional Innovation Index of China: 2017*,
https://doi.org/10.1007/978-981-13-1205-2_6

Table 6.1 Jiangsu innovation capability comprehensive value

	2017 comprehensive value	2017 sub-index rank			
Name of indexes	Index value	Rank	Strength	Efficiency	Potential
Comprehensive index value	53.3	2	2	5	27
1 Knowledge creation	47.28	2	2	7	19
1.1 R&D input	37.01	4	2	5	21
1.2 Patent	63.05	2	1	7	16
1.3 Research paper	36.29	5	2	22	16
2 Knowledge acquisition	42.99	3	1	6	26
2.1 Science and technology cooperation	40.65	2	2	14	19
2.2 Technology transfer	30.12	5	2	20	19
2.3 Foreign-funded enterprise investment	54.4	2	1	4	24
3 Enterprise innovation	62.43	2	2	6	20
3.1 Enterprise R&D input	83.08	1	1	4	19
3.2 Design capability	42.44	3	2	9	23
3.3 Technology improvement	56.97	1	1	10	18
3.4 New product sales income	66.89	3	1	8	10
4 Innovation environment	46.88	3	2	10	31
4.1 Innovation infrastructure	43.88	2	1	19	28
4.2 Market environment	38.29	6	4	5	31
4.3 Employee quality	45.15	3	2	17	25
4.4 Financial environment	51.86	2	1	11	25
4.5 Entrepreneurship	55.21	2	2	14	25
5 Innovation performance	62.14	2	2	1	14
5.1 Macroeconomic	82.56	1	2	4	11
5.2 Industry structure	66.52	2	2	3	18
5.3 Industrial international competitiveness	47.97	3	2	8	12
5.4 Employment	51.43	2	2	2	21
5.5 Sustainable development and environment	62.21	20	28	5	7

new materials. However, Jiangsu's population density is relatively high in China; therefore, there is huge pressure on resources and environment, and the traditional momentum of economic growth is gradually declining. To reach a new stage, an innovation-driven development strategy is put forward that will enact more technology innovation projects and improve the innovation system of the region, with the goal to make innovation as the main driving engine for economic development.

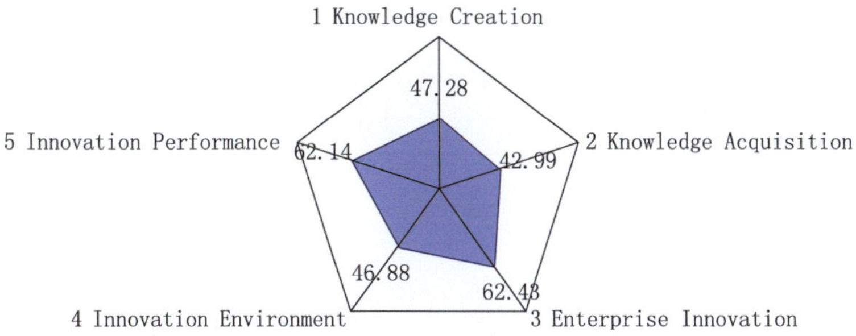

Fig. 6.1 Jiangsu innovation capability spider chart

Table 6.2 Jiangsu enterprises in "China Top 500 Enterprises List 2016" (top 20)

	Enterprise	Operation revenue (billion yuan)	Rank
1	Pacific Construction Group	459.06	19
2	Suning Holdings Group	350.29	35
3	Hengli Group	212.08	65
4	Shagang Group	205.82	69
5	Yueda Group	101.81	145
6	Xuzhou Construction Machinery Group	73.94	189
7	Nantong Sanjian Holding	68.83	196
8	Hailan Group	67.18	202
9	Wuxi Industry Development Group	60.52	225
10	Hodo Group	50.31	265
11	Jiangyin Chengxing Industrial Group	48.86	273
12	Hengtong Group	48.03	279
13	Jiangsu Nantong No.2 Construction Engineering (Group)	44.59	294
14	Nanjing Iron and Steel	44.57	295
15	Jiangsu Sanfangxiang Group	43.58	301
16	Jiangsu Guotai International Group Guomao	43.22	305
17	Jiangsu Suzhong Construction Group	42.19	311
18	Xuzhou Coal Mining Group	41.07	320
19	Jiangsu High Hope International Group Corporation	39.14	334
20	Jiangsu Huaxi Group	38.87	336

Source: China Enterprise Confederation 2017

Jiangsu is one of the provinces with lots of good universities and students. The number of academicians is only less than that of Beijing and Shanghai. It is the first province with more than 3 million patent applications. It has a high growth of intellectual property in China. There are more than 14 patents per 10,000 people in 2015, which is doubled compared with 2010. By the end of 2016, Jiangsu has 2800 national and provincial high-tech R&D laboratories and centers, more than

Table 6.3 2016 Jiangsu enterprises in the "China Top 1000 Innovative Enterprises List 2016"

	Enterprise	Patent application (Nr)
1	Wuxi Little Swan	561
2	Jiangsu Province Metallurgical Design Institute	387
3	NARI Technology	312
4	Bozhon Precision Industry Technology	282
5	NARI Group Corporation	279
6	NR Electric	218
7	Kunshan Govisionox Optoelectronics	202
8	Xuzhou Construction Machinery	195
9	Shoto Group	192
10	Suzhou Canadian Solar	191
11	Nanjing CIGU Limited Corporation	172
12	InfoVision Optoelectronics Kunshan	167
13	Centec Networks (Suzhou)	150
14	Jiangsu Haipeng Special Vehicles	150
15	Wuxi Jiangnan Cable	145
16	Xuzhou Construction Machinery Group	117
17	Nanjing Kangni Mechanical & Electrical	117
18	Jiangsu ZTT	108
19	A.O. Smith (China) Water Heater	105
20	China Energy Engineering Group Jiangsu Power Design Institute	105

300 various R&D institutes, and 6 provincial industrial innovation centers including Suzhou Nano, Suzhou Medical Engineering, Changzhou Intelligence, Nanjing Laser, Wuxi Environmental Protection, and Nanjing Future Network. Figures 6.2 and 6.3 show R&D investment level and patent number in last ten years in Jiansu Province.

6.2 The Upgrading of Traditional Industry in Jiangsu

Jiangsu is a big manufacturing region, characterized by traditional industry. It is near Shanghai and with good infrastructure for manufacturing. Before 1978, central government had made huge investment there, including the Jinglin Chemical Engineering, Sinopec Yangzi Petrochemical Company, Xuzhou Engineering, etc. According to the list of "Top-500 China's Companies in 2016" released by the China Enterprise Confederation and the China Enterprise Directors Association, 44 companies of Jiangsu are listed. Most of them are engineering technology and efficiency-driven industries, involving engineering equipment, automotive and spare parts, manufacturing, metallurgical processing, chemical fiber manufacturing, etc. In the list of "Top-500 China's Private Companies in 2015" released by the China Private Enterprise Association and Enterprise Research Center of Chinese Academy

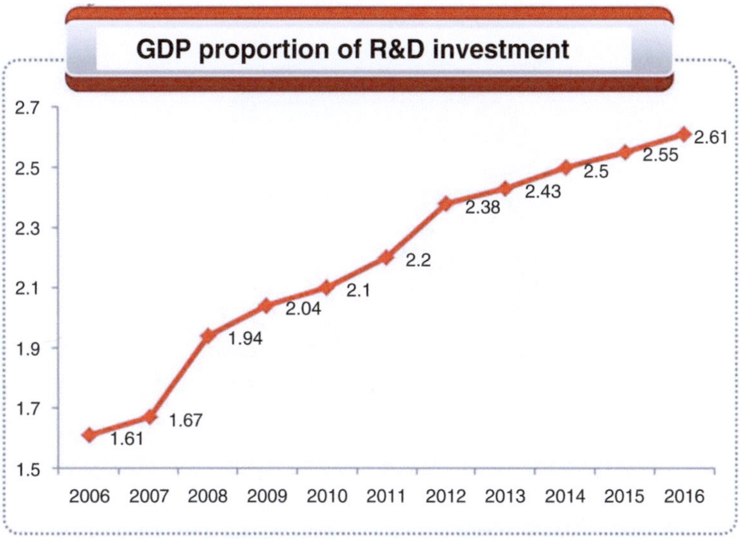

Fig. 6.2 GERD/GDP in Jiangsu. (*Source*: Statistics Bureau of Jiangsu 2007–2017)

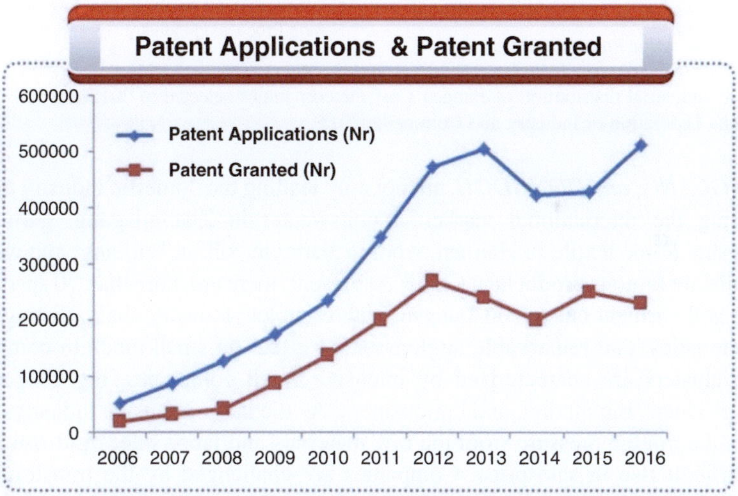

Fig. 6.3 Patent applications and patent granted. (*Source*: Statistics Bureau of Jiangsu 2007–2017)

of Management Science, Jiangsu takes 91 seats, also mainly involving engineering technology and efficiency-driven industries (see details in Fig. 6.4).

Textile and garment industry is used to be the key one for its economy with great international competitiveness. Among them, 117 companies in Jiangsu were listed as the top-500 China's textile industries 2014–2015, which is nearly one quarter of the total. Famous provincial brands, like *HENGLI, SHENGHONG, BOSIDENG,*

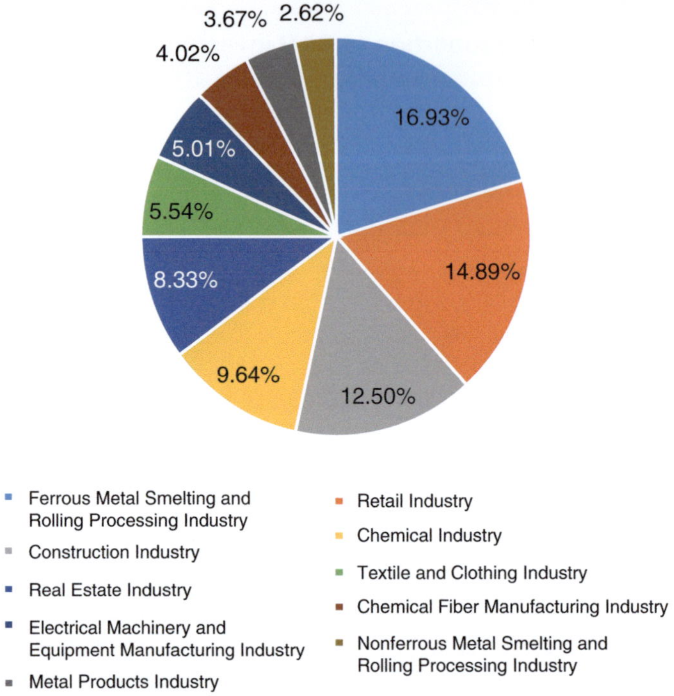

Fig. 6.4 Industrial distribution of Jiangsu's private companies selected in 2015 top 500. Source: All-China Federation of Industry and Commerce 2015

YANGGUANG, and HONGDOU, are not only leading the domestic industry but also exploring the international market. Wool/sweater in Zhangjiagang, garment in Changshu, home textile in Haimen, wool in Jiangyin, silk in Wujiang, and weaving in Wujin are famous products in China. At present, there are more than 70 specialized textile and garment clusters in Jiangsu, with complete industry chain, distinguished characteristics, and remarkable agglomeration effect on small-medium companies. These clusters are characterized by intensive small companies, big cooperation, small products, big market, and big business. As the labor-intensive industry, textile suffers the double pressure from the raw materials and labor side: rigid rise in cost and difficult rise in sale price. Companies are challenged by the transformation. Shenghong Group is a good example of how to make the transaction possible in the process (see Box).

Shenghong Group

The current competition of the garment manufacturing has changed from traditional one-dimension based on quantity to the multi dimension on technology, design, information, lean manufacturing, and so on. New technology and business models like customization and 3D printing keep emerging. "Market is not lack of purchasing power. Instead, personalized products are the real shortage. For the product differentiation, we should try our best to satisfy the personalized needs." This is the value of Shenghong Group, which helps Shenghong to occupy the leading position of the industry. The 20D144F of Shenghong is currently the finest fiber that can be industrialized in China. 9 kilometers of 20D144F weighs only 20 grams. The finest-look thread is made by 144 monofilaments. Moreover, Shenghong has achieved new breakthroughs in wrinkle resistance, flexibility, comfort and other aspects. Just for functional differentiation, the company has 10 kinds of fibers. No mention about other subdivided varieties. Due to these specialized fibers, demand is activated by supply, and brings vitality to the downstream industries.

Source: authors editing

Jiangsu's textile companies are also good at exploring the overseas market. In 2016, Jiangsu Yangguang Group signed an agreement with Ethiopia State Investment Commission on the wool project. 980 million US dollars are planned to invest on the production base in Ethiopia. Yangguang intends to build the overseas production base in the national development area of Ethiopia in Adama and move out its low-end manufacturing part to satisfy the need of transformation and upgrading. Yunfu Group, as another example, has built a three-dimensional, internationalized, and digitalized warehouse in Los Angeles, USA, based on its own developing status. By this strategy, the group has realized the 45-day quick reaction from taking orders to launching products, while most foreign trade companies in China need 90 to 120 days. Right now, Yunfu has used this warehouse in Los Angeles as the "Jiangsu public warehouse for cross-border e-commerce" and helps many companies to expand their overseas business.

The upgrading also happened in shipping industry. This is the traditional competitive industry of Jiangsu, with a number of national top brands like COSCO, Rongsheng Heavy Industries, New Yangtze, etc. Now it has been transformed toward marine engineering equipment, and now it is the current national leader of the marine industry. COSCO is renowned at home and abroad.

Pharmaceutical industry is another good example. It has been transformed into biomedical industry in clusters in Qidong City. To access the docking of the biomedical industry of Shanghai, Qidong uses Qidong High-tech Area as the platform of bioscience and Qidong Binjiang Pharmaceutical-chemical base as the platform of drug substance, gathering a number of leading companies of biomedicine, such as SHANGHUA, Bayer, Acebright, QZU, GAITIANLI, and DONGYUE. Brands like GAITIANLI, White & Black, and DaXi are well-known throughout the country. Due to the limited space of Zhangjiang, Shanghai, 70% new drugs need to be industrially produced outside Zhangjiang. Qidong Economic Developing Area is only an hour drive from Zhangjiang, so officials send employees to Zhangjiang to closely dock with this drug valley. By keeping exploring the industry development mode of "R&D in Zhangjiang and production in Qidong," Qidong has finally become the industrialization base and accelerator of Zhangjiang drug valley. Meanwhile, Qidong has further strengthened the cooperation with Shanghai Biomedicine Industry Area and outputs on firms, management, and talents with Shanghai Zhangjiang Biology Medical Company.

In 2000, Simcere established a new drug research center. In recent years, the annual R&D investment is accounted around 10% of sale with an increasing trend. In 2014, the company founded Baijiahui entrepreneurial community to share its R&D and pharmaceutical platform resources to global scientists and entrepreneurs, which started the open drug R&D mode in China. In 2015, Ministry of Science and Technology approved Simcere to establish the "translational medicine and innovative medicines," the state key laboratory. The company has been selected as "China's Most Innovative Pharmaceutical Enterprises" for many years.

In manufacturing industry, Xuzhou Construction Machinery Group Co., Ltd. (XCMG Group) provides another good example of how to deep their innovation capabilities. It has been the No. 1 construction machinery company in China for 28 years since its establishment. In the current downturn situation of construction machinery manufacturing, Xugong Group has all the way to improve counterattack development, working on smart manufacturing, and has completed the manufacturing transformation and upgrading by leaps and bounds. In 2017, XCMG ranked No.8 which is the only Chinese enterprise among the global top 50 construction machinery manufacturers' rankings.

The intelligent manufacturing model of XCMG takes intelligent equipment R&D and the integrated application of industrial Internet as starting point to improve the upgrading of intelligent production lines/modules to achieve R&D/manufacturing automation which turn out to be more lean, digital, and intelligent. In November 2016, XCMG and Alibaba jointly announced the release of the industry's first industrial cloud platform, which successfully filled the gap in the large-scale application of domestic big data in the industrial field. China's first industrial Internet of big data platform was led by independent research and development (Xinhua Daily 2017). In January 2018, Bank of Jiangsu and XCMG Group signed a cooperation agreement to innovate the use of thing of Internet big data technology to solve the pain points of corporate loans, transforming and upgrading the manufacturing industry with thing of Internet finance to create a new ecology of online Internet (CNR 2018). The development strategy of XCMG Group may profoundly affect the manufacturing industry in Jiangsu and even the entire manufacturing industry in China.

The emerging of new manufacturing industry in Jiangsu is another way to ensure the industrial transformation. With solid base of manufacturing, Jiangsu quickly catches the new wave of new industry. Photovoltaic industry is the typical one. It started from zero to one in Jiangsu. The earlier birds are Sentech, which was once listed in New York Nasdaq and now bankrupted, and Trina Solar. Now there is a large industry in Jiangsu and the successful one is Trina Solar.

Trina Solar Limited (TSL) was established in 1997 as one of the first PV companies in China and has grown into the largest PV module supplier and leading system integrator in the world. The cumulative shipment volume of the company's components exceeded 30GW, which is No.1 in the world and ranked No.19 among the "Top 500 Global New Energy Enterprises." The company has a layout of its downstream ecological chain globally, providing one-stop system integration solutions for customers with development, financing, design, construction, operation, and maintenance.

In 2008, through the active planning and layout of the government, Trina Solar Industrial Park was set up in Changzhou, Jiangsu Province, to form a regional PV industry cluster with Trina Solar as the leading industrial including equipment, accessories, and auxiliary materials from upstream to downstream.

Under the driving of Trina Solar, the supporting industrial chain has been gradually improved, and the industrial cluster effect has been constantly highlighted. As a result, a number of "vertically integrated" industrial platforms have been formed in the area of photovoltaic production. There are more than 40 photovoltaic manufacturing backbone enterprises including GCL, Youze technology Co. Ltd., AMAX, Baofeng New Energy, Monocrystal, Yourbest, YuBang Photovoltaic, Changzhou Giantion Optoelectronics, JiangSu JiuDing Solar Energy System Co., Ltd., Tokyo rope, and Komatsu NTC. A National Key Laboratory of R&D platform of Photovoltaic Technology Center has been built to serve as the platform of PV International Testing Center, the main personnel training platform and an information platform for conference and exhibitions related to photovoltaic.

The rapid development of Trina Solar Industrial Park is mainly due to the following aspects: first are the geographical advantages of Changzhou high-tech zone east next to Shanghai and west to Nanjing, surrounded by the three cities of Suzhou, Wuxi, and Zhenjiang; the second is due to the strong regional economic base, and the increase of the main economic indicators is higher than the provincial average; third, thanks to the superior industrial production and manufacturing base, Changzhou is a city with earlier realization of photovoltaic industrialization and solar energy industry clusters; fourth, it has established a number of new scientific research institutions; and fifth, the government has promulgated some certain policies to encourage the development of photovoltaic industry. All these have well promoted the integration of PV companies into the global competition (Feng 2013).

6.3 TVE: The Institutional Innovation in Jiangsu

Entering 1980s, an important institutional innovation has emerged in Jiangsu: the so-called township and village enterprises (TVEs). Jiangsu is the birthland of China's township company. They are not private but also not state-owned; they are collectively owned. Those TVEs played a very important role in Jiangsu and China's industrialization at an early stage. Later on, some of them have been transformed into state-owned companies, some into private ones.

The TVE-based economy in county level in Jiangsu has great advantages:

First is its large scale. Jiangsu's county economy occupies the half of the whole provincial economy. The economic scale in the south of Jiangsu is equal to that of some capital cities.

Second, they are very innovative. Jiangsu owns 60 innovative counties and 117 innovative towns. Among them, the innovation index of Kunshan, Changshu, and Zhangjiagang ranks top 3 nationwide.

Third, these counties provide good environment for investment and S&T. For example, according to the "investment environment and risk investigation of Mainland China" released by Taiwan Electric Power Association, Kunshan could be the first choice to invest (Jiang 2016).

However, with the increasingly fierce market competition, the original market of Jiangsu's township companies is shrinking. This tough situation leads the companies to understand that only the technology innovation could create the initiative in the market competition. Through introducing foreign advanced equipment, strengthening R&D of innovative products, hiring talents, reforming management mode, and other measures, the township companies in Jiangsu finished the restructuring before 2000 and maintained the dominant position in the provincial development. Then, with the further development of the township companies, a series of problems have been revealed, such as the lack of scale economy due to scattered layout, lagged urbanization, land waste, and environmental pollution. Therefore, since the mid-1990s, Jiangsu's township companies have gradually concentrated and formed the clusters. By the end of 2010, there were 400 township industrial clusters in Jiangsu with more than 1 billion yuan revenues. They offer effective platforms for township companies to deepen the regional economic cooperation.

Now, private firms are more important for the economic development of Jiangsu, producing more than 50% of the total provincial GDP with a steady increase. In 2016, the added value of the private economy is about 55% of the provincial GDP and 57% contribution rate to the GDP growth (Jiangsu Federation of industry and Commerce 2016). Meanwhile, the private companies in Jiangsu are active in entrepreneurships. In 2016, the number of new registered private companies has reached 500,000 and 2.229 million companies in total (Jiangsu Government 2017), which is ranked as 2nd in the whole country just behind Guangdong. Private investment takes 70% of the total provincial investment. It has greatly contributed to the economic development and become the main body of Jiangsu's innovation-driven transformation (Song 2016).

6.4 From FDI to High-Tech Zone

In 1990s, Jiangsu entered a second wave of high economic growth, triggered by intensive FDI: first from investment of Taiwan and Hong Kong, then Japan, and later, from multinationals of European and the USA. They come one by one here to stay at the special zones: Economic and Technology Development Zone (later merged with high-tech zone) and high-tech zone. Most of the FDI is in name of manufacturing as Jiangsu has great advantage in suppliers, friendly business environment, close to Shanghai for better life for foreign business people, efficient public support, and good transportation. The most important of them are General Electric, Philips, Sharp, Coca-Cola, Siemens, Samsung, Bosch, Fujitsu, and other foreign finance and insurance institutions like HSBC, Standard Chartered Bank, Mitsubishi UFJ, AIG, and AEGON Insurance Group (Wang 2001; An and Zhang 2007).

In November 2014, there are 18 national high-tech industrial developing areas in Jiangsu, including Nanjing, Suzhou, Wuxi, Changzhou, Kunshan, Jiangyin, Wujin, and Zhenjiang. There are also 48 province level high-tech zones in Jiangsu.

In Fig. 6.5, we can see the high-tech zones have asserted a very high effect on growth: with only 4.8% of land, these zones produced 40.6% of the total export value and 25% gross regional domestic product.

According to the Statistical Yearbook of Jiangsu Province, it has been the province that attracted the most FDI in China since 2003 (except in 2015); see Fig. 6.6.

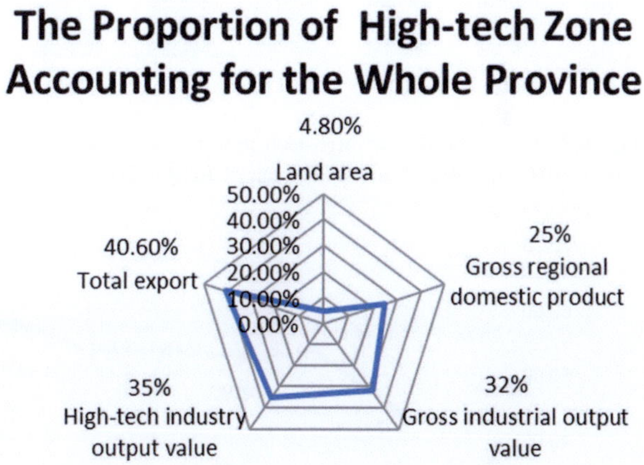

Fig. 6.5 The proportion of high-tech zone accounting for the whole province. (*Source*: Statistics Bureau of Jiangsu 2015)

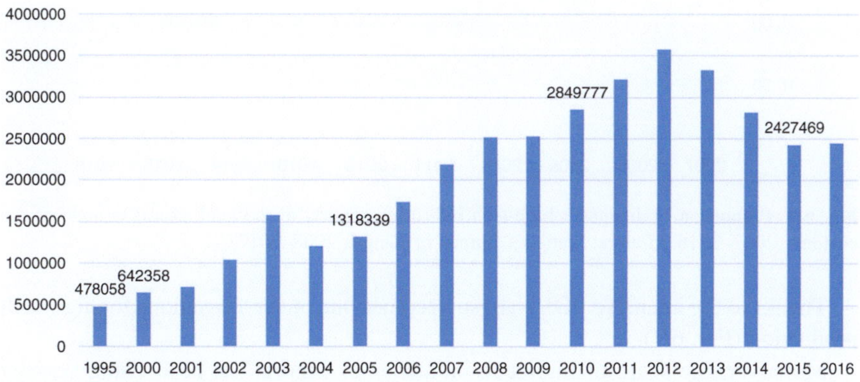

Fig. 6.6 FDI in Jiangsu 1995–2016 (10,000 dollars). (*Source*: Statistics Bureau of Jiangsu 2001–2017)

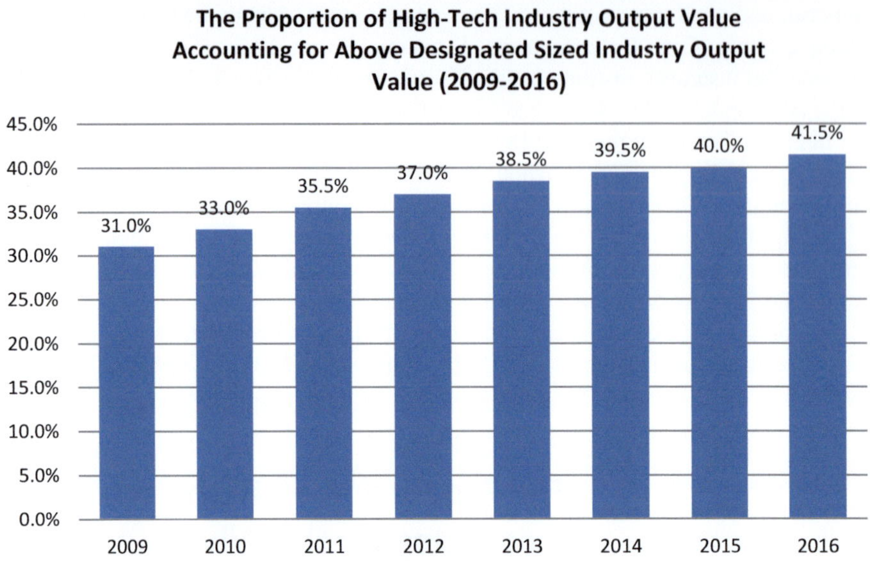

Fig. 6.7 The proportion of high-tech industry output in total of large enterprise industry value (2009–2016). (*Source*: Statistics Bureau of Jiangsu 2010–2017)

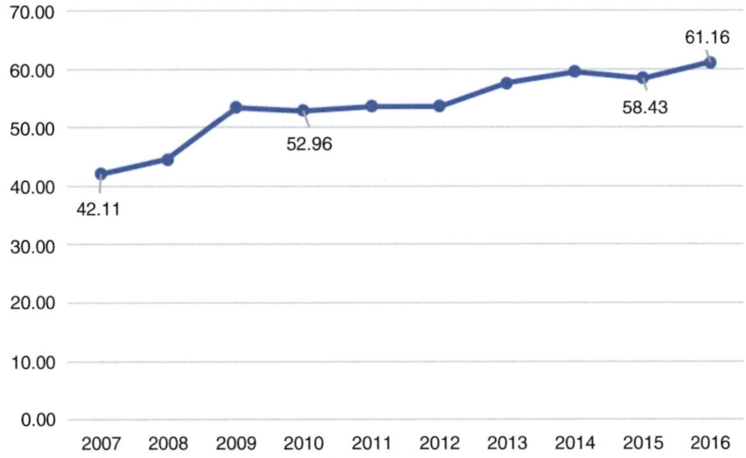

Fig. 6.8 Proportion of Jiangsu's high-tech industries exports in the total export volume of the customs, 2007–2016. (*Source*: Statistics Bureau of Jiangsu 2008–2017)

Thanks to the intensive FDI, Jiangsu becomes one of the top regions for high-tech production (Fig. 6.7).

Moreover, the share of high-tech industries in exports has been rising year by year, exceeding 60% in 2016 (Fig. 6.8).

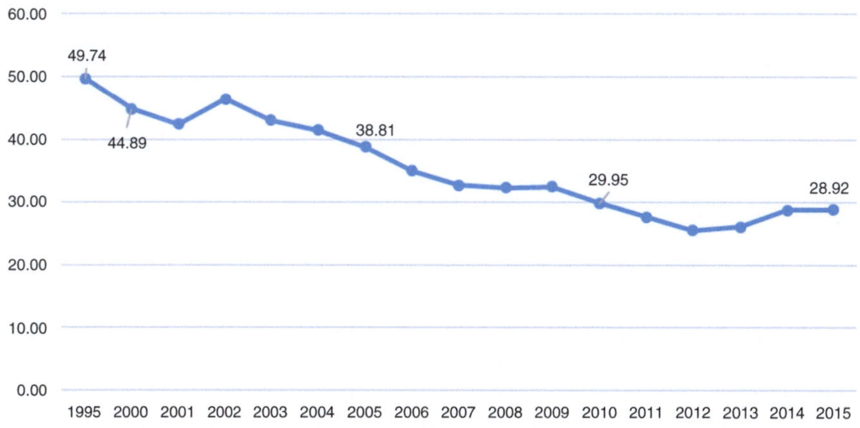

Fig. 6.9 The share of Juzhou FDI in Jiangsu Province. (*Source*: Statistics Bureau of Jiangsu 2000–2016)

Suzhou High-Tech Park

Take Suzhou as an example. Figure 6.9 *shows the proportion of Suzhou's FDI in Jiangsu Province. It can be seen that in the 1990s and early 2000s, the share of FDI in Suzhou was as high as 40% of Jiangsu.*

Multinational corporations have played a significant role in Jiangsu's economic growth. According to the statistics of Department of Commerce of Jiangsu Province (2004), 91 out of Fortune 500 companies have invested in Suzhou by March of 2004, and their investment projects are generally large in scale. The investment of multinational enterprises is mainly focused on high-tech industries, especially information technology industries. The investment distribution is relatively concentrated in central area of Suzhou, Suzhou Industrial Park, and the New District of Suzhou (Hua 2004). The multinational subsidiaries within the cluster maintain close contact with the parent companies outside the region, which makes Suzhou Industrial Park a highly open industrial cluster. There is not only a product exchange with the outside world; what's more, there is an active and free flow of knowledge, technology, and talents within and outside the region which has attracted many students who got doctor degrees in the USA to return to home countries to start up their own business (Ren 2005).

Multinationals' investment has strengthened industrial agglomeration in the area of technology spillover, talent flow, innovation incubation, industrial chain, innovation chain and value chain, etc., and the short-term impact is higher than the long-term impact (Cao and Xi 2017). Suzhou Industrial Park is an important cooperation project between governments of China and Singapore. It is one of the most fast-growing development zones in China with strong international competitiveness. Its major economic indicators reached an average annual increase of 30%. In 2016, the total GDP from the park achieved the 215 billion yuan, and its R&D investment accounted for 3.36% of the GDP. These made the park as the highest performance one among the national development zones in China (Suzhou Industrial Park Administrative Committee 2017).

But around 2003–2006, criticism about the Suzhou development model appeared; it has been claimed that the model too much relied on foreign investment and less on indigenous innovation (Du 2005; Zhang 2005a, b). There is less spillover effect from these high-tech investments. Even today, the local official document said the park is mainly composed of vertically integrated large foreign capital enterprises, which are mainly associated with processing and assembly. Those large multinationals operate as the manufacturing base, lacking core business and independent R&D capability, so it is difficult for them to radiate the overall innovation capability of the park.

Hu and Jefferson (2002) studied the impact of FDI on technological innovation in China's manufacturing industry. The results show that foreign investment has a significant positive spillover effect on China's manufacturing industry. By studying the provincial data from 1995 to 2000, Cheung and Lin (2004) showed that "the spillover effect of FDI has a positive influence on the number of patent applications for three types domestically (invention patents, utility model patents and design patents)." According to Ye and Wang (2014), FDI has a positive spillover effect on technological innovation in Jiangsu Province. The level of economic development, infrastructure, and human capital have contributed to the spillover effect of FDI and have achieved remarkable results.

Saltz (1992) analyzed the relationship between FDI and economic growth in the developing countries by using a combination of theoretical analysis and empirical analysis and concluded the negative correlation between FDI and economic growth in developing countries. That is to say that it was the introduction of FDI that may hinder the economic development of a country. Zhang (2005a, b) found that foreign capital did not show any significant spillover of productivity to the domestic capital sector and foreign capital played a negative competitive role in the technical efficiency of the domestic capital sector. Wang and Li (2003) think that in order to maintain the monopoly advantage, transnational corporations will choose to transfer technology internally. At the same time, due to the fact that our current absorption mechanism and capability are not strong, FDI has no obvious influence on the spillover effect of technological progress in our country.

In this background, Suzhou has experienced a growth model transition: from FDI driven to innovation driven. Entering 2000s, they begin to attract university and research institutes to come here, the most important of which are Suzhou Institute of Nano-Tech and Nano-Bionics of CAS, Suzhou Institute of Biomedical Engineering and Technology of CAS, and Institute of Medical Equipment of CAS. Many universities come here to open their subcamp-like graduate school or institute, for example,

Xi'an Jiaotong-Liverpool University (2006), Duke Kunshan University (2012), Oxford Suzhou Centre for Advanced Research (2016), Renmin University of China, University of Science and Technology of China, Xi'an Jiaotong University, etc.

A good example is: in October 2015, Suzhou industrial area also got the approval to become the first national open innovation pilot. In the same month, the Innovative Biologics located in the industrial area signed three cooperation agreements with Eli Lilly Group in the USA for the global development of the tumor-immune bispecific antibody drugs. The Innovative Biologics will receive more than 1 billion US dollars from Eli Lilly Group on this R&D project before any economic outcomes. Private township companies, with large quantity and advanced industrial chain, rely on the innovation to drive the economic development (Gu 2016).

But the problem is still there: The colleges and universities have not really been integrated into the local industry and culture, and the relationship between university research institute and its main department is constrained by institutional system. Moreover, another aspect that should be considered is the lack of cooperative innovation among different universities. Therefore, the gap between university and enterprise is obvious. But this is not just the case of Suzhou. This is a generous phenomenon in China.

Following the local entrepreneurship call, more and more SMEs have been set up there, but, at present, there are a large number of self-owned brands of emerging industries, such as nanotechnology, biopharmaceutical, cloud computing, etc. However, the majority of these enterprises is in development phase or in the stage of small-scale production, lacking of advantages in scale, growth speed, technology innovation, industry standard formulation, and international competition participation, and has not formed well-known leading enterprises or listed enterprises with leading advantages and greater influences in the domestic and overseas.

In recent years, some foreign-funded enterprises have started to evacuate from Suzhou. The dilemma faced by Suzhou model is not only due to the soaring cost of real estate, hydropower, transportation, raw materials, and other costs in China but also to the manufacturing structure (Zhang 2017). For example, since 2016, in the Suzhou Industrial Park, some foreign companies have successively closed the production department, including German-owned Nokia Communications Suzhou Branch. American-owned Express Semiconductor (Suzhou) Co., Ltd. has been merged, and the printer project of Samsung Electronics Co., Ltd. (South Korea) has moved to Weihai, Shandong Province. Samsung Motor (Suzhou) Co., Ltd. moved to Tianjin. The equity of Japan-owned Suzhou Mining Electronics Co., Ltd. was acquired. The relocation and evacuation of other large and medium-sized foreign-owned enterprises caused mass labor employee issues in January 2017 when Seagate, the US hard drive manufacturer, closed its Suzhou factory and laid off more than 2000 people. And similar happened in January 2018, when Nitto Denko (Suzhou) Co., Ltd. closed their optical division in Suzhou.

6.5 Policy and Strategy for Future

The economic openness of Jiangsu is high. As the front region of China's reform and opening, it entered the international market earlier and deeper, which becomes part of the local business culture and has trained the innovation and entrepreneurship awareness. Jiangsu government has implemented an innovation and entrepreneurship plan, to create good environment for business, and built various service platforms, which encourages entrepreneurship. The public entrepreneurship index of Jiangsu ranks the top nationally.

By the end of 2015, the innovation and entrepreneurship plan of Jiangsu supported and introduced 3127 talents, 236 teams, and 2598 PhDs. 40% of the introduced talents work for companies. There are more than 1300 new companies established by introduced talents, and 183 of them have become high-tech companies, while more than 80 of them achieve 100 million yuan sales. Jiangsu encourages the no border innovation. It has held many technology entrepreneurship competitions to integrate innovation and entrepreneurship factors and create innovation atmosphere. Companies with the excellent performance in the competitions could enjoy corresponding support policies (Gu 2016).

There are also many existing shortages. So far, Jiangsu still has a big gap with global leading innovation areas, and only Nanjing and Suzhou are relatively strong in the urban innovation globally. Lacking of world's famous companies and brand is also a shortage. There are 106 China's companies in the "Fortune Global 500 Companies in 2016," but only 2 from Jiangsu; in Forbes "World's Most Innovative Companies in 2015," 6 mainland companies are listed without Jiangsu; and in the "World's Top-100 most valuable brands in 2015," there are no Jiangsu companies. Therefore, Jiangsu needs to pay attention to these shortages in the future.

The government puts all efforts to promote the industry-university-research cooperation and the transfer of technology outcomes. Its S&T expenditure and regional R&D expenditure are all national lead; see Figs. 5.7 and 5.8. The government takes the benefits of innovation activities to train, find, and attract talents, including high-level introduction planning of innovation and entrepreneurship talents, double-thousand talent project of technology innovation and entrepreneurship, technology entrepreneur plan, youth doctoral programs of companies, etc. As a result of it, Jiangsu has also established enterprise academician workstations, graduate workstations, postdoctoral workstations, and enterprise technology centers, as well as promoted the innovation platforms like S&T entrepreneurship zone, university S&T zone, software zone, entrepreneurship service center, and various labs and R&D centers. In addition to that, the government has released the "Outline of Jiangsu Intellectual Property Strategy" to protect the intellectual properties (Fig. 6.10).

In August 2016, Jiangsu released "Jiangsu Implementation Plan of National Innovation-Driven Development Strategy." It clearly points out to promote the "six changes": the change of development mode from quantity-oriented rough

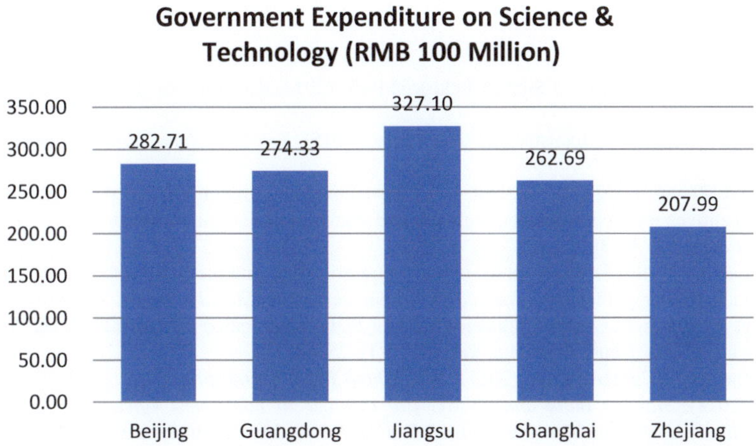

Fig. 6.10 2015 comparison of five provinces' government S&T expenditure (*Source*: NBS 2016)

expansion to quality-oriented sustainable development; the change of development factors from traditional to innovative; the change of industrial division from the low end to the medium-high end in the value chain; the change of innovation capability from "track and parallel" to "parallel and lead"; the change of resource allocation from R&D as the main part to the integration of industrial chain, innovation chain, and capital chain; and the change of innovative group from small scientist group only to the interaction with public innovation and entrepreneurship.

In order to address the new innovation goal, Jiangsu also set up some innovation system reforms:

Firstly, Jiangsu set up the provincial level Industrial Technology Research Institute to provide the industrial generic technology. The institute aims to conduct projects that university and research institutes are unwilling to do or single company cannot do. All these institutes have become the vanguard of the technology system reform and the most active power of innovation and entrepreneurship.

Secondly, Nanjing city was selected by the Ministry of Science and Technology as the pilot city of national S&T system reform in 2009. In 2010, it was also listed in the first batch of national innovative pilot cities. In 2012, Jiangsu Science and Technology Department, Education Department, and Nanjing government jointly issued the "breakthrough policy": "Nine Principle of Science and Technology." More than 20 universities and colleges developed or modified their related regulations based on it. More than 1000 college teachers own 1 billion yuan registered capital to start the entrepreneurships in Nanjing; Suzhou, Wuxi, Changzhou, and other places use new science city or original development/industry zones to attract universities and academy of sciences to come and cooperate.

References

Administrative Committee of Suzhou Industrial Park. (2017). Derived from http://www.sipac.gov.cn/

All-China Federation of Industry and Commerce. (2015). *Top-500 China's private companies in 2015*.

An, L., & Zhang, E. (2007). The entry of transnational corporation and the promotion of local competitiveness– achievements, problems and strategies of Jiangsu. *Forum of World Economics & Politics, 02*, 35–39.

Cao, Z., & Xi, Y. (2017). Study on the impact of transnational enterprises in Suzhou Industrial Park on industrial agglomeration. *Journal of Tianjin University (Social Science), 19*(05), 407–414.

Cheung, K., & Lin, P. (2004). Spillover effects of FDI on innovation in China: Evidence from the provincial data. *China Economic Review, 15*(1), 25–44.

China Enterprise Confederation. (2017). *2016 China Top 500 enterprises list*.

CNR. (2018). *Jiangsu Bank and XCMG are together to inject new kinetic energy into the development of manufacturing industry*. Jan 31. Derived from http://js.cnr.cn/2011jsfw/syyw/20180113/t20180113_524097908.shtml

Department of Commerce of Jiangsu Province. (2004). *Ninety-one of the world's top 500 enterprises have invested in Suzhou*. Apr 07. Derived from http://swt.jiangsu.gov.cn/art/2004/4/7/art_12624_4473526.html

Du, W. (2005). Suzhou mode': Reflection and affirmation. *Economics*, (10), 92–94.

Feng, J. (2013). *The research on Changzhou Tianhe photovoltaic Industrial Park development: Based on the perspective of industrial clusters*. Nanjing University of Science and Technology.

Gu, Y. (2016). Jiangsu: The secret of innovation ability tops of the country. *China Strategic Emerging Industry, 3*, 56–59.

Hu, A., & Jefferson, G. H. (2002). FDI impact and spillover: Evidence from China's electronic and textile industries. *The World Economy, 25*(8), 1063–1076.

Hua, J. (2004). The studies of FDI flow mechanism and the characteristics of transnational Enterprise investing in Suzhou. *Studies in International Technology & Economy, 7*(03), 45–49.

Jiang, Y. (2016). The important grasp of Jiangsu innovation-driven strategy implements. *Weishi* (12), 44–47.

Jiangsu Federation of industry and Commerce. (2016). *Research and analysis of private enterprises in Jiangsu province*.

Jiangsu Government. (2017). *The development of private economy*. Apr 10. Derived from http://www.jiangsu.gov.cn/col/col31388/index.html

Ren, S. (2005). The empirical analysis of the interaction between Suzhou industrial cluster and transnational enterprises. *China Soft Science Magazine* (01), 99–106.

Saltz, I. (1992). The negative correlation between foreign direct investment and economic growth in the third world: Theory and evidence. *Rivista Internationale di scienze Economich e commerciali, 7*(39), 617–633.

Song, X. (2016). *The quantity and quality of the private investment in Jiangsu province have been improved, showing a flourishing development*. JSxinhua Net. May 23. Derived from http://www.sohu.com/a/76663663_381581

Statistics Bureau of Jiangsu (SBJ). (2001–2017). *Jiangsu Statistical Yearbook 2001–2017*. Beijing: China Statistics Press.

Wang, X. (2001). The main motivation and strategies of investment in Jiangsu. *Reform and Opening*, (05), 17–18.

Wang, Z., & Li, Z. (2003). Study on the impact of foreign capital on the production efficiency of Chinese industrial enterprises. *Management World (Chinese)* (04), 17–25.

Xinhua Daily. (2017). *XCMG (Xuzhou Construction Machinery Group): Develop intelligent manufacturing to get achievements*. May 19. Derived from http://xh.xhby.net/mp3/pc/c/201705/19/c324261.html

Ye, J., & Wang, J. (2014). Research on the impact of FDI on local technological innovation–demonstration based on the panel data of Jiangsu province. *International Trade Journal*, (1), 131–138.

Zhang, H. (2005a). Human capital absorption, foreign technology diffusion and China's economic growth. *Studies in science of science (Chinese), 23*(01), 64–70.

Zhang, Y. (2005b). The reflection of 'Suzhou mode' and the choice of regional development path. *Shanghai Econmic Review*, (05), 35–41.

Zhang, Y. (2017). *Why the Ministry of Commerce come to declare for the phenomenon that foreign companies have closed Suzhou factories?* Jan 23. Derived from http://js.people.com.cn/n2/2017/0123/c360299-29636532-2.html

Chapter 7
Beijing: From High-Tech to Business Model Innovation

Beijing is a metropolis, the capital of China, as well as the center for political, cultural, and international exchanges and science, technology, and innovation of China. At the end of 2015, the population of Beijing reached 21.705 million, and the annual GRP reached 2301.459 billion yuan, with an increase of 7.89% compared to the previous year. GDP per capita reached 106,497 yuan, ranking the 2nd in the country. The main business income in the high-tech industry was 399.709 billion yuan, accounting for 17.37% of the GDP (Beijing Municipal Bureau of Statistics 2016).

Now, Beijing becomes the center of business model innovation based on Internet technology.

7.1 Profile of Beijing Innovation

In 2017, Beijing's innovation capability ranks 3rd in China, the same as the previous year. As to the sub-indexes, knowledge creation ranks 1st; knowledge acquisition and innovation environment both rank 2nd; and innovation performance index ranks 3rd, five places higher than last year; and enterprise innovation capability is relatively weak, ranking 4th (Table 7.1, Fig. 7.1).

According to "China Top 500 Enterprises List 2016" published by China Enterprise Confederation, there are 101 enterprises in the list that are located in Beijing, which are mainly large state-owned enterprises like State Grid, CNPC, Sinopec Group, etc. This phenomenon clearly reflects the economic feature of central enterprises headquarters in Beijing (Table 7.2).

In terms of patent applications, Beijing's top 20 patent enterprises are mainly from high-tech industries, like software and information technology service, aerospace, power, etc., obviously different from the top 500 enterprises list (Table 7.3).

Beijing can generally be considered an innovative city driven by science, with a large amount of science and technology agents, especially in the field of Internet Plus

X. Liu et al., *Regional Innovation Index of China: 2017*,
https://doi.org/10.1007/978-981-13-1205-2_7

Table 7.1 Beijing's innovation capability comprehensive value

	2017 comprehensive value	2017 sub-index rank			
Name of indexes	Index value	Rank	Strength	Efficiency	Potential
Comprehensive index value	52.56	3	3	1	26
1 Knowledge creation	74.10	1	1	1	23
1.1 R&D input	78.95	1	1	1	17
1.2 Patent	66.22	1	2	2	21
1.3 Research paper	80.14	1	1	1	18
2 Knowledge acquisition	47.42	2	4	2	4
2.1 Science and technology cooperation	57.72	1	1	5	16
2.2 Technology transfer	31.62	4	4	6	25
2.3 Foreign-funded enterprise investment	51.54	3	4	2	2
3 Enterprise innovation	46.30	4	10	1	26
3.1 Enterprise R&D input	47.46	7	14	3	25
3.2 Design capability	51.65	2	7	1	19
3.3 Technology improvement	49.84	4	8	1	19
3.4 New product sales income	32.98	10	12	7	25
4 Innovation environment	51.05	2	4	1	27
4.1 Innovation infrastructure	33.76	3	13	2	31
4.2 Market environment	66.70	2	2	2	23
4.3 Employee quality	52.05	1	7	1	5
4.4 Financial environment	61.25	1	3	1	23
4.5 Entrepreneurship	41.47	3	10	1	27
5 innovation performance	49.97	3	4	4	28
5.1 Macroeconomic	61.28	5	13	2	26
5.2 Industry structure	38.69	4	7	4	29
5.3 Industrial international competitiveness	32.62	9	11	5	26
5.4 Employment	29.97	7	3	18	31
5.5 Sustainable development and environment	87.30	1	2	2	2

industry, where start-ups are continuously emerging. Thanks to its geographical advantages, Beijing enjoys high starting point and is equipped with abundant financial resources and talented people. At the same time, Beijing is confronted with unavoidable problems, mainly environmental ones, including traffic congestion, overpopulation, resource pressure, and other problems that are common in large cities.

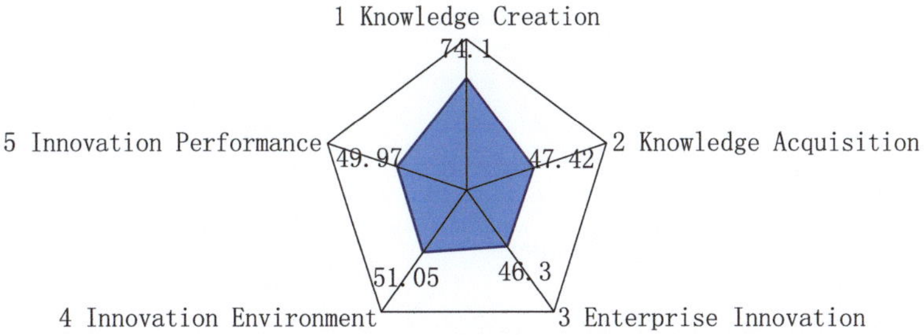

Fig. 7.1 Beijing's innovation capability spider chart

Table 7.2 Beijing enterprises in "China Top 500 Enterprises List 2016" (top 20)

	Enterprise	Operation revenue (billion yuan)	Rank
1	State Grid	2071.35	1
2	China National Petroleum	1880.74	2
3	China Petrochemical	1849.78	3
4	Industrial and Commercial Bank	1079.84	4
5	China Construction Bank	925.88	5
6	China State Construction Engineering	880.58	6
7	Agricultural Bank of China	833.77	7
8	Bank of China	769.22	8
9	China Mobile Communications	670.93	10
10	China Life Insurance (Group)	636.44	12
11	China Railway Group	624.89	13
12	China Railway Construction	601.11	14
13	China Development Bank	595.32	15
14	China Resources (Holdings)	481.22	17
15	China South Industries Group	440.42	20
16	China Post Group	437.63	21
17	China National Offshore Oil	426.08	22
18	China Communications Construction	425.86	23
19	The People's Insurance Company (Group)	406.01	24
20	China National Cereals, Oils and Foodstuffs	405.44	25

Source: China Enterprise Confederation 2017

In general, the innovation-driven development pattern in Beijing has been almost formed with unique innovation elements and gradually increasing development power. The economic characteristics of large SOEs' headquarters in Beijing are apparent, and the public institutions in Beijing are the main contributors to its innovation. As a scientific and technological innovation center, Beijing will play a supporting role in optimizing and upgrading industrial structure as well as technological innovation in the surrounding regions. The innovation radiation will be gradually achieved. The future development of Beijing-Tianjin-Hebei cooperation

Table 7.3 Beijing enterprises in "China Top 1000 Enterprises in Innovation Capability List 2016" (top 20)

	Enterprise	2016 invention patent application number
1	BOE Technology Group	3569
2	Lenovo (Beijing)	1142
3	China Electric Power Research Institute	918
4	Kingsoft	813
5	Baidu Online Network Technology (Beijing)	743
6	Leshi Internet Information & Technology	734
7	Beijing Shenwu Environment & Energy Technology	686
8	China Academy of Launch Vehicle Technology	560
9	Kingsoft Security	546
10	Beijing BOE Optoelectronics Technology	546
11	Inspur (Beijing) Electronic Information Industry	484
12	Beijing New Energy Automotive	463
13	Beijing BOE Display Technology	409
14	Yimin Juyuan (Beijing) Media Technology	333
15	China Institute of Water Resources and Hydropower Research	254
16	China National Offshore Oil Corporation Energy Development	238
17	China Academy of Railway Sciences	211
18	Beiqi Foton Motor	196
19	Global United Technology	161
20	China Academy of Aerospace Aerodynamics	157

Source: Renmin University of China 2017

and the construction of Beijing Municipal Administration Deputy Center and the Xiong'an District will provide Beijing with a broader space for development.

7.2 Zhongguancun Park: The Place for High-Tech Industry

Innovation in Beijing starts from its story of Zhongguancun. Tsinghua University is located there, as well as Beijing University and the Chinese Academy of Sciences, among others. This area is academically highly advanced in China and, even in the world, with a long-lasting tradition in education and research.

However, for a long time, outcomes of research were not published and have not been utilized for societal development and innovation. Thus, Professor Chunxian Chen from CAS who visited Silicon Valley for three times set the first technology-based start-up in Zhongguancun between 1978 and 1980, based on the "Silicon Valley model." Following Mr. Chen, many researchers found that S&T results could be commercialized by means of spin-off and licensing. They began to increasingly set up their start-ups in the Zhongguancun area. Though in that time, this seemed to be illegal, but finally in May 1988, the State Council supported the new business, claiming that S&T projects could be used for commercial purpose. This led Zhongguancun become the first high-tech zone in China. Lenovo is one of the most prominent start-ups from Chinese Academy of Sciences and later as the top manufacturer of personnel computers in the world. Now, in Zhongguancun high-tech zone, more than 20,000 technology companies have been located in 2016, and every day, there are 66 new companies settled there.

Now, the entrepreneurial gene gradually was formed here, and Zhongguancun became a model of industry-university-research triple helix and the symptom of high-tech innovation for other regions.

Zhongguancun's advantage for high-tech industry came from the rich resources: there are 40 universities, such as Beijing University and Tsinghua University, located in Zhongguancun; more than 200 national/provincial research institutes as the representative of the Chinese Academy of Sciences; the quantity and quality of national research infrastructures ranks the top in China; this is the world's most densely populated area of talents and research resources. By the end of 2016, there are more than 30,000 returnees working in Zhongguancun area (Fangdi Institute 2017).

Later, Zhongguancun has expanded to Beijing's other district and county, such as Chaoyang, Fengtai, etc. This made Zhongguancun having a sub-park system around it. The attracting effect for high-tech zone is that the companies in there can enjoy some tax cut and other subsidies. General enterprises have to pay a 25% of income tax, but high-tech enterprises only 15% of income tax, i.e., a 10% income tax cut. But there are criteria before a company can be qualified as high tech, including R&D expenditures and number of patents and R&D staff.

This cluster effect of Beijing Zhongguancun is outstanding. The six high-end industrial zones have already become the great engine of the economy of Beijing. Zhongguancun S&T Zone, Financial Street, Economic and Technology Development Zone, CBD, Airport Economic Zone, and Olympic Central Zone are important drivers to promote the high-end, high efficiency, high radiation development of the capital economy. In 2016, the added value of the six high-end industrial zones took 48.3% of the GDP of Beijing, with 59.9% economic contribution rate and 66.4% above-scale technology companies (Beijing Municipal Bureau of Statistics 2017a, b). Now, Zhongguancun has become the fertile land for China's entrepreneurs and companies to gather, integrate, develop, and grow. Many unicorn (fast-growing) companies are born here. In 2016, there are about 65 unicorn companies from Zhongguancun, about half of total China, just second to Silicon Valley. Among them are Xiaomi (market value, 46 billion USD), DiDi taxi (33.8 billion USD), and Meituan (18 billion USD) (Fangdi Institute 2017). They are the so-called unicorn companies in Beijing (Fig. 7.2).

Lenovo Group

Lenovo is one of the pilot groups in China's technological innovation and also the leader of China's information industry. Lenovo began as a spin-off from the Institute

Fig. 7.2 Unicorn company in Beijing

of Computing Technology, a research institute under the Chinese Academy of Sciences, in 1984 with an initial capital investment of RMB200,000 (Xie and White 2004), by Founder Liu Chuanzhi together with ten scientists and technicians.

Unlike other start-ups, Lenovo came with the institution background, which is an active explorer and practitioner of Chinese technology reform. Mr. Liu Chuanzhi broke the shackles of the old system and the traditional concepts of the planning economy at that time. He led Lenovo to have found out of a road of high-tech industrialization with Lenovo characteristics and made successful implementation of the joint-stock reform of Legend Holdings. Under his promotion, Lenovo has established property and incentive mechanisms which may maximize the long-term development and the subjective initiatives of the employees (Liu 2004).

Lenovo is a company which went out from Zhongguancun, Haidian District, with the help of "Zhongguancun electronic street" originated in the early 1980s and the environmental advantage of the approval to establish the experimental zone for the development of high-tech industry in Zhongguancun. This situation provides a huge market opportunity for Lenovo. At the same time, the success of Lenovo has also led to the development of a large number of national IT enterprises in Zhongguancun and even in the whole domestic country.

Today, Lenovo has become a large enterprise group with diversified development of the information industry and an innovative international technology company, producing products like server, desktop computer, notebook computer, smart TVs, printers, handheld computer, mobile phone, computer motherboard, all-in-one PC, and other goods. From 1996, sales of Lenovo computer rank the top in China's domestic market. In 2005, it bought IBM PC business department and became the world's largest PC manufacturer in 2013; in 2014, it completed the acquisition of Motorola mobile, and in 2017, it ranked No. 226 on the list of Fortune 500, with about 60,000 staff around the world.

Baidu is the result of second wave of innovative entrepreneurship in Zhongguancun area. By the end of 1999, having seen the huge potential of the Internet search engine service industry in China, Robbin Li (Li Yanhong) quitted his well-paid job in the Silicon Valley in the USA and decided to return to his home country with his search engine patented technology. Baidu was founded in Zhongguancun in January of 2000. Baidu had a very rapid development due to the highly skilled human resources and environmental advantages of Zhongguancun.

Baidu is a high-tech company that continues to innovate, with the mission of "making the complex world simpler with technology." Many people put the label of "Chinese Google" to Baidu. In 2010, Google quit China which made Baidu the biggest beneficiary. Because of this, Baidu eliminated the most threatening competitor (Xu 2010). Now, Baidu has become the largest Chinese search engine in the world and the largest Chinese website. Its technology R&D team has obtained the most advanced search engine technology in the world, making Baidu a high-tech enterprise with the core technology of the world's top science in China. Through the business model and continuous innovation on products and technologies, Baidu has played a positive role in promoting the integration and development of the industries like finance, medical treatment, education, automobile, and life services of the real economy with depth linkage of the Internet. Its open connection to traditional industries has made positive influence on the promotion of economic innovation and the transformation of economic development mode.

As the first-class Internet company in China, Baidu is in the leading level of search, artificial intelligence, cloud computing, big data, and other technical fields worldwide. By the end of 2015, Baidu, after Google, has made business restructure to separate the mature business and new business so as to put more time and energy on innovative businesses like the Internet finance, unmanned vehicles, artificial intelligence, etc. What's more, it put more emphasis on the concerns of the layout and construction of content ecology, service ecology, and financial ecology on a strategic level (Tan 2016). Take the unmanned vehicle as an example; Baidu not only has a strong R&D strength and investment input as a strong support; it obtains natural advantages to develop automatic driving technology. It has not only the advanced big data and artificial intelligence technology, as the largest Internet map service provider; it also obtains the leading network of road data collection and exploring capability in China (Xiang 2016).

In the selection of the 50 most intelligent companies in the world in 2016, Baidu ranked the 2nd highest than other technology companies. In 2017, on the list of the ten major technology breakthroughs that were published by MIT, Baidu was on the top of the list as the leading company with face recognition technology which won all the awards of artificial intelligence.

7.3 Industrial Transformation and Business Model Innovation

Since the founding of the People's Republic of China, the economic development of Beijing has gone through a transformation from service economy via industry economy to a service economy. From 1949 to 1978, Beijing had the transition from consumption-oriented to the heavy chemical industry. From 1978 to 1990, Beijing began the structural adjustment for high-tech industry. From the1990s on, Beijing has become more and more service-oriented. From the 2000s until now, Beijing has entered the stage of high-end economy, more and more focused on business model innovation.

Details of industry structure are shown in Fig. 7.3.

Comparing to other regions in China, Beijing has already entered the transition of decline of its traditional manufacturing industry to emerging industry. The modern service industry of Beijing is in the leading position in China. The economic development of the capital is deeply restricted by the resource environment, and traditional industry and manufacturing are no longer suitable to develop in Beijing.

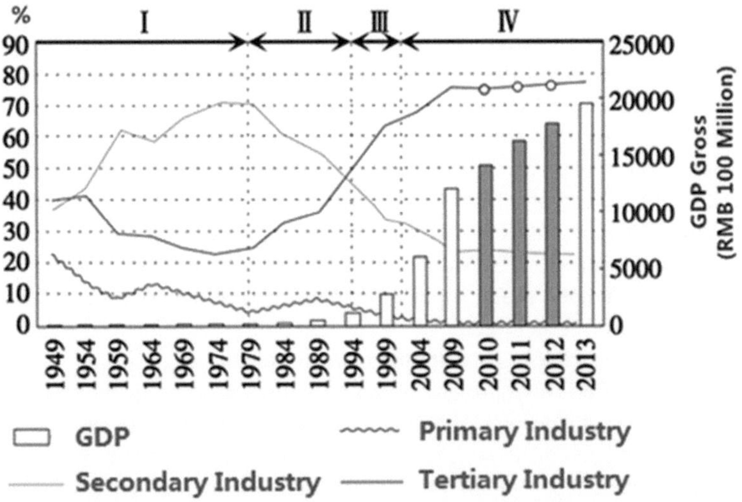

Fig. 7.3 1949–2013 Beijing economic development trend. (*Source*: NBS 2010; Beijing Bureau of Statistics 2014)

The productive service industry with low consumption and high added value should become the top priority of Beijing's industrial development. Therefore, the city should keep enhancing the development concept of "Beijing service-based innovation" and build the high-end service center rooted in Beijing and from there spread to the world.

The following measures have been taken: first, accelerate the development of the service industry, increase its proportion to the industrial structure, optimize the industrial development environment, and promote the brand of Beijing service; second, optimize the internal structure of the service industry, accelerate the development of modern and productive service industries, and enhance the global influence of "Beijing service" by the format innovation from the global demands; and third, accelerate the high-end talent pool and respective training, and activate the potential of the industrial development relying on regional infrastructures and the following industrial factors:

1. Finance. Beijing is the place where major domestic financial institutions set their headquarters and foreign financial institutions set branches and representative offices. It completely satisfies the requirements to build an international financial center and the national financial control center. So, Beijing has a great advantage to develop the financial sector.
2. Telecom industry and business service industry. Beijing is the national political center, international communication center, and aggregation area of the national major economic management and research institutions. It is why Beijing should also be the "digital Beijing" and the national information exchange center. Correspondingly, Beijing has great development advantages of its telecom industry, network service industry, e-commerce industry, consulting industry, advertising industry, exhibition industry, occupation introduction industry, software industry, notarization industry, accounting industry, lawyer industry, design industry, and so on.
3. Education, training, healthcare, cultural media, tourism, environmental management, scientific research, and technology services. Beijing is the national cultural center with high level of education and scientific research as 50% of academicians are working in Beijing.
4. Real estate industry and logistics industry. Beijing is a large consuming city. It is well developed in its road network and owns the important national railway hub and aviation port. The car ownership per capita ranks the top in the country, while the residents are relatively concentrated in the downtown area of Beijing. The daily urban mobile population makes people have quite demand on the real estate industry, business services, logistics, environmental management, automotive service industry, etc. The percentage of Beijing's service industry is 30% higher than the national average and 20% higher than the large-medium cities, such as Shanghai and Guangzhou.

Following this trend, Beijing became an important hub for business model innovation in China. Initially, Beijing paid much attention to the traditional industry. With the development of Zhongguancun and more and more technology, Internet companies were set up there, and meanwhile many returnees came back to set up their own companies and brought back advanced business philosophy and close communication with high-tech areas like in Silicon Valley. Highly skilled talents in hardware and software engineering made this area an effective cluster on the development of the Internet industry. The innovation of business model does not need to rely on too many substantive resources. Therefore, business model innovation is constantly accelerating in the region of Beijing. Nowadays, there are a large number of IT technology enterprises grow quite fast due to their business model innovation in Beijing, such as Xiaomi, DiDi, OFO, Mobike, and so on. Some main examples are reflected on in the following:

Beijing Xiaomi Technology Co., Ltd. was established in April 2010. It is a mobile IT company which focuses on the development of intelligent hardware and electronic products. "Just for fans" is the product concept of Xiaomi. It has created the mode of using Internet to develop mobile phone operating system and encourage enthusiasts to participate in the development and improvement of the system. Xiaomi, after Apple, Samsung, and Huawei, has become the fourth company that obtains the technology and capability to develop its own mobile chips.

To be able to achieve the speed of Xiaomi, the root to make the Xiaomi company create fast-growing is its talent strategy[1]. Since the setup of the company, Lei Jun set up the strategy plan for what the new company will be: one is to learn from Tongrentang (a famous Chinese traditional medicine company) to produce products with real materials with faith; the second is to learn from Haidilao (a famous cafeteria) to provide over-expected service and gain word of mouth reputation; and the third is to learn from Walmart and Costco to gain low margin but high efficiency (Lei 2015). They share their interests with their employees and share as much interest as possible. Xiaomi has carried out the plan of employee stock ownership and investment plan since it was founded.

Xiaomi is now in dominant position in the smartphone market. From the perspective of technological innovation and product innovation, Xiaomi really has nothing to do, but it creates a business model directly communicating with users, which is totally different from the business model of Apple's mobile phone. The key of Xiaomi is to iterate quickly and establish a test version. Since Xiaomi began to update the software every week, users may often find a lot of new functions. These functions may be quite tiny and small, but the company builds a relationship of "accept and give" with users which is very unique (Feng 2015). The business model innovation of Xiaomi is mainly rooted in the following three aspects: (1) Xiaomi

[1]During the first half year after Xiaomi was founded, Lei Jun spent 80% of the time to look after talents and formed a super luxurious co-founder team later. Lei Jun is the chairman of the Kingsoft software and a famous angel investor. Lin Bin was the vice president of Google Research Institute; Hong Feng was a senior engineer in Google; Huang Jiangji was the chief engineer of Microsoft Engineering Institute; Li Wanqiang was the design director for human-computer interaction design in Kingsoft and general manager of Kingsoft Power Word; Zhou Guangping was the chief engineer of Motorola R&D center in Beijing; and Liu De was an industrial designer who graduated from Art Center, one of the world's top design schools.

uses the production and manufacturing model of "independent design and production outsourcing," (2) a combination use of B2C e-commerce model and "light assets" model, and (3) adaptive use of "hunger marketing" and "network marketing" (Wang 2016).

DiDi taxi is another big company success related to business model innovation.

In July 2012, Beijing Small Orange Technology Co., Ltd. was founded and launched the taxi APP DiDi, with followed investment made by Tencent, Alibaba, Apple, and other big Internet companies. The sharing economy represented by DiDi trips originated from overseas country. The business model of DiDi trip changed the traditional way of taking trips in China completely, which made trips with intelligent, low cost, light asset, and diversified features. The company is committed to be a sharing economy practice with response to the Internet innovation strategy in China, working with different communities and industry partners to complement each other, using deep learning technology which is driven by big data so as to solve trips and environmental challenges in China and create social value by enhancing the user experience, building efficiency, and sustaining ecological new mobile travel. Through sharing, carpool, and the complementation of part-time and full-time drivers, it effectively promotes the optimal allocation of social vehicle resources and solves the problem of mobile difficulties in trips.

Today, DiDi has become the largest one-stop trip platform in the world with diversified services. It provides comprehensive trip services in more than 400 cities for nearly 300 million users with taxis, tailored taxis, DiDi express cars, carpools, buses, driving services for enterprise clients, etc. In 2015, DiDi was selected as the Global Growth Companies in Davos.

Since the beginning of the second half year of 2016, bike sharing has sprung up and began to blossom in the big cities of China, which has met the huge demand of people to solve the problem of traveling in short distance. Among all, two companies are the most prominent ones which are OFO and Mobike. In 2014, OFO was founded by five graduates from Peking University. It is the earliest established bicycle-sharing company, which pioneered the bicycle-sharing mode. It is the world's first non-pile sharing platform.

In January 2015, Mobike was established, creating the first intelligent global bicycle-sharing model and development of its independent research on intelligent patent lock integrating GPS and communication module with the use of a new Internet of Things technology. Bikes can be located, and users can find and use the bikes near them whenever and wherever they need through the APP in their smartphone. From the business model, the operating profit of the company is determined by the income and the cost. The products of the two companies are not very different, but the market is different. At the beginning, colleges and universities are the main battlefields. As long as the consumer groups of teachers and students are verified, the first order per day can be free of charge. Second order will be charged 0.5 yuan/hour. Consumers who are not teacher or student need to pay 199 yuan as deposit and be charged 1 yuan/hour. Mobike charges 299 yuan as deposit and 2 yuan/hour for using its bikes. The latest Mobike lite bike charges 1 yuan/hour. Both companies are able to secure a quick return of the deposit as soon as in a second to consumers (Feng 2017).

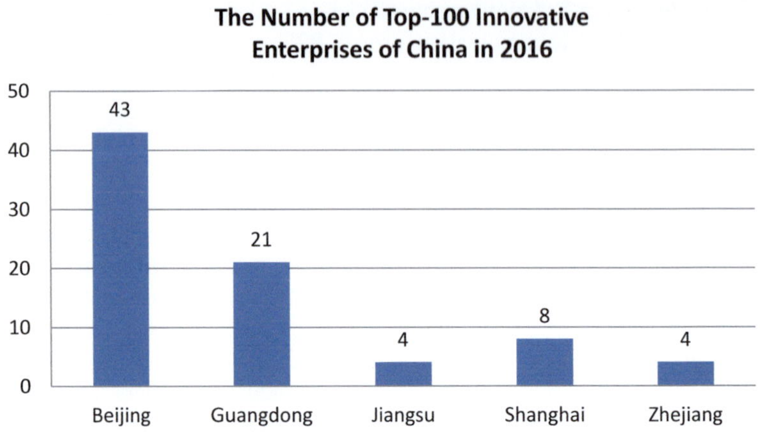

Fig. 7.4 Quantity comparison of the top 100 innovative companies of Mainland China in 2016 in 5 regions. (*Source*: Clarivate Analytics 2016)

Now, there are many innovative companies in Beijing which form a strong entrepreneurial atmosphere. Figure 7.4 above shows the top 100 innovative companies of Mainland China in 2016. The selected companies are geographically concentrated, coming from 15 provinces (headquarter location), mainly in the Eastern China. Forty-three of them are located in Beijing, almost half of the total quantity of innovative companies.

With the emergence of graphene, 3D printing, artificial intelligence, Internet industry, and other new technologies, new processes, new modes, and new formats, Beijing continues to adjust and correct its direction and path selection of the industrial upgrading. Thus, Beijing will continue to be a frontier region in innovation of the digital economy in China.

7.4 Capital Effect: Cluster of SOEs Headquarter

As a capital, many large companies like to stay at Beijing, which makes Beijing having the obvious characteristic of the headquarter economy. Various central government-owned companies and institutions such as China Mobile, four major national banks, Sinopetroleum, and Sinochemical are all located here, producing another cluster effect. The headquarter economy means a lot for the regional development, such as the tax effect, industrial multiplier effect, consumption effect, employment effect, and social capital effect. A batch of headquarters of both domestic and foreign companies can improve the visibility and credibility, promote the government's service quality, optimize the business environment, improve the city infrastructure and living environment, promote multicultural integration and interaction, and accelerate the internationalization. The China's top 500 companies

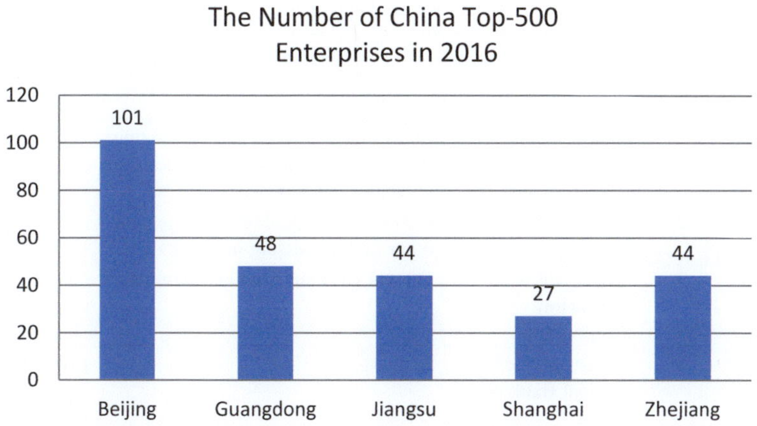

Fig. 7.5 The distribution of China's top 500 enterprises in 2016 in 6 provinces. (*Source*: China Enterprise Federation 2017)

in 2016 are distributed in 6 provinces (Beijing, Tianjin, Jiangsu, Zhejiang, Guangdong, and Shanghai), as shown in Fig. 7.5. Beijing with 101 companies ranks 1st.

By the end of 2008, the headquarters in Beijing, which is 0.3% of the total Beijing's institutions, had controlled 61.3% of the assets of the whole city, accounting for 69.3% of the total profits of Beijing's second and third industries. The characteristics of the headquarter economy also cause the assets of Beijing's second and third industries take 31%. Just the financial industry with the headquarters in Beijing has the assets as high as 42 trillion and 700 billion yuan (Tu and Zhao 2010).

In the selected top 500 companies of Beijing, the large state-owned enterprises occupying are represented most often. As shown in Table 7.4, the top 10 companies in terms of revenues in Beijing are all state-owned.

The state-owned enterprises are one of the important parts of the economic development of Beijing. The cluster effect of the headquarters of state-owned enterprises has played an important role in the economic development. From the macroperspective, state-owned enterprises can overcome the institutional arrangement of "market failure," as the "macroeconomic stabilizer" in period of economic transition by raising the marginal output of capital, improving the capital allocation efficiency, and promoting dynamic TFP growth or using other ways to boost economic growth (Xu, Zhang 2015).

At the same time, state-owned enterprises can also become the center of "technology imitation, technology diffusion and technology catching up" (Liu 2001) and play a central role for external spillover effects (Zhao 2017). In addition, state-owned enterprises have strong foundation, which can play a role for technical demonstration for the development of other enterprises which also have strong investment capability and the ability to absorb talents.

Beijing also has a rich VC pool and other kinds of capital inputs for science-based innovation. In the aspect of investment factor, in 2015, the angel venture and VC/PE investment reached 525.49 billion yuan, which is five times the investment in 2010.

Table 7.4 Top 10 companies in Beijing selected in China's top 500 in 2016

Location	Rank	Top 10 in Beijing	Industry	Operating income (unit: 10,000 yuan)
Beijing	1	State Grid	Energy (power, heat, gas, etc.) supply, development, emission reduction, and recycling services	207,134,945
Beijing	2	SINOPEC	Petrochemical products, coking, and other fuel processing industries	204,727,191
Beijing	3	China National Petroleum Corporation	Oil and gas extraction and production	201,675,666
Beijing	4	Industrial Commercial Bank of China	Banking	107,983,800
Beijing	5	China Construction Bank	Banking	92,587,800
Beijing	6	China Construction Engineering Corp	Construction	88,057,713
Beijing	7	Agricultural Bank of China	Banking	83,376,600
Beijing	8	Bank of China	Banking	76,922,200
Beijing	10	China Mobile	Telecommunications, mail, express, and other services	67,092,753
Beijing	12	China Life Property & Casualty Insurance Group	Insurance	63,644,438

Source: China Enterprise Federation 2017

In the aspect of innovation environment, the R&D intensity of Beijing is ranked top nationally with the rate of 6.01%, 3.94% higher than the average. Informatization index achieves 131.3 with the highest rate of Internet penetration all over the country (Dong 2016).

In the same time, in the first half of 2016, Beijing has established three new national intelligent manufacturing demonstration projects; as a result, Beijing has 18 national intelligent manufacturing projects, which ranks the top in the whole country. National Power Battery Innovation Center has been built in Beijing as the first national manufacturing innovation center.

7.5 S&T and Innovation Performance

With a high density of universities, public research institutes, high-tech enterprises, and graduate students, Beijing has a very high performance in S&T and innovation.

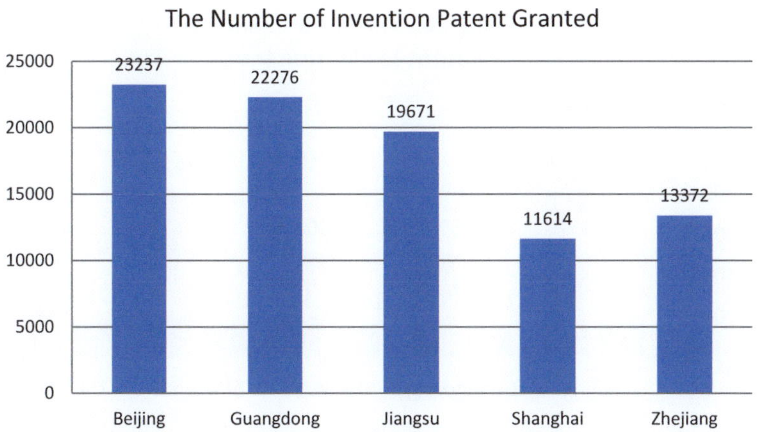

Fig. 7.6 Comparison of six provinces' invention patents in 2015. (*Source*: NBS 2016)

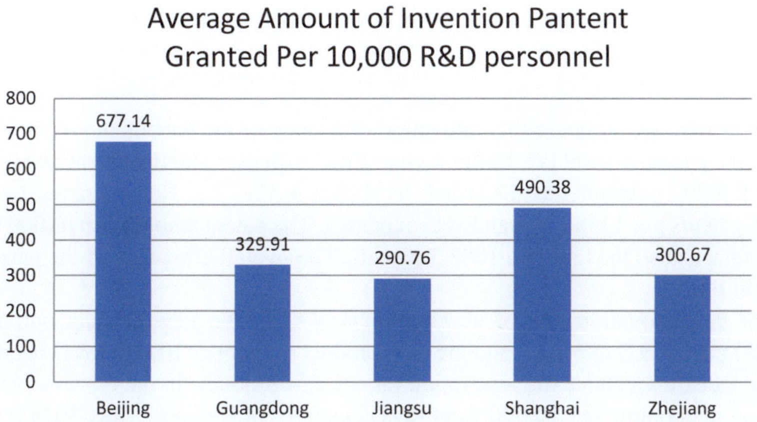

Fig. 7.7 Amount of five provinces' invention patent authorizations per 10,000 R&D personnel in 2015. (*Source*: NBS 2016)

In 2016, the quantity of patent applications and authorizations is 189,129 and 100,578, respectively, with an increase of 21% and 7%, respectively. The quantity of patent applications and authorizations for the invention is 104,643 and 40,602, respectively, which increased by 17.7% and 15%, respectively (Beijing Municipal Bureau of Statistics 2017b).

Figure 7.6 illustrates the comparison of the invention patents in six provinces. It shows that Beijing has the highest number of invention patent authorizations (23,237). Meanwhile, comparing the amount of invention patent authorizations per 10,000 R&D personnel, Beijing is still on the top of six provinces as Fig. 7.7. It demonstrates the strong technology capability of Beijing.

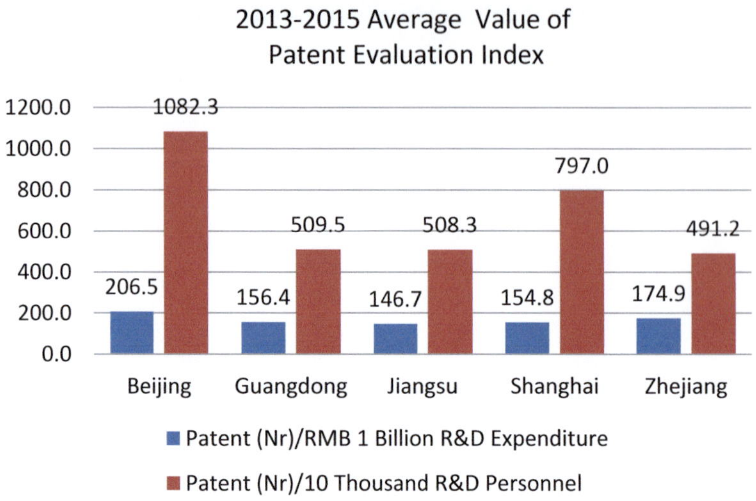

Fig. 7.8 2013–2015 average value of patent evaluation index in five provinces. (*Source*: NBS 2014–2016)

Moreover, we compare the innovation efficiency of the patents in six provinces, using the patent amount per 1 billion yuan R&D expenses and the patent amount per 10,000 R&D personnel as the index, as shown in Fig. 7.8. Beijing ranks 1st with 206.5 patents per 1 billion yuan R&D expenses. The patent amount per 10,000 R&D personnel is also the top with 1082.3 patents. The overall efficiency of the patents is high in Beijing.

For the innovation rate of new products, we use the new product output per 10,000 yuan R&D expenses and the new product output per 10,000 R&D personnel as the index to evaluate the innovation rate of new products in the 6 provinces. The new product output per 10,000 yuan R&D expenses of Beijing is only 30,000 yuan, and the new product output per 10,000 R&D personnel is only 15.7 billion yuan, all the bottom in the 6 provinces. It means the new product output efficiency of Beijing is low, as shown in Fig. 7.9.

We also use the number of international papers per 1 billion yuan R&D expenses and the number of international papers per 10,000 R&D personnel as the index to evaluate the innovation rate of the papers in the 6 provinces. Beijing ranks 1st for both two evaluation indexes, as 561.1 papers and 2941.3 papers, respectively. See Fig. 7.10.

Beijing is also a leading city in terms of VC and incubation. In 2015, the number of venture capital investment in Beijing ranked 1st in China, with investment amount of RMB 43.093 billion, accounting for one third of the total national investment. More than 10,000 "angel investors" are active in Zhongguancun area, accounting for

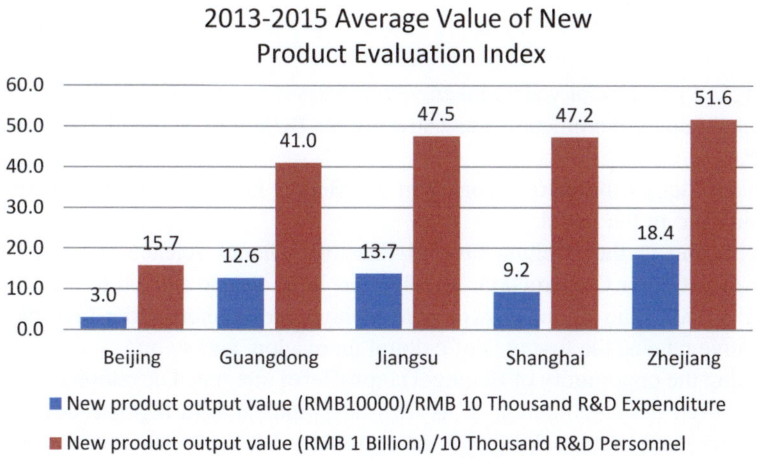

Fig. 7.9 2013–2015 average value of new product evaluation index in five provinces. (*Source*: NBS 2014–2016)

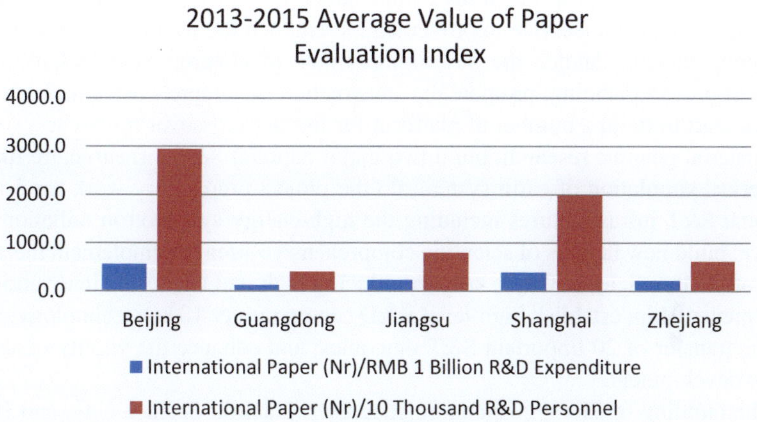

Fig. 7.10 2013–2015 average value of paper evaluation index in six provinces. (*Source*: NBS 2014–2016)

80% of the whole country. Entrepreneurial incubation, entrepreneurial finance, and entrepreneurial business services are developing rapidly. By the end of 2016, there were more than 150 science and technology business incubator and university science and technology parks in Beijing and more than 200 co-working spaces. Among them, 125 institutions were selected for the state-level co-working space. There are nearly 16,000 national high-tech enterprises, 6.13 times as many as in 2005 (Liu 2017).

7.6 Policy for Future

Beijing is a city lack of water and energy; it faces challenge of air pollution. So, the current Beijing government made a strategy for Beijing to be the global innovation center.

This strategy is based on a condition that Beijing has the richest human resource in China and in the world.

According to the "Beijing Overall Plan of National Science and Technology Innovation Center Construction," Beijing has been clearly defined to be the leader of nationwide innovation, growth pole of high-end economy, the first choice of innovative talents, the first area of cultural innovation, and so on.

Taking the opportunity of Beijing-Tianjin-Hebei integrated development, Beijing has achieved the combination of its own development with regional development and the national strategy. Beijing is the region that S&T drives innovation. Massive S&T are developed in Beijing, especially the IT Internet industry.

In the future, Beijing will further expand and optimize the development space of Zhongguancun Science City; actively set up national laboratories and major projects; implement series of projects on S&T innovation/outcome transfer/innovation entrepreneurship platform/service improvement; strengthen the leading and guiding role of Zhongguancun; launch the development plan of Huairou Science City; refine kinds of project planning; promote the construction of comprehensive national S&T center; start to build a number of platform for frontier crossover researches, such as the material genome research; build two major national S&T infrastructure like the numerical simulation of earth system; do the project preparatory work for 3 major national S&T infrastructures including the high-energy synchrotron radiation light source; build new heights of scientific comprehensive strength; implement the action plan of "Future Science City"; keep introducing high-end talents for innovation and entrepreneurship; establish high-level R&D center; master 10 key technology; speed up the transfer of 20 important S&T outcomes; and enhance the vitality of science city's development.

The building of Zhongguancun Science City, Huairou Science City, and Future Science City, and Beijing (Yizhuang) Economic and Technology Development Area is called three cities and one area (see Fig. 7.11).

The strategy of Zhongguancun Science City is to focus on radical S&T breakthrough, radical technology and international standards, innovation centers, and hub for cluster of innovative enterprise with global influence. The new Huairou Science City is to become the national major scientific and technological infrastructure and advanced science and technology cross-disciplinary platforms, as the park of world-class original innovation. The new Future Science City will be the main platform of national science and technology innovation center for large SOEs, a collaborative innovation center and a global leading technology innovation park. Lastly, the technology development area or zone will be served as the space for technology transfer and commercialization. It should become an innovative industrial cluster and "Made in China 2025" innovation demonstration zone.

Fig. 7.11 "Three cities and one area" based on relevant information

References

Beijing Fangdi Institute Economic Development. (2017). *Zhongguancun index 2017*.

Beijing Municipal Bureau of Statistics (BMBS). (2014). *Beijing Statistical Yearbook 2014*. Beijing: China Statistics Press.

BMBS (2016) Beijing Statistical Yearbook 2016. China Statistics Press, Beijing.

BMBS. (2017a). Six high-end industrial functional zones create half the total economic volume of the city. *Beijing Daily*. Aug 30. Derived from http://news.youth.cn/jsxw/201708/t20170817_10528492.htm

BMBS. (2017b). *Statistical bulletin of the national economic and social development 2017*. China Enterprise Confederation (2017) 2016 China top 500 enterprises list.

China Enterprise Confederation. (2017). *China top 500 enterprises list*.

Clarivate Analytics. (2016). *Top-100 innovative companies of mainland China in 2016*.

Dong, X. (2016). The contribution rate of S&T progress in Beijing exceeds 60%, R&D input intensity ranked first in China. *China Business Network*. Sep 06. Derived from http://www.yicai.com/news/5090169.html

Feng, S.. (2015). *Business Model Innovation in China*. Investment and Cooperation. Dec 14. Derived from http://news.hexun.com/2015-12-14/181181349.html

Feng, R. (2017). Discussion on the development of business model under the background of sharing economy—Take the OfO and Mobike as examples. *Journal of Kaifeng Institute of Education, 37*(5), 258–259.

Lei, J., (2015) Xiaomi learned the essence of these four companies. Manager (02): 71–77.

Liu, Y. (2001) On the macroscopic efficiency of state-owned enterprises—Theory and its verification. *Social Sciences in China* (5), 69–81.

Liu, N.. (2004). Creating Lenovo – President Liu Chuanzhi. *Franchise Magazine* (10), 24–27.

Liu, H. (2017). *Beijing's economic development doubles in quality and efficiency: GDP per capita reached 115,000 yuan*. Qianlong.com. Jun 19. Derived from http://finance.qianlong.com/2017/0619/1780895.shtml

National Bureau of Statistics of China. (2010). *China compendium of statistics 19–2008*. Beijing: China Statistics Press.

National Bureau of Statistics of China. (2014–2016). *China's statics yearbook 2014–2016*. Beijing: China Statistics Press.

Renmin University of China. (2017). *China's top 1000 enterprises in innovation capability list 2016*.

Tan, X. (2016), Baidu restructuring of business structure, established "Baidu search company", which is to imitate Google? *PE Daily*. Aug 13. Derived from http://news.pedaily.cn/201604/20160413395673.shtml

Tu, L., & Zhao, H. (2010). Beijing's headquarter economy has obvious features. *Beijing Daily*. Jan 22. Derived from http://news.163.com/10/0122/07/5TK8NJ9S000120GR.html

Wang, T. (2016). Analysis of the impact of business model innovation on corporate working capital management—Take Xiaomi technology as an example. *Commercial Accounting*, (11), 64–65.

Xiang, Y. (2016). Baidu: Driverless enters the world's first camp. *Science and Technology Daily*. Mar 01: 001.

Xie, W., & White, S. (2004). Sequential learning in a Chinese spin-off: The case of Lenovo Group Limited. *R&D Management, 34*(4), 407–422.

Xu, H. (2010). Google withdraws from mainland China, Baidu follows the trend. *Securities Times*. Mar 24: A04.

Xu, Z., & Zhang, W. (2015). SOEs reform impact on economic growth. *Economic Research, 50*(4), 122–135.

Zhao, Q. (2017). Does state-owned enterprise really inefficient? Based on the perspective of regional innovation efficiency spillover effect. *Science of Science and Management of S.& T, 38*(3), 107–116.

Chapter 8
Shanghai: From Industrial City to Global Innovation Center

8.1 Profile of Shanghai Innovation

In 2017, Shanghai's synthetic index of innovation capability ranked 4th in the country for 9 consecutive years. In terms of knowledge acquisition, Shanghai ranks 1st. For knowledge creation, innovation environment, and enterprise innovation, Shanghai ranks 3rd, 4th, and 5th, respectively, whereas all remained stable in comparison to the previous year. Innovation performance ranks 4th, one place higher than the year before (Table 8.1, Fig. 8.1).

In the list of large enterprises in China's top 500 enterprises 2016, there are 27 enterprises from Shanghai, most of which are state-owned enterprises and no private companies in top 10 (Table 8.2).

In China Top 1000 enterprises in innovation capability list, there are 98 enterprises from Shanghai in 2016. Top 20 enterprises are mainly from electronic apparatus and machine building, computer telecommunication, and other electronic device manufacturing as well as special equipment manufacturing (Table 8.3).

According to the data from "2016 Shanghai National Economic and Social Development Bulletin," in 2016, the expenditures of the whole year research and experimental development (R&D) are 103 billion yuan, which accounted for 3.80% of Shanghai's GRP and accounted for 6.6% of the nationwide total R&D expenditures. Published scientific papers and scientific books accounted for 6.1% and 6.7% of the total national quantity, respectively, which are higher than most of the other regions (Shanghai Bureau of Statistics, 2017).

The number of patents can in some way reflect the strength of patent application awareness and principle innovative capability. In 2016, there are 119,937 patent applications in Shanghai, which showed an increase of 19.9% compared to the previous year, accounting for 3.5% of the total national patent applications, of which 54,339 patent applications were filed, showing an increase of 15.7%. The number of patents granted for the whole year was 64,230, up by 5.9%. Among them, the number of invention patent grants was 20,086, up by 14.1%. The number of

X. Liu et al., *Regional Innovation Index of China: 2017*,
https://doi.org/10.1007/978-981-13-1205-2_8

Table 8.1 Shanghai innovation capability comprehensive index

Name of indexes	2017 comprehensive value	2017 sub-index rank			
	Index value	Rank	Strength	Efficiency	Potential
Comprehensive value	44.81	4	6	2	28
1 Knowledge creation	42.21	3	5	2	18
1.1 R&D input	44.09	2	4	2	9
1.2 Patent	40.1	7	6	5	19
1.3 Research paper	42.65	3	3	6	27
2 Knowledge acquisition	54.06	1	3	1	10
2.1 Science and technology cooperation	33.59	6	3	16	13
2.2 Technology transfer	33.15	3	3	7	27
2.3 Foreign-funded enterprise investment	85.09	1	2	1	6
3 Enterprise innovation	42.83	5	5	2	27
3.1 Enterprise R&D input	49.24	6	6	5	24
3.2 Design capability	30.24	6	6	4	29
3.3 Technology improvement	51.55	2	5	3	13
3.4 New product sales income	43.36	6	5	2	26
4 Innovation environment	38.73	4	6	2	29
4.1 Innovation infrastructure	27.2	7	9	4	29
4.2 Market environment	67.79	1	3	1	20
4.3 Employee quality	34.92	6	12	3	22
4.4 Financial environment	32.35	6	5	6	24
4.5 Entrepreneurship	31.42	12	7	6	29
5 Innovation performance	49.9	4	5	2	21
5.1 Macroeconomic	61.23	6	12	3	25
5.2 Industry structure	43.27	3	5	2	23
5.3 Industrial international competitiveness	45.59	5	3	6	18
5.4 Employment	21.6	18	25	4	24
5.5 Sustainable development and environment	77.82	4	5	4	8

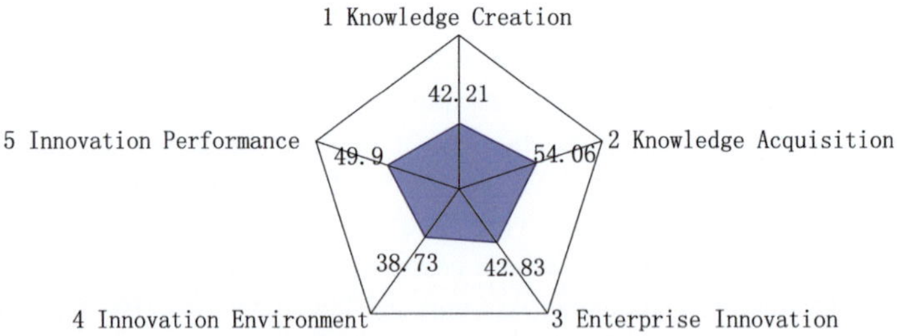

Fig. 8.1 Shanghai innovation capability

Table 8.2 Shanghai enterprises in the China's top 500 enterprises list 2016

	Enterprise	Operation revenue (billion yuan)	Rank
1	SAIC Motor Corporation	670.45	11
2	Bank of Communications	357.99	33
3	Shanghai Pudong Development Bank	264.13	49
4	CEFC China Energy	263.15	50
5	China Pacific Insurance (Group)	247.20	52
6	Baosteel Group	230.06	57
7	Bright Food (Group)	147.56	105
8	Shanghai Tobacco Group	131.83	115
9	Shanghai Construction Group	125.43	119
10	Bailian (Group)	119.84	121
11	Shanghai East Best and Lansheng International (Group)	118.94	122
12	Shanghai Pharmaceuticals Holding	105.52	133
13	China Eastern Air	97.07	155
14	Shanghai Electric Group	96.52	156
15	Shanghai Junhe Group	66.38	205
16	Shanghai Huayi (Group)	61.39	223
17	Shanghai Urban Construction (Group)	55.04	242
18	Shanghai Fosun High-Tech (Group)	53.79	247
19	Shangtex (Group)	46.07	290
20	Shanghai Yongda Holding (Group)	44.61	293

Source: China Enterprise Confederation, 2017

accepted PCT international patent application in the whole year was 1560, an increase of 47.2% to the previous year. By the end of 2017, there were 85,049 valid patents (Shanghai Bureau of Statistics, 2017).

8.2 Shanghai: From an Industrial City to a Global S&T and Innovation Center

Shanghai used to be the main industrial city in China for a long time. By 1978, Shanghai's GRP is about the same level as of Hong Kong. By 2016, Shanghai is about 415 billion USD, and HK is 321 billion USD.

By 1949, Shanghai contributed more than half of the national GDP. With many foreign and domestic industrial companies, Shanghai was the most industrialized city in China in that time. Entering the 1950s, many Shanghai skilled workers went to other regions such as northeast and west of China to engage in emerging industries located in these regions. Shanghai used to dominate a number of industry sectors, among them textile, steelmaking, chemical engineering, automobile, and so on.

Table 8.3 Enterprises in the "China top 1000 enterprises in innovation capability" list

	Enterprise	2016 Invention patent application number
1	Shanghai Phicomm Communication Technology	1931
2	China Construction Eighth Engineering Division	1078
3	Shanghai Siemens Medical Equipment	688
4	Shanghai Belling	664
5	Shanghai Huali Microelectronics	551
6	Shanghai Guao Electronic Technology	504
7	Hudong Heavy Machinery	482
8	SETP	434
9	Tianma Microelectronics	399
10	Saic Motor	397
11	Shanghai Municipal Engineering Design Institute (Group)	395
12	Shanghai United Imaging Healthcare	391
13	Shanghai Mechanized Construction Group	363
14	Opple Lighting service	352
15	Hua Hong Semiconductor	341
16	Shanghai Koito Automotive Lamp	314
17	China Union	278
18	Shanghai Nuclear Engineering Research and Design Institute	274
19	Shanghai Engineering Research Center for Nanotechnology	259
20	Shanghai Droi Tech	243

Source: Renmin University of China, 2017

Currently, combined with the development of Shanghai industrial foundation and advantages, six industries were identified as being highly developed, including electronic information product manufacturing, automobile manufacturing, petrochemical and fine chemical manufacturing, high-quality steel manufacturing, complete sets of equipment manufacturing, and biological medicine manufacturing. In 2016, the industrial output value of the six industries totaled 211 billion yuan, accounting for 67.6% of the total industrial output value of Shanghai (Shanghai Bureau of Statistics, 2017). However, we can observe different economic characteristics in these industries.

Auto manufacturing is a competitive industry with a powerful momentum of development in Shanghai. Auto manufacturing developed fastest, with the output value of 578.158 billion yuan, accounting for 17.5% of the gross value of Shanghai's industrial output, increasing 12.6% over the previous year.

The biomedical manufacturing industry ranked second with an output value of 95.9 billion yuan and a growth rate above 5%, which is also a strategic emerging industry for the future development of Shanghai, but it is still in the early stages of developments at a relatively small scale.

Since the new century, the Shanghai's automobile manufacturing industry is more developed. The change of people consumption concept and the upgrading of consumption structure have promoted the rapid development of industry and enterprise. Especially the industrial transformation caused by the deep integration of "Internet Plus" and manufacturing industry is forming new modes of production, industrial morphology, business models, and economic growth points.

As a large and open city, Shanghai used to host scaled investment and love imported technology from abroad for mass production. So, they gave large multinationals much favorite policy for a long time; in this way, SME was less motivated (Lu and Chang, 2014). Here there are many large-scale, super-scale state-owned companies, and foreign-funded companies by large multinational companies, of which the ownership structure is 25% central, 5% local state-owned, 25% foreign-funded, and 25% private (Wei, 2016). For a long time, these companies have different degrees of monopoly in the market due to various reasons. In 2015, the gross industrial output value of Shanghai state-owned companies was 36.8% of the total: 37.3% in 2014 and 37.3% in 2013. By analyzing the statistical data of Shanghai in recent 10 years, we can figure out that this proportion has always been maintained at 35% and above.

Most of the companies in Shanghai are medium-to-large state-owned companies in the traditional industries. Under the planning economy for a long period, many of them are quite weak in terms of indigenous innovation. Even after the 30-year reform, they have still not become competitive subjects for the tech innovation. Influenced by historical factors and mechanisms, the state-owned companies in Shanghai are facing a series of problems, including lagging production capacity, sunk costs, inflexible operation, employee settle, etc. If just following the old traditional development mode, companies will be easily trapped by the technological lock-ins and have difficulty getting out. The only way to regenerate the industrial life is to accelerate the innovation transformation. With the increasingly fierce market competition, deepening reform, and higher degree of marketization, the state-owned companies in Shanghai have implemented the reform in the forefront of the country. The previous phenomenon, which the companies existed and developed relying on administration and monopoly resources, has gradually disappeared. The innovation transformation could facilitate the state-owned companies' long-term development in the market competition. Moreover, some state-owned companies have already achieved the largest scale in the domestic industry under the existing mechanisms and technical conditions. The marginal profit of production factors, such as capital and labor force, is declining. To maintain the sustainable and positive development of the companies, innovation on mechanisms and technology is needed (Cai, 2014).

8.3 From FDI to Open Innovation

Shanghai is an immigrant city with a high degree of internationalization. Shanghai-style culture has a strong and comprehensive innovation capability, creating an atmosphere of entrepreneurship, enabling individuals to develop.

In 1843, Shanghai was announced to be a treaty port, but at that time, there were only hundreds of foreign traders, far from being an international city. Between 1920 and 1940, the population of Shanghai reached about 3 million, including nearly 100 thousand residents from foreign countries, such as England, the USA, France, Germany, Italy, Portugal, Japan, India, etc. These foreigners worked, lived, and engaged in trading business in Shanghai, co-working with natives, which laid the foundation for Shanghai to become an international city (Yang, 1993).

The historical factors have greatly influenced the development of Shanghai's globalization and have more standardized policies and stronger attractions for the foreign people. The increase of the Shanghai's resident foreign population is closely related to the development of foreign capital economy. Shanghai has attracted a large number of foreign enterprises, and foreign capital economy has already become one of the important motivations to promote the development of Shanghai economy.

In 2016, the amount of foreign direct investment in Shanghai was \$18.5 billion, a 3.5 times increase from the \$2.4 billion in 2000 (Beijing Bureau of Statistics, 2017). The proportion of permanent resident foreign population has increased rapidly, increasing from 0.28% in 2000 to 0.41% in 2016. The resident foreign population ranks first in China. The development of foreign-funded enterprises in Shanghai has become a significant reason for attracting foreigners to live in Shanghai, and the staff and their family members of foreign-funded enterprises have become the most important components of resident foreigners. According to the data from Sixth National Population Census of Shanghai in 2010, half of the foreign residents in Shanghai come from Japan, the USA, and Korea. Europe and America area mainly include the USA, France, Germany, Canada, etc.; also the amount of residents from African countries shows an upward trend. Most of foreign residents have a higher level of education, 71.03% of which have got college or above degrees. The foreign residents are normally highly educated and working-aged (15- to 64-year-old). However, in terms of the proportion, there is still a large gap compared with other typical international metropolises, where foreigners can make up more than 10% of the total population.

In view of the increasing foreign population, Shanghai also issued a series of exit-entry policies to make it easier for foreign residents to work in Shanghai. Shanghai is promoting the integration of foreign population into the city with a more open mind. These policies aim to attract overseas high-end talents, gather and support their innovative entrepreneurship, and improve and optimize the soft environment of science and technology innovation center. Meanwhile, many universities in Shanghai, such as Fudan University, Shanghai Jiaotong University, Tongji University, and East China Normal University, use various measures to attract overseas students, which urgently prompted the local government to pay more attention to the introduction of international talents (Zhu, 2016).

Relying on transnational corporations and joint venture, such as Shanghai Bell, Shanghai has rapidly improved the capabilities of domestic companies in IT industry and gained advanced foreign managerial experience, technological process, and access to international markets. For example, while developing large-scale manufacturing in Shanghai, Transcend Information Inc. is equipped with the development of large-scale research and development facilities, as well as the situation that the manufacturing industry is the focus of Shanghai's development.

In 2016, the total amount of foreign direct investment reached 18.514 billion dollars, of which the actual amount of the tertiary industry is 16.335 billion dollars, accounting for 88.2% of the actual use of foreign capital. There were 168 countries and regions investing in Shanghai. The number of regional headquarters of transnational corporations located in Shanghai reached 580; moreover, 330 investment companies and 411 foreign-funded R&D centers were also located in Shanghai.

8.4 Global S&T and Innovation Center

If local and FDI helped Shanghai finish the industrialization, Shanghai wants to solid its resources on science-based innovation and aims to be the global innovation center beyond the catching-up process.

Since 2010, Shanghai tries to transform its status as manufacturing city into a global S&T and innovation. This is the result of the national strategy for China to become the S&T Power and Innovation Power country by 2050. And Shanghai and Beijing have to be the leading region for that goal.

Shanghai's science and education resources are second, just behind Beijing[1].

In 2015, scientific research institutions totaled 206 in Shanghai, with 11,910 published science and technology papers, 37% of which were published abroad. Accepted patent applications recorded 5605. Especially invention patent application amounted to 4444, and the proportion is as high as 79.3%. In the same year, scientific research institutions in colleges and universities totaled 227, with 59,957 published academic papers, including 30,774 published abroad, accounting for more than 50%, and 35,481 of them (nearly 60%) were recorded by SCI, EI, and ISTP.

[1]By the end of 2015, the total number of professional personnel in Shanghai had exceeded 4.73 million, and there were 67 universities, 44 national key laboratories, and 11 municipal key laboratories. In Shanghai, 177 academicians of the Chinese Academy of Sciences or Chinese Academy of Engineering and 771 overseas high-level personnel were selected into the "National Thousand Talents Plan"; 676 overseas high-level personnel were selected into the "Shanghai Thousand Talents Plan"; 2787 students studying abroad were selected into the "Shanghai Pujiang Talents Plan"; 371 people were selected as national candidates for "Million Talents Program"; 1186 people were selected into the "Leading Talents Plan"; and 1021 personnel were selected into "Shanghai's Chief Technologist Thousand Talents Plan."

The number of patent applications was 9455, of which 7687 were invention patent applications, accounting for 81.3%. The number of patents granted was 6116, of which 4387 were invention patent granted, accounting for 72%. The number of patent application and patents granted from research institutions and universities has increased year by year, especially invention patent, normally accounting for 70%–80% (Beijing Bureau of Statistics, 2016). These data show that Shanghai's universities and research institutions have a good performance, and Shanghai put much emphasis to the intellectual property protection of scientific and technological achievements. In addition, under the situation of collaborative innovation becoming a new trend of today's innovation, the number of cooperative patent applications also showed an upward trend, which reflected the innovation level with external cooperation of Shanghai's universities and research institutions, showing a great significance in integrating innovation resources and improving innovation efficiency.

A number of mega-science projects have also been initiated in Shanghai, greatly enhancing its technological innovation capacity. Examples are Shanghai Supercomputing Center, protein scientific research facilities, Synchrotron Radiation Facility, 65 meters radio telescope, brain artificial intelligence, giant aircraft project, lithography machine project, etching machine project, and other advanced research projects.

Universities and research institutes in Shanghai and their top scientific research level and professional talents have made Shanghai to build up a solid industrial and technological foundation. A large number of returnees tended to choose Shanghai as their foothold. By the end of 2015, more than 130,000 returnees had been working or started up business in Shanghai, and more than 4900 companies had been set up in Shanghai.

Furthermore, Shanghai's universities and research institutes have attracted a large number of skilled talents and gathered rich scientific and technological innovation resources, meanwhile driving the development of intermediary organizations related to science and technology innovation services, e.g., technology trading market, property right trading market, entrepreneurship center, public service platform, etc. Shanghai has formed an effective carrier and a huge network of scientific and technological innovation. As the center of the Yangtze River delta urban agglomeration, Shanghai has the ability of cooperating with domestic and foreign organizations and sharing resources of various kinds of science and technology innovation.

In the future, Shanghai will focus on the following key activities in order to sustain its transformation:

1. Focusing on science driven industry

The biomedical industry is a typical science-driven industry. In the "2014–2017 Shanghai biological medicine industry development plan," issued by Shanghai, it said it will continue to be more developed as the manufacturing, business, and innovative R&D center of biomedical high-end products in the Asia-Pacific region. The aim is to realize a leapfrog development of Shanghai's biological medicine industry.

In recent years, a large number of the leading international biomedical industry enterprises have emerged in Shanghai, such as Roche, Siemens Healthineers, Boehringer Ingelheim, Microport, United Imagine, Tofflon Technology, Kaibao Pharmaceutical, CP Guojian Pham, etc. Competitive biological pharmaceutical products cover a range of fields including tumor, cardiovascular, diabetes, neurological diseases, antibody drugs, genetic engineering drugs, small molecule chemical medicine, digital medical imaging, minimally invasive interventional instrument, orthopedics, rapid diagnostic reagents, etc. Numerous pharmaceutical companies, research institutes, and various service platforms in Shanghai have formed a R&D innovation industrial cluster composed ((of "2 universities, on institute and one academy, 18 public service platforms and more than 40 centers." Now, there are two medical instrument clusters – the international medical zone and Zhangjiang east zone – and a cluster of high-end medical services in the international medical zone. Twelve globally well-known multinational pharmaceutical enterprises have set up regional headquarters or R&D center in Zhangjiang, including some domestic topping pharmaceutical enterprises, such as Shanghai Hengrui Pharmaceutical Co., Ltd.; Shanghai Fudan Zhangjiang Biomedical Co., Ltd.; Hutchison Whampoa (Shanghai) Co., Ltd.; and Shanghai CP Guojian Pharmaceutical Co., Ltd.

2. More funds for S&T

In 2016, Shanghai's R&D reached 104.9 billion yuan, a 10% increase than in 2015; its GERD/GDP also reached a highest level of 3.72% (Fig. 8.2).

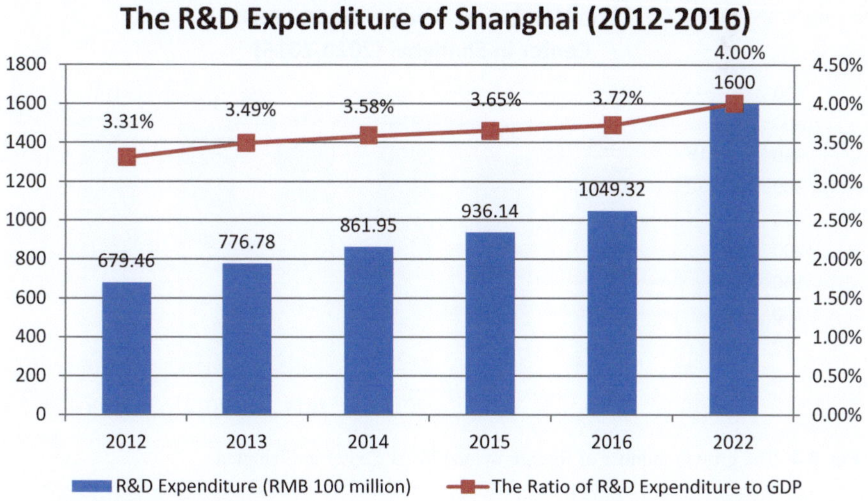

Fig. 8.2 GERD/GDP 2012–2016. Note: The value of 2022 is a predicted one. (*Source*: Shanghai Bureau of Statistics, 2013–2017)

3. Strengthening the position as a hub for R&D center for multinationals

Shanghai used to be the center for foreign R&D, as they have good infrastructure for R&D. Aiming to be the global S&T and innovation center, they will like to be a better place for global R&D center. From Fig. 8.3, we can see the emergence of foreign workspace in Shanghai, and the number of foreign R&D is growing in recent years (Fig. 8.4).

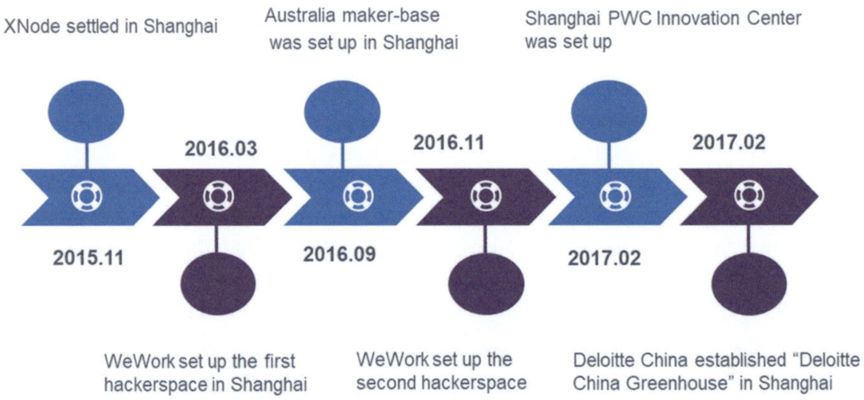

Fig. 8.3 The emergence of foreign workspace in Shanghai

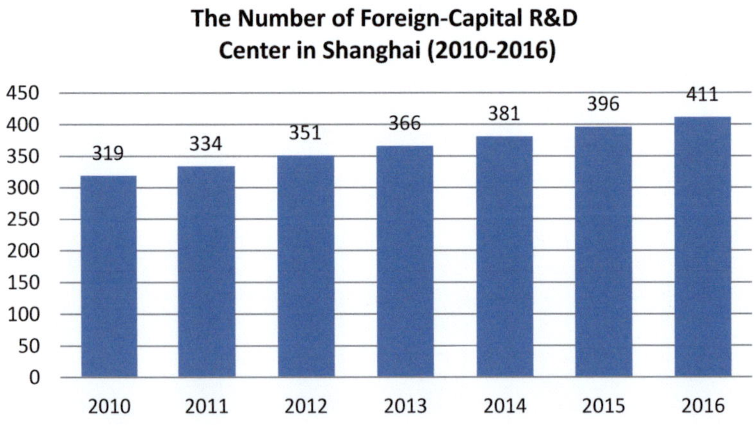

Fig. 8.4 The growth number of foreign-owned R&D Center in Shanghai

8.5 Construction of Special Functional Zones

Entering post-industrialization stage, Shanghai has actively established and developed many special functional zones as new model of development, including special high-tech zones, trade zones, and so on. Each zone is operated independently and is located in the periphery of Shanghai. Main examples are discussed in the following:

8.5.1 Shanghai Zhangjiang High-Tech Industrial Development Area

A successful technology zone is not built on the imitation. It needs excellent creation and targeted construction. Shanghai Zhangjiang High-Tech Zone satisfies this point and has become the symbol of Shanghai technology innovation and the direction of its technology innovation transformation. Since its establishment in 1992, Shanghai Zhangjiang High-Tech Zone has been internationally famous for its reputation as "The Silicon and Medicine Valley in China." Until 2011, it has formed "Zhangjiang National Indigenous Innovation Demonstration Zone" with 13 zones, including Core Zone, Caohejing, Zhabei Zone, Qingpu Zone, Jiading Zone, Jinqiao Zone, Zizhu Zone, Yangpu Zone, Xuhui Zone, Changning Zone, Hongkou Zone, Songjiang Zone, and Minxing Zone. In 2014, it extended to 22 zones. Zhangjiang High-Tech Zone is a successful project with more than 70,000 tech companies, 2million employees, 1400 R&D institutions, 700 multinational corporations' R&D centers and regional headquarters, 300 public service platforms, 44 colleges and universities, and 80% high-end talents of the whole city. It has formed large industrial clusters including biomedicine, information technology, energy saving, environmental protection, high-end equipment manufacturing, new materials, new energy, new energy vehicles, cultural-tech integration, and modern service industry. It has played an important role of agglomeration, demonstration, guidance, and replication, as one of the most intensive areas of innovation resources in China.

Zhangjiang gains comprehensive advantages from its technology innovation resources and industrial clusters. By sufficiently taking advantage of both international and domestic resources and markets, Zhangjiang Zone has extended its innovation development to the Yangtze River Delta, the Yangtze River Economic Belt, and the Belt and Road Initiative. Zhangjiang implements the "government guiding + marketized operation" mode, following international standard and experiences of international advanced innovation centers. For example, in the field of integrated circuit, it has studied from Hsinchu, Taiwan, to explore the pilot of bonded industrial chain; it has also learned from Bangalore, India, to build the Pudong national software base and the software service outsourcing base; in the field of biomedicine and information industry, Zhangjiang has used the mode of Silicon Valley and Boston, America, to introduce returnees, support company-centered innovation and entrepreneurship, and allocate large-scale venture capital funds; and in the field of S&T center and academic city construction, it has combined the innovation research centers of universities and innovation departments of multinational institutions for advanced S&T researches, referring to the US National Laboratory and Japan's Tsukuba mode. Now, Zhangjiang High-Tech Zone has become the innovation highland for forefront researches and international incubator cooperation. It has achieved its leading role in technology innovation, system innovation, industrial clusters, urban renewal, talent development, and many other aspects.

8.5.2 Yangpu "National Innovative City Pilot": An Innovation Base Around Universities

In 2010, Yangpu was authorized as the "National Innovative City Pilot" and became the agglomeration area of universities, scientific research institutions, and science and tech valleys of Shanghai. It has implemented the innovation development idea that "integrates three types of zones and develops interactively": to promote the technology innovation and output transfer through integrating the resources from university zones, technology zones, and public community zones. Yangpu also became the national innovation-driven demonstration zone in 2016.

The "university zone" is responsible for knowledge creation and talent cultivation, providing human capital and intellectual support; "technology zone" takes the responsibility of technology incubation and innovation products, providing carriers for university-industry-research cooperation, innovation and entrepreneurship, and employment; and "public community zone" provides the public services to universities and tech valleys, building living space and ecological space for living, communication, and recreation.

The universities in Yangpu have formed deep cooperation projects such as "one circle, one corridor, one valley, one zone." The "one circle" is the update of Tongji knowledge economy circle; "one corridor" is to promote the innovative corridor of Fudan; "one valley" is to build the global financial valley; "one zone" is to accelerate the Terahertz Industrial Zone of Shanghai University of Technology. Yangpu takes advantage of the interaction of university-tech-community zones, positioning as knowledge innovation area, and becomes the central area of Shanghai with innovation characteristic. Promoting the close cooperation of universities, tech zones, and communities could create the sustainable productive forces and a good ecological environment for the economic development that benefit the innovation. The integration and expansion of various resources could achieve higher value. The interactive development of universities and cities is the important issue for the urban development. If the universities cannot effectively interact with research institutions, companies, and governments, it will cause the knowledge spillover and the decline of resource utilization rate.

Based on universities' resources, Yangpu chooses the marketization thought and innovation mechanism to allocate the international innovation resources from the global view. It also supports the organization system of multidisciplinary and multi-subject collaborative innovation and develops strategic emerging industries relying on the new Internet economy. Yangpu keeps promoting the upgrade of traditional industries and establishes transformation funds in various fields, such as industry, finance, and talent. It has firstly built an incubator service chain with four levels, before entrepreneurship – entrepreneurship emerging – incubator – accelerator, to boost the entrepreneurship and the innovation of old industrial areas.

8.5.3 *Shanghai Free Trade Pilot*

As an internationalized and open city, Shanghai established the Shanghai Free Trade Pilot in 2013, which covered four custom supervision areas: Shanghai Waigaoqiao Bonded Zone, Waigaoqiao Bonded Logistics Zone, Yangshan Bonded Port Zone, and Shanghai Pudong Airport Bonded Zone. The Free Trade Pilot could facilitate the allocation of international industries, promote the transformation and upgrade of domestic industries, gain international competitive advantages, and offer the high-level platform for the connection of international finance, business, and services. By the end of April 2016, there are 35,000 new companies set up in the pilot, within 5500 foreign-funded companies. More than half of the foreign-funded companies in Shanghai settled in the free trade zone. The number of foreign-funded companies in the zone increased from 5% to 20%. Now, the free trade zone has created 25% of Shanghai's GRP in its 2% land area.

The system innovation is the core of the development of the free trade zone. Shanghai free trade zone takes the companies as the subjects to transform the government functions and launches a series of measures for system innovation in four fields, including investment management, trade regulation, finance innovation, and process supervision. According to the practices, the system innovation of the free trade zone has supported the companies' innovation in three aspects:

One is to help the companies to reduce the trade expenses, increase the productivity and factor fluidity. It further increases the efficiency of resource allocation, which is fully shown in various innovation cases of financial services and specialized service platforms.

Second is to incent the innovation of market subjects. It could benefit the technological update of companies, which is already reflected in the innovation cases of R&D services, overseas equity investment, and tax reporting.

The third is to guide the market subject to construct a reasonable organization structure to achieve the sale merit, which is used in the cases of capital management and supply chain management (Bai et al., 2015).

The greatest advantage of the innovation reform pilot in the free trade zone is the market vitality stimulated by the created regime dividends. The foreign investment regime has changed from the traditional "positive list" to the "negative list." It is not only a change of a word but also a great change of the government administration. From restricting the companies "what could be done" to regulate them "what cannot be done," the change of administration lets the market play a bigger role and be consistent with China's reform and internationalization. The state-owned companies will have more investment choices and good legal environment; the entrepreneurial companies will also get better starting points. The innovation of the trade supervision leads to the "Enter before Declaration" system. It directly improves the efficiency of the custom clearance, which reduces average 10% costs of logistics and warehousing due to original 2–3-day waiting time. This complete change of traditional custom declaration process facilitates the emergency orders and consequently seizes the domestic and foreign market. In the free trade zone, the market access threshold is destroyed, which causes the rationalization of emerging industries' investments. All these innovations in the free trade zone have gradually changed the foreign trade mode of China.

8.6 Policy for Future

Shanghai government has taken great steps to become a global innovation center, based on its traditional resources, such as human resources, good infrastructure, and globalization.

8.6.1 Attention on Private and Small-Medium Companies

To build the technology innovation center of Shanghai, efforts of private and small-medium companies are needed in the future. In recent years, with a series of preferential policies by the Shanghai government, as well as the increasing opening degree of the market, the development environment of small-medium companies in Shanghai is getting better and better. But overall, the current development mode of Shanghai is still led by large companies. Relatively, the innovation activities of private and small-medium companies have been limited. The innovation activities will bring positive impacts to Shanghai's development. The business environment for the small-medium companies, especially for those without state-owned sectors, has difficulty in getting the support from the national finance. The financial services hardly offer the large-scale systematic support to the technology innovation activities of small-medium companies (Jiang, 2010). This situation has begun to change, which is good.

The private companies are also an important part of social economic development. Comparing the top 50 state-owned companies to the top 50 private companies in Shanghai, we found out that the R&D expenses of state-owned companies are more than that of private companies, as well as the quantity of invention patents. However, the private companies own much higher growth rate of the invention patents. Moreover, state-owned companies have larger scale of technical personnel than that of private companies. The strength of technical personnel represents the scientific and technological strength of the company to a certain degree, which is directly related to the technology innovation capacity and invention patents. Thus, it further influences the development and performance of the companies. At the same time, the scale and proportion of highly educated personnel in state-owned companies is much larger than those of private companies. The talent attraction and incentives of private companies do not get enough awareness, which restrict the companies' innovation (Chang and Hang, 2015).

According to the "The Innovation Capacity Survey of Shanghai's Small-Medium Companies" by Shanghai Democratic National Construction Association (Xu, 2016), the innovation weaknesses of Shanghai's small-medium companies mainly include "limited policy role," "lack of R&D talents," "lack of innovation projects," "difficult risk support," and so on. These factors also impact the innovation environment for small-medium companies to develop. The current innovation planning and policies of Shanghai for existing small-medium companies are weak, on aspects of its innovation environment, talent attraction and training, VC contact, innovation and entrepreneurship support, etc. On the side of innovation investment, the proportion of added innovation expense to the company's total income is relatively small; on the side of innovation output, the proportion of new product sale to the total sale is relatively small, too; and on the side of innovation resources, the informatization level of small-medium companies is in general or even lower. These companies are facing the terrible situation of R&D talent shortage. 84% of them do not have any experts named in the "Thousand Talents Program," 80% of them do not have government-identified experts or leaders, and only 12% of these companies have one to five leading figures and experts. For those technical small-medium companies, their unbalanced development reflects in several points: first, the technology ability is irregular, as most of them develop their technologies following the hot issues, such as electronic information and biomedicine; second, the patents are concentrated in few companies, and 50% companies develop without any own patents. What's more, there is a huge gap on the patent quantity among companies with different patent structures.

8.6.2 Building Innovation Ecosystem

The large company priority strategy of Shanghai causes the gap between large and small companies, which also limits the innovative spirit of entrepreneurs in Shanghai. The companies' R&D expenses and technology activities are little, and most of them lack the core technology and independent intellectual property rights. Compared to the companies in developed countries and advanced regions of China, companies in Shanghai have not really become focused on R&D, innovation, and output application. The leading role of companies in the innovation activities is insufficient. In addition, the leading companies in domestic emerging industries seldom set up their headquarters in Shanghai. Baidu chooses Beijing, Alibaba chooses Hangzhou, and Tencent and Huawei choose Shenzhen.

In the future, the goal to build the internationalized city, global innovation center, and four centers of traditional industries will offer a well regional economic environment for Shanghai's state-owned companies, as well as various advantages like technology, talent, and brand. A series of incentive policies of innovation and transformation also provide strong guidance and support. The increasing degree of opening up has accelerated the internationalization of state-owned companies, which help them to get more opportunities to engage into key projects (Cai, 2014). However, for private and small-medium companies, the living environment greatly impacts them to develop innovation ability. Shanghai should further enhance its awareness of innovation and entrepreneurship, improve the level of entrepreneurship, and focus on the innovation and transformation of existing companies as well, including more preferential policies and development planning, relieving companies' pressure, and attracting more tech-innovative talents. Shanghai needs small-medium companies to build the global technology innovation center. Take Germany as an example. Its success is built based on thousands of small-medium companies with core technology. To support national industries, relying solely on large companies is not enough.

To Be an Important Manufacturing and Innovation Center

Shanghai city is an ideal developmental platform for multinational corporations because it has the global resources needed for today's innovation. However, Shanghai's overwhelming superiority in traditional manufacturing leads to challenges in industrial transformation and new innovative enterprise development. First, Shanghai's digital economy industry is not as developed as Shenzhen and Beijing. Second, the new growth point of entrepreneurial driven is relatively slow. Third, in the process of scientific driving innovation, Shanghai has encountered the problem of the slow development of some important biological industries.

References

Bai, S., Bai, D., & Geng, Z. (2015). Case study of Chinese enterprises on innovation drive. *Science & Technology Progress and Policy, 32*(22), 88–92.

Cai, W. (2014). Research on the innovation and transformation of state-owned Enterprises in Shanghai. *Journal of Shanghai Economic Management College, 12*(6), 1–9.

Chang, Y., & Hang, A. (2015). Analysis and suggestions on the innovation capabilities of private-owned Enterprises in Shanghai. *Shanghai Enterprise, 9*, 68–70.

China Enterprise Confederation. (2017). 2016 China Top 500 Enterprises List.

Jiang, Y. (2010). *Research on model of Shanghai science and technology innovation development*. Shanghai: Shanghai Jiao Tong University.

Lu, & Chang. (2014). The evolution of "made in China" to "make in China": Research summary of innovation ability and its enlightenment to Shanghai. *Productivity Research*, (08), 128–130.

Renmin University of China. (2017). China's top 1000 enterprises in innovation capability list 2016.

Shanghai Bureau of Statistics. (2013–2017). Shanghai Statistical Yearbook 2013–2017. Beijing: China Statistics Press.

Shanghai Bureau of Statistics. (2017). Statistical bulletin of the national economic and social development 2016.

Wei, D. (2016). Innovation development in Shanghai suffered four major bottlenecks. Aug 31. Derived from http://sz.house.qq.com/a/20160831/035957.htm

Xu, Z. (2016). What is the shortcoming of SME innovation in Shanghai? Shanghai Committee of China Democratic National Construction Association first released the "Report on the Innovation Capability of SMEs in Shanghai." *United Times*. Sep 30: 002.

Yang, X. (1993). From open ports to international cities: Shanghai and Hong Kong – historical experience can be learned from Shenzhen. *Practice and Theory of Special Zones*, (3), 33–34.

Zhu, B. (2016). *Study on the integration of Shanghai's foreign population in cities*. East China Normal University.

Chapter 9
Zhejiang Province: Role of Entrepreneurship and Industry Cluster

9.1 Profile of Zhejiang Innovation

In 2017, Zhejiang's innovation capability ranks 5th in China, the same as the previous year.

Looking at the sub-indexes, Zhejiang ranks 3rd for enterprise innovation, 5th for knowledge creation, 6th for innovation environment, 8th for knowledge acquisition, and 9th for innovation performance (Table 9.1). Relatively, enterprise innovation capability in Zhejiang innovation system is higher than in other dimensions (Fig. 9.1).

There are 44 enterprises from Zhejiang in the "China's top 500 enterprises list 2016," most of them are private enterprise and ranking up middle in the list, including Wuchan Group, Geely, and Alibaba (Table 9.2).

In the China's top 1000 innovative enterprises list, there are 90 enterprises from Zhejiang. Top 20 are mainly from industries like electric apparatus and equipment manufacturing, computer, telecommunication, and other electronic device manufacturing, software, and information technology service (Table 9.3).

Main actors of the Zhejiang regional innovation system are SMEs. They show a quite good entrepreneurial behavior and can quickly identify customer needs, rapidly occupy resources, and therefore develop a large number of special industries. Alibaba has originated in Zhejiang, nowadays one of the largest Internet companies in the world.

In recent years, Zhejiang Province has promoted the construction of special industrial towns. This new model has played an important role in gathering talents, promoting innovation, and expanding effective investment.

According to the statistics of National Economic and Social Development Statistical Bulletin in Zhejiang Province in 2016, the annual research and development (R&D) expenditure in the whole society is 113 billion RMB, equivalent to 2.43% of local GDP in 2016, a level similar to the developed country level in EU.

X. Liu et al., *Regional Innovation Index of China: 2017*,
https://doi.org/10.1007/978-981-13-1205-2_9

Table 9.1 Zhejiang innovation capability comprehensive index

	2017 Comprehensive value	2017 Sub-index rank			
Name of indexes	Index value	Rank	Strength	Efficiency	Potential
Comprehensive value	37.66	5	5	7	25
1 Knowledge creation	34.63	5	4	12	16
1.1 R&D input	31.55	6	5	6	14
1.2 Patent	45.64	6	4	9	10
1.3 Research paper	18.78	23	7	29	29
2 Knowledge acquisition	22.86	8	5	11	24
2.1 Science and technology cooperation	30.15	10	11	8	17
2.2 Technology transfer	12.6	16	7	30	30
2.3 Foreign-funded enterprise investment	25.08	6	5	6	15
3 Enterprise innovation	50.07	3	3	5	21
3.1 Enterprise R&D input	70.77	3	4	2	12
3.2 Design capability	17.97	13	5	24	24
3.3 Technology improvement	41.02	6	3	13	29
3.4 New product sales income	76.22	1	3	1	8
4 Innovation environment	34.97	6	5	7	26
4.1 Innovation infrastructure	30.43	6	5	5	25
4.2 Market environment	45.61	5	5	6	19
4.3 Employee quality	30.58	9	6	20	30
4.4 Financial environment	29.22	7	6	9	17
4.5 Entrepreneurship	38.99	6	4	11	22
5 Innovation performance	38.9	9	6	10	18
5.1 Macroeconomic	60.89	7	4	5	19
5.2 Industry structure	27.69	10	4	9	24
5.3 Industrial international competitiveness	10.51	21	8	25	10
5.4 Employment	30.32	6	6	7	17
5.5 Sustainable development and environment	65.12	15	22	10	17

9.2 Entrepreneurship: From Lagging Behind to Catch Up

Zhejiang Province, located south of Shanghai, had little central government investment before 1978 as it is close with Taiwan. But Zheijing emerges as a powerful region after 1978 with fast industrialization, especially in light industry with cluster advantage. As a catching-up region, with limited government investment, Zhejiang's

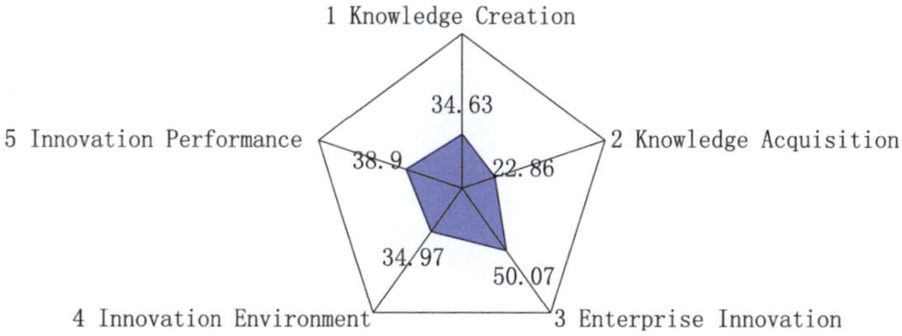

Fig. 9.1 Zhejiang innovation capability

Table 9.2 Zhejiang enterprises in the China's top 500 enterprises list 2016

	Enterprise	Operation revenue (billion yuan)	Rank
1	Wuchan Zhongda	182.57	87
2	Zhejiang Geely Holding	165.30	99
3	Haigliang	140.16	108
4	Wanxiang	115.36	125
5	Alibaba	101.14	148
6	China Guangxia	89.71	162
7	Qingsha	87.69	164
8	RiseSun Enterprises	80.61	172
9	Zhejiang Xinghe	80.20	174
10	Hengyi Petrochemical	79.41	177
11	China Tobacco Zhejiang Industrial	78.61	178
12	Hangzhou Iron & Steel	77.70	179
13	Chilwee	74.98	186
14	Tianneng Battery	72.39	191
15	Zhejiang Energy	67.95	201
16	Youngor	62.36	219
17	Zhongtian	61.97	220
18	Aux	59.61	229
19	Yin Yi	58.84	231
20	Hangzhou Turbine	57.01	235

Source: China Enterprise Confederation 2017

development model is to build an entrepreneurial region around low- to medium-tech industries. Many cities win the competition from other regions by specializing in some products, such as lighters (Wenzhou), textile (Shaoxing), and shoes (Wenzhou).

Though mainly low-tech, companies in Zhejiang feature a core competence of marketing and distribution. This gives Alibaba the power to development. Alibaba is born and located in Hangzhou, the capital of Zhejiang. Ma Yun, the founder of Alibaba, graduated from Hangzhou Normal University, in China, which is a low-below average one. In some way, Ma Yun only extends the capability to the

Table 9.3 Enterprises in the "China top 1000 Enterprises in innovation capability" list

Rank	Name	Invention patent application 2016
1	Zhejiang Supor	827
2	Fotile	821
3	Zhejiang Geely Automobile Research Institute	708
4	New H3C Group Telecommunication Technology	595
5	Zhengtai (Chint)	458
6	Uniview Technology	450
7	Aux	417
8	NetEase (Hangzhou)	341
9	Huadian Electric Power Research Institute	341
10	Huada Technology	323
11	Hangzhou Hikvision Digital Technology	312
12	Hangzhou DPtech Technologies	258
13	Zhejiang Aishida Electric	237
14	Zhejiang Yat Electrical Appliance	170
15	Jack Sewing Machine	169
16	Hengdian Group DMEGC Magnetics	148
17	Xilinmen Furniture	144
18	Zhejiang Roomeye Energy-Saving Technology	127
19	Hangzhou Honyar Electrical	123
20	Tianneng Battery	82

Source: Renmin University of China 2017

online business. When he set up Alibaba at Hangzhou, he knows the e-commerce business is what the million SMEs in Zhejiang needed. It provides the platform for both B2B and B2C business for SME.

The root of entrepreneurship in Zhejiang can date back to Ming and Qing dynasty; in that time, Zhejiang and Shanghai were the early capitalist regions with lots of textile and other industries in China. Many of new industry in that time were imported or learned from European country; it is called westernization (Yangwu) movement. So, Shanghai-Zhejiang-Jiangsu areas have richer entrepreneurial culture than other parts of China. In 2015, among the over 40,000 industrial enterprises on above industries in Zhejiang Province, only 750 of them are state-owned and state-holding enterprises.

The entrepreneurial culture also owned a fact that Zhejiang lacks natural resources. 70% of Zhejiang's land is mountains, 10% of water, and 20% of plains, and so, they have long been devoting to optimize productivity in within limited space. From black economy and town economy to industrial zone, development zone, and highly innovative zone and to cluster area and science town, Zhejiang tries every effort to maximize the utilization of limited resources.

In the early 1980s, many Zhejiang business people were the early bird of doing private business with great venture. At the early stages of the reform and opening up, Zhejiang people dared to be the first, breaking the original mode of production and

Table 9.4 Industrial sector breakdown in Yiwu City 2007

Sector	Number of firms	Turnover (billion RMB)
Textile	1700	5
Clothing	1750	3.5
Jewelry and ornaments	1000	4
Hosiery and socks	1400	10
Bands	430	0.3
Zip	300	5
Printing and packaging	750	3
Arts and crafts products	700	2.5
Cosmetics	150	1

Source: Bellandi and Lombardi 2012

embarking on the road of adventure and also creating a group of world-known Zhejiang merchants. At the same time, as they did not have much capital and technology in heavy industry, they focused on consumer- or family-used products for new venture. The local government has made great efforts to support innovation and entrepreneurship in those industries, not only to simplify the examination and approval and value-added services but also pay attention to the cultivation of market-oriented incubator and the introduction of industrial funds into the back end.

Take Yiwu City as an example; ever since the "barter chicken feather for candy" started in Yiwu City over 30 years ago, the private economy was blooming here for small commercial products. Yiwu now is the world's largest wholesale center for small products; they have low-value and daily used products. Yiwu wholesale market has around 70,000 booths. There are accessories, Christmas products, all kinds of hardware, and many other products. Most of these products are not made in Yiwu but made in cities closed by. About the booth owner, some are factory and others are middlemen, and it needs some experience to distinguish them.

Now, Yiwu has become the world's largest commodity distribution center, being recognized by the United Nations and the World Bank and other international authoritative institutions as the world's largest market. Table 9.4 can show us how the Yiwu developed in daily used products.

Wenzhou is origin of the first individual industrial and commercial household in China after the reform and opening up. It has the first joint-stock company registered in Wenling and Geely which later merged Volvo. Geely also is the first Chinese private company among the world's top 500 companies. Besides, Wanxiang has acquired more than 20 industries in the USA, and the average annual net assets abroad is more than 80%, far higher than that of American counterparts. Alibaba created the world's largest e-commerce platform, becoming a world leading company of big data.

Following the Internet revolution, second-generation Zhejiang entrepreneurs are born. This new generation, with Alibaba as frontrunner, comes from Alibaba staff, overseas talents, university students, and Zhejiang business people, with Ma being the representative, pioneered in entrepreneurship and innovation, making Zhejiang one of the Chinese economic developed areas. With their sharp market sense and innovative spirit, they are active both at home and abroad, becoming the source of vitality in Zhejiang.

Today, the number of shared workspace, the heat of entrepreneurship, and the award-winning in innovation and entrepreneurship competitions in Zhejiang are among the top in all provinces. In order to better protect the grassroots innovation business groups, the government is expanding its supports in shared workspace and technology services, to create a fair competitive environment for innovation and strengthen the building of innovation platform, the integration of technology and finance, the scientific and technological innovation system, as well as the protection of intellectual property rights. In December 2015, *Opinions of Zhejiang Provincial Government on the Implementation of Shared Work Space Innovation* was promulgated, another significant policy support in promoting Zhejiang as a leader in innovation and entrepreneurship.

9.3 Building Special Cluster: From Low Tech to High Tech

Porter (1998a, b) defined the industrial cluster as the geographic concentration of interconnected companies, specialized suppliers, service providers, and firms in related industries and associated institutions. Zhejiang can be considered as a typical industrial cluster region in terms of Porter's definition. Towns with distinctive industry, strong cultural atmosphere, and ecological environment, with both tourism and community features, are strategies in Zhejiang for accelerating regional innovation development under the new economic norm and also the effective path to promote supply and demand structural reform in new urbanizations. A very interesting lesson is that the special industrial cluster as way of industrialization in rural area is not the design of the government and vice versa; it is the autonomy result or bottom-up effect. Different regions learn from each other and get their industrial specialization.

Building special industrial towns with characteristics has become Zhejiang's key way to improve regional innovation and industrial transformation and upgrading. At present, there are three types of town construction. For the first type, enterprises take the main role, and the government provides services, responsible for the town's positioning, planning, infrastructure, and approval, guiding private enterprises to build towns with characteristics.

The second type is the cooperation between government and enterprises, and the government makes plans and cooperates with big enterprises to cultivate large industries. The third type is government construction plus marketing, with the government setting up state-owned enterprises and national investment promotion according to industry positioning.

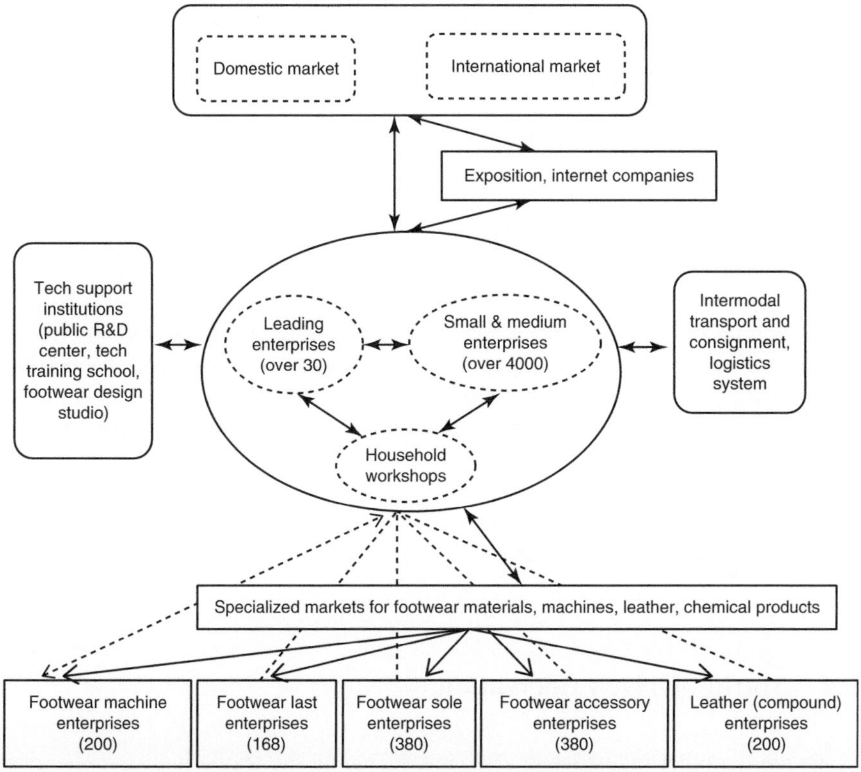

Fig. 9.2 Wenzhou shoe cluster. (*Source*: Huang et al. 2008, P411)

For instance, after 1978, Wenzhou private shoemaking business developed rapidly. By the end of 1981, there were 99 shoemaking factories in Wenzhou, but by 2006, Wenzhou had become the most important footwear production base in China (Huang et al. 2008) (Fig. 9.2).

Their model of special industrial town later became a national best practice for the whole country. In 2016, after experts' review and recognition by the National Development and Reform Commission and the Ministry of Finance, 127 towns were identified as the first batch of towns with Chinese features.

Towns with features conform to the law of industrial structure evolution and become an important starting point to solve the shortage of effective supply in Zhejiang. Shaoxing textile, Datang sock industry, Shengzhou tie, Haining leather, and other economies are contributing greatly to Zhejiang's developing from a province with limited resource into a big manufacturing province with strong economy and market.

Entering the 2010s, government likes to promote the knowledge level of special town. The new special town is more closely related to the seven strategic new industries and historical industries. It focuses on the most basic and advantageous special industries and avoids homogeneous competition. For example, the Cloud

Town and Dream Town feature on information economy with the former emphasizing on big data and cloud computing and the latter focusing on "Internet Entrepreneurship and Venture Capital." Special towns also require investment projects to attach importance to quality and closely integrate with real economy transformation, paying attention to cutting-edge technology, emerging industries, high-end equipment, and advanced manufacturing. As of November 2016, the first batch of 37 special towns attracted over 3300 enterprises and 13,000 talents, including a large number of young talents, bringing in new investment, new projects, and new tax revenues. They are also targeting high-end industries, introducing into leading teams with strong innovation and enterprises with high potentials of development, along with encouraging the new college graduates who were born after the 1990s, the senior executives of big companies, technicians, and returnees to become entrepreneurs, particularly for providing a start-up platform for the young entrepreneurs with dreams, passions, and creativity yet without capital, experience, or supports.

The building of special towns complies with the development law and trend in Zhejiang Province, once again lifting the spirit of "dare to be the first, especially to start a business" among Zhejiang people.

9.4 Market-Driven Innovation

There are ten pillar industries in Zhejiang, including the textile industry; electrical machinery and equipment manufacturing; common equipment manufacturing industry; chemical fiber manufacturing industry; transportation equipment manufacturing; pharmaceutical manufacturing industry; metal industry; plastic industry; textile and garment, shoes, and hat manufacturing; and handicrafts and other manufacturing, the majority of which is efficiency-driven.

Though an SME-dominated region, the manufacturing industry in Zhejiang is export oriented. In 2016, Zhejiang's total exports amounted to RMB 1766.6 billion, accounting for 12.8% of the country's total exports, and private enterprises exported RMB 1338 billion, accounting for 75.7% (Zhejiang Provincial Bureau of Statistics 2017). According to the statistics from Hangzhou Customs, in 2014, seven categories of traditional labor-intensive products (including textiles, clothing, bags, footwear, toys, furniture, plastic products) were approximately USD 104.97 billion, accounting for 38.4% of the province's total exports and 21.7% of total national exports. The exports of electromechanical products were about USD 112.62 billion, accounting for 41.2% of the province's total exports and 8.6% of total national exports. Therefore, the manufacturing industry has a huge effect on the exports in Zhejiang Province and is one of the main driving forces of economic growth in Zhejiang Province. This proves that SME in Zhejiang has efficiency advantage by clustering and specializing in some regions.

Fig. 9.3 E-commerce enterprises in five provinces and cities in 2015 (Source: NBS 2016)

In the era of Internet, Zhejiang enterprises quickly use Internet for product marketing. So, the information industry in Zhejiang Province is well developed, among which the development of e-commerce is the most prominent.

Based on the constant expansion of the Internet market, the number of online shops in Zhejiang reached 1.47 million in 2015. At present, there are more than 1000 e-commerce platform websites in Zhejiang, accounting for more than 20% of that of China, ranking the first in the country. In terms of the strength of the platform, nationally, about 85% of the country's online sales, 70% of cross-border e-commerce (including B2B), and more than 60% of e-commerce transactions are made on Zhejiang platforms (Wang and Zhang 2015). At the same time, the number of enterprises with e-commerce transactions in Zhejiang is in the lead in China; see Fig. 9.3.

Case: Alibaba

Alibaba Group was established in 1999 by 18 people led by Jack Ma, a former English teacher from Hangzhou, China. From the outset, the company's founders shared a belief that the Internet would level the playing field by enabling small enterprises to leverage innovation and technology to grow and compete more effectively in the domestic and global economies. Since launching its first website helping small Chinese exporters, manufacturers and entrepreneurs to sell internationally, Alibaba Group has grown into a global leader in online and mobile commerce. Today the company and its related companies operate leading wholesale and retail online marketplaces as well as businesses in cloud computing, digital media and entertainment, innovation initiatives and others. In the quarter ended December 31, 2017: Alibaba's Revenue was RMB83,028 million (US$12,761 million), an increase of 56% year-over-year.

At the same time, Zhejiang is a region making small products manufaturing as its core capability, but it is also an export-oriented business. Its trade is dominated by general trades. This feature has become an important aspect of Zhejiang, making it different from other major provinces of foreign trade, including Guangdong, Shanghai, and Jiangsu. The reason why general trades flourish in Zhejiang is that the products of state-owned foreign trade enterprises and private enterprises are innovative, with low cost, diverse, and competitive, directly selling at the international market through trade transactions and obtaining higher added value. The raw materials of exported traditional agricultural products and textile, technology, and mechanical and electrical products do not need to be imported and can be produced locally. Domestic materials can be used in general trades, and the development directly or indirectly leads to a large number of domestic enterprises to export products, making foreign trade enterprises and local enterprises more and more close, and gradually a number of large and small industrial clusters are formed in Zhejiang Province.

This feature of private economy in Zhejiang is clearly different from that of more developed places such as Guangdong and Jiangsu in exports. Entrepreneurs in Zhejiang are bold to "going out" in the economic globalization, building effective "production and sales network" in many parts of the world. However, the location feature of Zhejiang indeed limits its development of processing trade with Guangdong and Fujian in the south and Shandong and Jiangsu in the north, making Zhejiang weak in the transfer of manufacturing in Hong Kong, Taiwan, Japan, and South Korea. Meanwhile, Zhejiang lags behind in introducing foreign capital, and the number of bonded zones, export processing zones, and national development zone is relatively small, resulting in a relatively small number of foreign-funded enterprises on processing trades. Therefore, being unable to utilize processing trade to carry out international trade, Zhejiang has to rely on OEM to undertake foreign OEM business on production and export, making the processing trade taking up an extremely low percentage. On the other hand, it also reflects the fact that Zhejiang makes little use of foreign capital and is unbalanced in trade structure. The processing trade is developing from scratch, from small to large, but it is still a shortage in Zhejiang's foreign trade, in which the wholly foreign-owned enterprises and Sino-foreign joint ventures take the lead, and foreign investment enterprises is the largest component. This situation to some extent limits the development of private enterprises engaged in processing trade, becoming another main reason for development limitations of processing trade in Zhejiang. In addition, Zhejiang private processing trade enterprises started late, due to lack of international experience, talents, and other factors, resulting in the current situation of Zhejiang processing trade imbalance. In the 1990s, trades in Zhejiang are mainly made among state-owned enterprises. After 2010, private economy in Zhejiang has become its largest export channel, becoming the most active economic area in foreign trade. Since the 1980s, the structure of Zhejiang's export commodities has changed a great deal. The proportion of primary product exports in total exports in Zhejiang was decreasing, while exports of manufactured goods including the export of electromechanical products were increasing, becoming a major component and the main driving force of Zhejiang exports. After 2010, the export growth rate of high-tech products is significantly faster than that of mechanical and electrical products, and the export of high-tech products will become a new force in Zhejiang trade. However, the largest proportion of exports of mechanical and electrical products is the main force of industrial manufactured goods in Zhejiang.

Maintaining trade advantages and accelerating the development of competitiveness are the main driving force for Zhejiang change from resources driven to innovation driven, achieving the leap from "made in Zhejiang" to "created in Zhejiang," which is also the major issue concerning the overall development in China.

9.5 Policy for Future

In Zhejiang, the government created a fertile market for rapid growth. According to the practice result of Zhejiang economic growth and innovation of small- and medium-sized enterprise activity, it shows that the government has set up a good platform for the market economy, stimulating the vitality of market cells. The balance between government intervention and market equilibrium can promote economic growth and small and microenterprises to stimulate sustained burst of vitality. For example, starting from 2013, in accordance with the requirements of simplifying procedures, Zhejiang made efforts in clearing, reducing, and controlling rights of provincial departments as most as 60%. It also released a list of power of various provincial departments on the Internet. Through the four clearances and one Internet, Zhejiang sets up the system architecture of the government to fulfill the duty and strives to become a province with least approval procedures, highest efficiency, and optimal investment environment (Su 2016). Zhejiang, which has a higher degree of export-oriented economy, is one of the provinces most influenced by the international financial crisis and also a province fastest recovery.

For a long time, market plays a more important role in promoting innovation than the government in Zhejiang, with the tradition of "inaction." Especially in the last century of the 1980s, the environment was rather tolerant of the things that the environment was not allowed, and yet the business community preferred to do by silent approval, supports, or guidance, allowing them to make mistakes. At the same time, the government advocated the strategies of no argument, no showing off, no discouragement, to promote the self-development of small- and medium-sized enterprises, because the government is little involved, which means that the market plays a big role. However, when the enterprise develops to a certain stage and needs to cross the provinces and cities to participate in bigger competition, the government's support is significant. In particular, Zhejiang mostly consists of small- and medium-sized enterprises, and when a relatively large private enterprise competes with a big state-owned enterprise, it needs the appropriate support of the government, since there is the phenomenon of enterprise helping each other in Zhejiang. With a large number of private enterprises, an individual enterprise's credit is not enough for bank loaning, and it is very common for enterprises to save each other in Zhejiang, while some innocent enterprises in a good momentum of development suffer from this and fail (Cai 2015). Therefore, there should be proper balance between the government's role and the freedom of the market.

Now, Zhejiang government, according to the "13th Five-Year Plan on Technology Innovation," sets up a new strategy by July 2016, which aims to be an Internet Plus as the core of the information economy and enter into the high-end global value chain.

They will build a global technology innovation highland of "Internet Plus" by taking the west side of Hangzhou as the main corridor or platform, so as to achieve the provincial's goal of changing to an innovative province. Alibaba will play a great role in this special region.

To implement the plan, a big initiative is to set a fund of 10–30 million RMB to provide supports for companies in Zhejiang with setting up for national engineering laboratory or major technological fundamental research and a two billion fund for transferring the provincial technology achievements and providing supports for banks and financial institutions with their promotion in building entrepreneur investment and loans.

Meanwhile, some new policies are made to fully develop talent potential, increase talent innovation autonomy, strengthen financial and science and technology performance reward, as well as reform science and technology project fund management.

One of the special financing innovations is to build a fund in 50 million RMB called "Zhejiang Angel Dream Fund" in the Dream Town, focusing on supporting high-quality start-ups. At the same time, some Internet banking, wealth management institutions, and fund industry are rapidly growing and cooperating with governmental poverty alleviation funds, better taking advantage of the role of markets in selecting projects and cost control (Zhu and Wang 2015).

References

Bellandi, M., & Lombardi, S. (2012). Specialized markets and Chinese industrial clusters: The experience of Zhejiang Province. *China Economic Review, 23*, 626–638.

Cai, R. (2015). What is lacking in Zhejiang: Big business and big government. *China Economic Herald, 06*, B01.

China Enterprise Confederation (2017) 2016 China Top 500 enterprises list.

Huang, Z., Zhang, X., & Zhu, Y. (2008). The role of clustering in rural industrialization: A case study of the footwear industry in Wenzhou. *China Economic Review, 19*, 409–420.

National Bureau of Statistics of China. (2016). China's statics yearbook. In *Beijing*. China Statistics Press.

Porter, M. E. (1998a). Clusters and the new economies of competition. *Harvard Business Review, 76*(6), 77–90.

Porter, M. E. (1998b). *The competitiveness advantage of nations*. New York: The Free Press.

Renmin University of China (2017) China's top 1000 enterprises in innovation capability list 2016.

Su, J., (2016) Entrepreneurial and innovative Zhejiang Genes. *Zhejiang Daily*. Aug 11:001.

Wang, Z., & Zhang, M. (2015). Research on the ideas and countermeasures of Zhejiang SME's re-innovation advantage. *Zhejiang Economy*, (3), 21–22.

Zhejiang Provincial Bureau of Statistics. (2017). Statistical bulletin of the national economic and social development 2016.

Zhu, G., Wang, Z., (2015) "Change in Zhejiang" Accumulation to Breakthrough. *Zhejiang Today* (14), 34–35.

Part III
Regional Innovation Capability Report in Different Provinces (Autonomous Regions, Municipalities)

Chapter 10
Tianjin

At the end of 2015, the population of Tianjin reached 15.47 million, and the annual GDP reached 1653.819 billion yuan, with an increase of 5.16% compared to the previous year. GDP per capita reached 107,960 yuan, ranking 1st in the country. The ratio of the three-industrial structure in 2015 changed from previous year's 1.3:49.2:49.6 to 1.3:46.7:52.0, with service sector exceeding over 50% for the first time. The main business income in the high-tech industry is 423.38 billion yuan, accounting for 25.6% of the GDP.

In 2017, Tianjin's innovation capability ranks 7th in China, the same as the previous year, and has been in the same ranking for 10 years since 2008. As to the subindexes, knowledge creation, knowledge acquisition, enterprise innovation, innovation environment, and innovation performance, the rankings are, respectively, 11th, 6th, 8th, 9th, and 5th in the country. Among them, the ranking of knowledge creation capability increased, while innovation environment and innovation performance decreased slightly (Table 10.1, Fig. 10.1).

In 2016, there are 17 enterprises from Tianjin in the "China Top 500 Enterprises List 2016," most of which are large state-owned enterprises. Enterprises in the traditional industry like oil, chemicals, minerals, and machinery are ranking in the front places (Table 10.2).

There are 35 enterprises from Tianjin in the "China Top-1000 Innovative Enterprises List 2016" (Table 10.3). We selected ten enterprises with the largest number of patent applications in 2016 based on the data from patent search and analysis system published by SIPO. From the data, we can find out that the top 10 are mainly state-owned enterprises (SOE) from heavy industry such a petroleum, chemicals, etc., except for Tianjin Santroll Electric Automobile Technology, which is a private company engaging in electronic vehicle technology and accessory production. In emerging industry, only Opzoon Technology. Engaging in cloud computing and big data is in the **list.**

In general, Tianjin's innovation capability is developing stably. With its strong advantages in manufacturing and heavy industry, Tianjin is gradually forming eight dominant industries – aerospace, petrochemical, equipment manufacturing,

X. Liu et al., *Regional Innovation Index of China: 2017*,
https://doi.org/10.1007/978-981-13-1205-2_10

Table 10.1 Tianjin innovation capability comprehensive value

Name of indexes	2017 comprehensive value	2017 subindex rank			
	Index value	Rank	Strength	Efficiency	Potential
Comprehensive value	33.71	7	12	3	12
1 Knowledge creation	27.28	11	12	13	3
1.1 R&D input	38.17	3	10	3	2
1.2 Patent	18.03	24	15	26	29
1.3 Research paper	24.02	14	13	18	8
2 Knowledge acquisition	26.04	6	9	3	6
2.1 Science and technology cooperation	32.69	8	13	7	3
2.2 Technology transfer	13.68	14	12	11	21
2.3 Foreign-funded enterprise investment	30.33	5	9	3	8
3 Enterprise innovation	36.56	8	12	4	18
3.1 Enterprise R&D Input	55.52	5	11	1	7
3.2 Design capability	24.36	8	11	7	12
3.3 Technology improvement	23.84	24	15	14	23
3.4 New product sales income	39.12	8	9	5	23
4 Innovation environment	28.52	9	15	5	22
4.1 Innovation infrastructure	24.91	10	15	7	19
4.2 Market environment	52.82	4	6	4	6
4.3 Employee quality	26.64	15	24	4	7
4.4 Financial environment	12.95	20	16	20	22
4.5 Entrepreneurship	25.26	19	11	13	30
5 Innovation performance	47.22	5	9	3	15
5.1 Macroeconomic	64.39	4	19	1	4
5.2 Industry structure	30.07	8	13	5	20
5.3 Industrial international competitiveness	30.51	10	6	7	14
5.4 Employment	25.23	13	17	3	25
5.5 Sustainable development and environment	85.92	2	4	1	9

electronic information, biomedicine, new energy and new materials, defense technology, and light industry and textile – most of which belong to engineering and technology industry. In particular, the petrochemical industry is an important component in the development of Tianjin. In recent years, Tianjin has formulated the

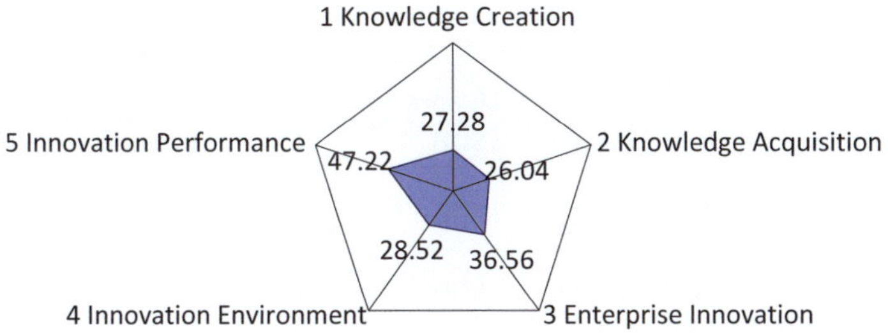

Fig. 10.1 Spider chart of Tianjin innovation capability

Table 10.2 Tianjin enterprises in the "China Top 500 Enterprises List 2016"

	Enterprise	Operation revenue (billion yuan)	Rank
1	Tianjin Material and Equipment Group	403.66	26
2	Tianjin Zhonghuan Semiconductor	121.52	120
3	Tianjin Benefo Equipment Manufacturing	118.66	123
4	Bohai Tianjin Chemical Refco Group	110.09	131
5	Tianjin Bohai Light Industry Investment Group	103.82	138
6	Tianjin Rockcheck Investment Holding Group	55.04	241
7	Teda Investment Holding	52.29	252
8	Tianjin Yishang Group	51.01	258
9	China Bohai Bank	43.74	299
10	Tianjin Textile Group (Holdings) Limited	38.82	337
11	Tianjin Baodi Agri and Tech Group	38.14	343
12	Tianjin Pharmaceuticals	37.35	348
13	Tianjin Port Group	35.08	368
14	TIENS Group	33.21	387
15	Bank of Tianjin	27.91	454
16	Tianjin Real Estate Group	27.35	464
17	Tianjin Youfa Steel Pipe Group	26.17	483

"big project, small giant, and building economy" development plan and achieved some achievements. Projects like FAW Volkswagen North China base, SMIC chip, etc. have promoted industry structure adjustment and improved the manufacturing core capacity. In 2016, there are up to 3900 small giant enterprises, with the output value accounting for more than 50% of the total industrial value. Meanwhile, Tianjin is actively expanding in the strategic emerging industry and gradually switching its industry structure from heavy industry to high-tech industry. However, shortages still exist in technology transfer, industry-university-research cooperation, and

Table 10.3 Tianjin enterprises in the "China Top-1000 Innovative Enterprises List 2016" (Top 10)

	Enterprise	Ownership	Patent (Nr)	Industry
1	CNPC Bohai Drilling Engineering Company Limited	SOE	176	Petrochemical
2	CNOOC Tianjin Chemical Research and Design Institute	SOE	133	Chemistry research
3	China Railway Third Survey and Design Institute Group	SOE	106	Railway survey and design
4	Opzoon Technology	State-owned holding	100	Cloud computing and artificial intelligence
5	China Oilfield Services Limited	SOE	73	Oil and gas extraction
6	Offshore Oil Engineering	State-owned holding	72	Petroleum engineering
7	Tianjin Santroll Electric Automobile Technology	Private	53	New energy/ electric vehicle
8	Tianjin Lishen Battery Joint-Stock	SOE	53	Lithium ion battery
9	China Tianchen Engineering Corporation	SOE	48	Petrochemical
10	Tianjin Optical Electrical Group	Stockholding	48	Optical, fax machines, special equipment

sensitivity in new emerging technology like big data and artificial intelligence. In the future, Tianjin should take advantage of the integration of Beijing-Tianjin-Hebei and strengthen its open innovation ability.

Chapter 11
Shanxi

At the end of 2015, the population of Shanxi province reached 36.64 million, and the annual GDP reached 1276.649 billion yuan, with an increase of 0.04% compared to the previous year. GDP per capita reached 34,919 yuan, ranking the 27th in the country. The ratio of the three-industrial structure in 2015 changed from previous year's 6.2:49.3:44.5 to 6.2:40.8:53.0, and the percentage of the third industry increased. The main business income in the high-tech industry is 86.471 billion yuan, accounting for 6.77% of the GDP.

In 2017, the ranking of innovation capacity in Shanxi province continues to decline, ranking the last in China. As to the subindexes, knowledge creation ranks the 27th and increases by four positions; knowledge acquisition ranks the 22nd and increases by one position; enterprises innovation ranks the 20th and decreases by ten positions than last year; innovation environment ranks the 27th and increases by two positions; and innovation performance ranks 30th and decreases by seven positions (Table 11.1, Fig. 11.1).

There are ten enterprises from Shanxi in the "China Top 500 Enterprises List 2016" and mainly in coal and steel industry. Companies such as Datong Coal Mine Group, Shanxi Coking Coal Group, Taiyuan Iron & Steel, etc. are in the list (Table 11.2).

As to innovation capability, there are six enterprises from Shanxi in the "China Top-1000 Innovative Enterprises List 2016," and all of them are state-owned enterprises mainly in coal, chemical, and steel manufacturing industry. It can be found out that both large and innovative enterprises in Shanxi are mostly in coal and chemical industry, which are closely related to the local nature resources (Table 11.3).

In general, the reason behind the decline of Shanxi's innovation capability is not only the decrease in one or two subindexes. The decrease in enterprise innovation and innovation performance is the reflection of economy decline. The vicious cycle started by the lack of innovation investment and resources and spread to industrial technology and further affected the innovation capability in the region. It is worth noting that this effect is still continuing. However, with the cut of the overcapacity

X. Liu et al., *Regional Innovation Index of China: 2017*,
https://doi.org/10.1007/978-981-13-1205-2_11

Table 11.1 Shanxi innovation capability comprehensive value

Name of indexes	2017 comprehensive value Index value	2017 subindex rank Rank	Strength	Efficiency	Potential
Comprehensive value	17.93	30	23	21	29
1 Knowledge creation	15.43	27	22	22	25
1.1 R&D input	9.97	27	22	23	24
1.2 Patent	18.21	23	21	18	24
1.3 Research paper	20.78	19	23	16	11
2 Knowledge acquisition	13.91	19	23	19	17
2.1 Science and technology cooperation	23.14	17	23	15	7
2.2 Technology transfer	10.57	22	23	15	22
2.3 Foreign-funded enterprise investment	4.19	25	21	23	21
3 Enterprise innovation	15.57	30	20	22	29
3.1 Enterprise R&D input	21.47	26	20	22	26
3.2 Design capability	8.12	28	23	26	26
3.3 Technology improvement	25.65	16	22	8	26
3.4 New product sales income	7.82	30	21	24	31
4 Innovation environment	20.12	27	23	20	25
4.1 Innovation infrastructure	17.77	27	20	16	24
4.2 Market environment	26.87	24	24	30	8
4.3 Employee quality	24.15	21	19	11	18
4.4 Financial environment	10.07	25	23	22	14
4.5 Entrepreneurship	21.74	22	24	15	26
5 Innovation performance	23.03	30	28	21	30
5.1 Macroeconomic	11.07	31	24	27	30
5.2 Industry structure	18.3	17	22	12	17
5.3 Industrial international competitiveness	24.61	11	17	9	11
5.4 Employment	13.85	30	24	21	30
5.5 Sustainable development and environment	47.3	29	26	28	16

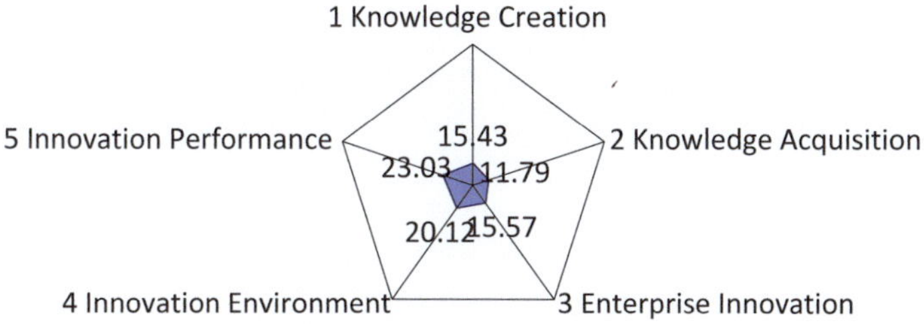

Fig. 11.1 Spider chart of Shanxi innovation capability

Table 11.2 Shanxi enterprises in the "China Top 500 Enterprises List 2016"

	Enterprise	Operation revenue (billion yuan)	Rank
1	Datong Coal Mine Group	200.68	71
2	Shanxi Coking Coal Group	195.06	78
3	Shanxi Lu'an Mining	180.00	89
4	Yangquan Coal Industry	177.91	90
5	Shanxi Jincheng Anthracite Mining Group	173.27	93
6	Taiyuan Iron & Steel	90.84	160
7	Shanxi Coal Imp. & Exp. Group	85.24	165
8	Jinneng Group	77.03	180
9	Shanxi Construction Engineering Group	34.07	373
10	Shanxi Provincial Guoxin Energy Development Group	29.17	433

Table 11.3 Shanxi enterprises in the "China Top-1000 Innovative Enterprises List 2016"

Rank	Enterprise	Ownership	Patent Amount	Industry
1	Shanxi Taigang Stainless Steel	SOE	109	Steel
2	Shanxi Xinhua Chemical	SOE	46	Chemistry
3	Taiyuan Heavy Industry	SOE	37	Machinery
4	CCETG Taiyuan Research Institute	State-owned holding	37	Coal
5	Changzhi Qinghua Machinery Factory	SOE	36	Machinery
6	Sedin Engineering	State-owned holding	25	Chemistry

and over-inventory in coal industry, it can be predicted that the supply side structural reform in Shanxi province can be helpful. In the future, industrial transformation will promote the overall innovation capacity.

Chapter 12
Hebei

At the end of 2015, the population of Hebei province reached 74.25 million, and the annual GDP reached 2980.611 billion yuan, with an increase of 1.31% compared to the previous year. GDP per capita reached 40,255 yuan, ranking the 19th in the country. The ratio of the three-industrial structure in 2015 changed from last year's 11.7:51.0:37.3 to 11.5:48.3:40.2. The main business income in the high-tech industry is 170.587 billion yuan, accounting for 5.72% of the GDP.

In 2017, Hebei's innovation capability ranks the 23rd in China, the same as the previous year, with a slight increase in value. As to the subindexes, the rankings of knowledge creation, enterprise innovation, and innovation environment rank, respectively, are the 21st, 17th, and 16th, with the increase of six, two, and four places, respectively, compared with last year, while the ranking of knowledge acquisition and innovation performance decreased five or six places, now ranking the 29th and the 28th (Table 12.1, Fig. 12.1).

It can be found out from the basic data that Hebei has made huge progresses in R&D input and patent applications with some indexes increasing by double. The development of technology incubations has been paid attention to, and innovation environment has been improved. However, the amount of foreign technology import showed reduction. As to innovation performance, employment in high-tech industry has increased, and urban unemployment rate has decreased.

According to the "China Top 500 Enterprises List 2016" published by China Enterprise Confederation, there are 24 enterprises from Hebei in the list, and most of them are in steel industry such as HBIS Group, Great Wall Motor, etc (Table 12.2). The Great Wall Motor is the only enterprise in the CA-2016 Top 100 Innovative Enterprises List in Hebei province.

When it comes to the patent number, companies with strong innovation in patents are mainly in material, energy, and other heavy industry, and some of them are in the biology and pharmacy industry (Table 12.3).

X. Liu et al., *Regional Innovation Index of China: 2017*,
https://doi.org/10.1007/978-981-13-1205-2_12

Table 12.1 Hebei innovation capability comprehensive value

Name of indexes	2017 comprehensive value	2017 subindex rank			
	Index value	Rank	Strength	Efficiency	Potential
Comprehensive value	20.50	23	16	31	13
1 Knowledge creation	18.47	21	17	27	7
1.1 R&D input	18.10	14	16	19	7
1.2 Patent	17.32	26	19	29	8
1.3 Research paper	21.53	18	16	20	13
2 Knowledge acquisition	9.84	29	20	30	21
2.1 Science and technology cooperation	14.60	30	16	31	26
2.2 Technology transfer	7.75	28	20	25	26
2.3 Foreign-funded enterprise investment	7.83	17	14	18	10
3 Enterprise innovation	21.64	17	14	25	12
3.1 Enterprise R&D Input	33.62	15	12	18	10
3.2 Design capability	10.65	25	16	28	3
3.3 Technology improvement	22.36	25	14	28	20
3.4 New product sales income	19.45	15	14	19	14
4 Innovation environment	23.79	16	12	27	11
4.1 Innovation infrastructure	23.97	13	7	21	11
4.2 Market environment	30.03	18	18	16	9
4.3 Employee quality	26.22	16	10	31	21
4.4 Financial environment	17.82	13	9	15	9
4.5 Entrepreneurship	20.90	26	17	29	13
5 Innovation performance	24.49	28	23	26	25
5.1 Macroeconomic	32.42	16	7	19	27
5.2 Industry structure	12.02	29	14	28	31
5.3 Industrial international competitiveness	6.25	28	20	28	24
5.4 Employment	18.42	21	23	19	14
5.5 Sustainable development and environment	53.36	26	29	21	6

In general, heavy industries including steel and machinery are still the main industries in Hebei province. Meanwhile, there are also innovative companies in high-tech industries such as pharmaceutical, new energy industry, etc. In recent

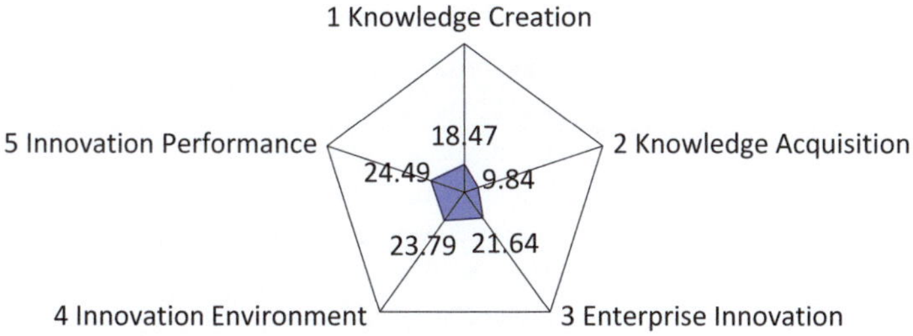

Fig. 12.1 Spider chart of Hebei innovation capability

Table 12.2 Hebei enterprises in the "China Top 500 Enterprises List 2016" (Top 20)

	Enterprise	Operation revenue (Billion Yuan)	Rank
1	HBIS Group	284.47	44
2	Jizhong Energy Group	237.66	55
3	Kailuan (Group) Limited Liability Corporation	171.69	94
4	Hebei Sinogiant Group (Xinhua United Metallurgical Holding Group)	67.97	200
5	Hebei Jinxi Steel Group	64.80	209
6	Hebei Jingye Group	57.57	234
7	Hebei Logistic Industry Group	55.01	243
8	Tangshan Reafon Steel	54.15	246
9	Jinan Steel Group	48.53	276
10	Shijiazhuang Northland Man Group	32.80	392
11	Hebei Construction Group	32.67	394
12	Hebei Puyang Iron and Steel Group	29.83	426
13	Hebei Xinjin Iron & Steel	29.76	427
14	Wuanshi Yuhua Steel	29.66	430
15	Hebei New Wuan Iron and Steel Group, An Iron & Steel	28.24	449
16	Wuan Mingfang Iron and Steel	27.97	451
17	Tangshan Ganglu Iron & Steel	27.94	452
18	Hebei construction Group	27.61	462
19	Great Wall Motor	76.03	183
20	Pangda Automobile Trade	56.38	237

years, Hebei province increased its R&D input and made some achievements in knowledge creation and innovation environment. With the development of Beijing-Tianjin-Hebei integration and the construction of Xiong'an District, Hebei will have more development opportunities in the future. To achieve better developments,

Table 12.3 Hebei enterprises in the "China Top-1000 Innovative Enterprises List 2016"

	Enterprise
1	CITIC Dicastal
2	Chengguang Biotech Group
3	Shijiazhuang Xinhua Energy Environmental Technology
4	Sinosteel Xingtai Machinery & Mill Roll
5	Great Wall Motor
6	Hebei Chenyang Industry & Trade Group
7	The 718th Research Institute of CSIC
8	Jihua 3052 professional suit
9	Yingli Solar
10	Shijiazhuang Chengzhi Yonghua Disaplay Material.
11	ChinaCoal Zhangjiakou Coal Mining Machinery
12	Juli Sling
13	Hebei Huijin Electromechanical
14	Shijiahzuang Yiling Pharmaceutical
15	China Lucky Group
16	JA Solar Holdings
17	Huida Sanitary Ware
18	BOSUN Tools
19	CRRC Tangshan

Hebei needs to put more efforts in fastening knowledge transfer and increasing absorption capacity, benefiting from the opportunities of industrial transfer to promote the transformation and upgrading of local industries.

Chapter 13
Inner Mongolia

At the end of 2015, the population of Inner Mongolia Autonomous Region reached 25.1104 million, and the annual GDP reached 1783.151 billion yuan, with an increase of 0.35% compared to the previous year. GDP per capita reached 71,101 yuan, ranking sixth in the country. The ratio of the three-industrial structure in 2015 changed from previous year's 9.2:51.3:39.5 to 9:51:40. The main business income in the high-tech industry is 39.432 billion yuan, accounting for 2.21% of the GDP.

In 2017, Inner Mongolia's innovation capability ranks the 29th in China, one place lower than that of last year, and the synthetic value is slightly higher than that of last year. As to the subindexes, knowledge creation and enterprise innovation have the same rankings as the previous year, the 31st and 28th, respectively. Knowledge acquisition ranks the 23rd, one place lower than that of last year. Innovation environment ranks the 22nd, six places higher than that of last year, and innovation performance ranks the 24th, four places lower than that of last year (Table 13.1, Fig. 13.1).

According to the "China Top 500 Enterprises List 2016" published by China Enterprise Confederation, there are four enterprises from Inner Mongolia in the list, and most of them are in traditional heavy industry, including Inner Mongolia Power Group, Yili Group, Yitai Group, and Baogang (Table 13.2).

According to patent data, there are only two enterprises from Inner Mongolia in the "China Top-1000 Innovative Enterprises List 2016," which are Inner Mongolia First Machinery Group and Norinco Group (Table 13.3).

In general, Inner Mongolia still has the industrial structure that mainly relies on traditional heavy industry and agricultural and livestock products. In recent years, Inner Mongolia Autonomous Region has been actively creating a favorable financial environment and paying attention to the expenditure of education and talent introduction. However, shortages still exist in enterprise innovation, R&D input, innovation environment, and development of private economy.

X. Liu et al., *Regional Innovation Index of China: 2017*,
https://doi.org/10.1007/978-981-13-1205-2_13

Table 13.1 Inner Mongolia innovation capability comprehensive value

Name of indexes	2017 comprehensive value	2017 subindex rank			
	Index value	Rank	Strength	Efficiency	Potential
Comprehensive value	18.32	29	24	29	19
1 Knowledge creation	11.91	31	26	31	10
1.1 R&D input	12.52	21	25	24	10
1.2 Patent	7.45	30	27	31	25
1.3 Research paper	19.61	22	27	26	2
2 Knowledge acquisition	11.63	23	22	17	28
2.1 Science and technology cooperation	13	31	27	30	20
2.2 Technology transfer	21.9	10	17	2	9
2.3 Foreign-funded enterprise investment	2.9	27	22	16	28
3 Enterprise innovation	16.32	28	22	27	16
3.1 Enterprise R&D input	27.62	20	19	21	17
3.2 Design capability	7.11	30	24	29	16
3.3 Technology improvement	21.64	27	24	25	9
3.4 New product sales income	7.86	29	22	30	18
4 Innovation environment	22.22	22	21	17	16
4.1 Innovation infrastructure	17.74	28	25	17	18
4.2 Market environment	29.21	19	14	29	17
4.3 Employee quality	23.23	23	21	14	8
4.4 Financial environment	23.59	11	17	4	7
4.5 Entrepreneurship	17.32	27	25	27	21
5 Innovation performance	25.75	24	27	19	23
5.1 Macroeconomic	43.07	9	16	6	24
5.2 Industry structure	7.31	31	23	31	30
5.3 Industrial international competitiveness	6.77	26	24	27	17
5.4 Employment	14.61	28	27	26	13
5.5 Sustainable development and environment	56.98	24	25	18	23

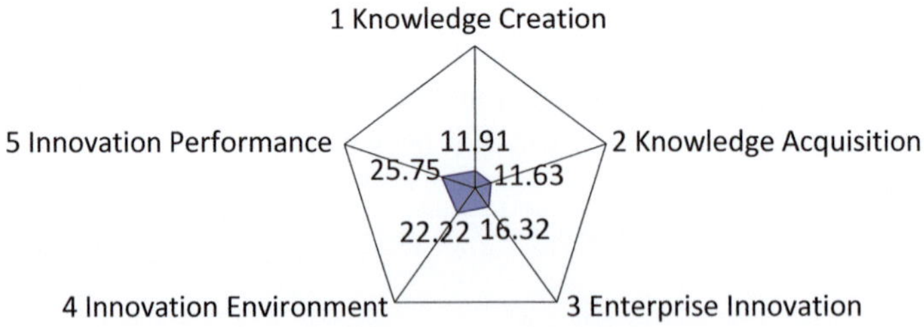

Fig. 13.1 Spider chart of Inner Mongolia innovation capability

Table 13.2 Inner Mongolia enterprises in the "China Top 500 Enterprises List 2016"

	Enterprise	Operation revenue (billion yuan)	Rank
1	Inner Mongolia Power	63.49	217
2	Inner Mongolia Yili Industrial Group	60.36	226
3	Inner Mongolia Yitai Group	32.44	396
4	Baotou Iron & Steel	32.29	397

Table 13.3 Inner Mongolia enterprises in the "China Top-1000 Innovative Enterprises List 2016"

	Enterprise	Patent application (Nr)
1	Inner Mongolia First Machinery Group	18
2	Inner Mongolia North Heavy Industries Group	16

Chapter 14
Liaoning

At the end of 2015, the population of Liaoning province reached 43.82 million; the annual GDP reached 2866.902 billion yuan, with an increase of 0.15% compared to the previous year. GDP per capita reached 65,354 yuan, ranking the ninth in the country. The ratio of the three-industrial structure in 2015 changed from previous year's 8.0:50.2:41.8 to 8.3:46.6:45.1. The main business income in the high-tech industry is 181.375 billion yuan, accounting for 6.33% of the GDP.

In 2017, Liaoning's innovation capability ranks the 17th in China, one place higher than of previous year. As to sub-indexes, knowledge acquisition, knowledge creation, innovation environment, innovation performance, and enterprise innovation rank the 10th, 14th, 20th, 21st, and 24th, respectively. Among them, only knowledge creation ranks seven places higher than the last year, and the other sub-indexes all decreased at different levels. The ranking of knowledge acquisition decreases two places, while the rankings of enterprise innovation, innovation environment, and innovation performance are three places lower compared to that of last year (Table 14.1, Fig. 14.1).

In 2016, there are nine enterprises from Liaoning province in the "China Top 500 Enterprises List 2016," including companies like Wanda Group, Dashang Group, Brilliance Auto, Ansteel, BX Steel, etc. (Table 14.2).

As to enterprise innovation ability, there are 15 companies from Liaoning in the "China Top-1000 Innovative Enterprises List 2016," 13 of which are state-owned (joint venture) from manufacturing, energy, and computer software development industry and 2 of which are private enterprises in machine manufacturing and industrial materials industry. Neusoft and MCC from information industry rank top regarding patent application (Table 14.3).

After comparison, it can be found that the traditional advantage industries in Liaoning province are mainly state-owned enterprises. Private enterprises still lag behind to a large extent in innovation capability compared to state-owned enterprise. It is worth noting that Liaoning has made great efforts in encouraging the development of new emerging industry in recent years and great achievements have been made in its high-tech area such as computing and pharmaceutical

X. Liu et al., *Regional Innovation Index of China: 2017*,
https://doi.org/10.1007/978-981-13-1205-2_14

Table 14.1 Liaoning innovation capability comprehensive value

Name of indexes	2017 comprehensive value	2017 sub-index rank			
	Index value	Rank	Strength	Efficiency	Potential
Comprehensive value	22.26	17	13	15	31
1 Knowledge creation	22.22	14	11	11	30
1.1 R&D input	13.59	20	13	13	28
1.2 Patent	24.99	13	13	12	23
1.3 Research paper	33.92	7	10	8	26
2 Knowledge acquisition	20.44	10	7	7	31
2.1 Science and technology cooperation	35.27	5	5	2	29
2.2 Technology transfer	8.68	25	13	26	31
2.3 Foreign-funded enterprise investment	18.14	9	6	7	29
3 Enterprise innovation	18.9	24	15	19	31
3.1 Enterprise R&D input	23.38	25	15	17	29
3.2 Design capability	11.55	24	15	21	30
3.3 Technology improvement	24.01	23	13	26	15
3.4 New product sales income	18.09	18	15	12	30
4 Innovation environment	22.47	20	11	19	30
4.1 Innovation infrastructure	21.85	16	12	14	30
4.2 Market environment	37.39	7	8	8	28
4.3 Employee quality	23.94	22	13	15	31
4.4 Financial environment	13.99	18	19	23	4
4.5 Entrepreneurship	15.2	30	14	24	31
5 Innovation performance	27.58	21	16	15	31
5.1 Macroeconomic	34.57	14	10	9	31
5.2 Industry structure	16.15	20	12	23	27
5.3 Industrial international competitiveness	9.09	22	16	23	20
5.4 Employment	16.75	26	19	20	27
5.5 Sustainable development and environment	61.35	22	23	11	24

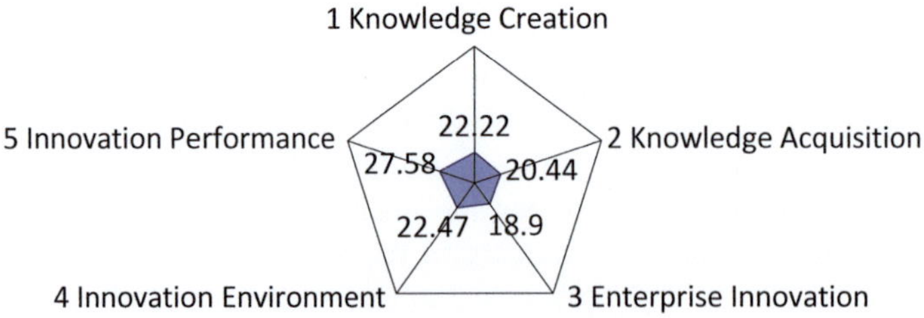

Fig. 14.1 Spider chart of Liaoning innovation capability

Table 14.2 Liaoning enterprises in the "China Top 500 Enterprises List 2016"

	Enterprise	Operation revenue (billion yuan)	Rank
1	Dalian Wanda Group	290.16	42
2	Dashang Group	200.44	72
3	Brilliance Auto Group	160.55	100
4	Angang Steel Company Limited	129.71	118
5	Benxi Steel Group Corporation	77.00	181
6	Jiachen Group	37.86	346
7	Liaoning Rilin Group	28.99	438
8	Dalian Fujia Group	26.39	480
9	Lingyuan Iron & Steel Group	26.79	476

Table 14.3 Liaoning enterprises in the "China Top-1000 Innovative Enterprises List 2016"

Enterprise	Ownership	Industry
Neusoft Corporation	SOE	Software and IT service
Dalian Huarui Heavy Industry Group	SOE	Machine manufacturing/ metal/ architecture
SANY Heavy Equipment	Private	Machine manufacturing
Shenyang Research Institute of Chemical Industry	SOE	Chemistry/ consultant
Neusoft Medical Systems	SOE	Armarium/ computer/ consultant
Shenyang Aluminum & Magnesium Engineering & Research Institute	SOE	Metal/ architecture
Dalian Shipbuilding Industry	SOE	Ship/ oceaneering
CNR Dalian Locomotive & Rolling Stock	SOE	Machine manufacturing
Shenyang SIASUN Robot Automation	SOE	Machine manufacturing
China Zhongwang Holdings Limited	Private	Architecture/ industry/ material
Sunrise Power	SOE	Energy
CFHI Dalian Design and Research Institute	SOE	Machinery
North Engineering & Technology Corporation, MCC	SOE	Mineral/ architecture/ engineering
Shenyang Aerospace Xinguang Group	SOE	Machine manufacturing/ energy/ manufacturing
Shenyang Piotech Co., Ltd.	SOE	Material/ machine manufacturing

industry and so on. The output value of the third industry continued to grow. In general, Liaoning's innovation capability and innovation environment still need to be improved. The government should put more efforts into innovation and entrepreneurship, actively promoting industry-university-research cooperation to further increase enterprise innovation vitality.

Chapter 15
Jilin

At the end of 2015, the population of Jilin province reached 27.53 million; the annual GDP reached 1406.313 billion yuan, with an increase of 1.88% compared to the previous year. GDP per capita reached 51,086 yuan, ranking the 12th in the country. The ratio of the three-industrial structure in 2015 changed from previous year's 11.0:52.8:36.2 to 11.2:51.4:37.4. The main business income in the high-tech industry is 184.847 billion yuan, accounting for 13.14% of the GDP.

In 2017, Jilin province's innovation capability ranks the 28th in China, one place lower than last year. As to sub-indexes, knowledge creation, knowledge acquisition, enterprise innovation, innovation environment, and innovation performance rank, respectively, the 23rd, 20th, 25th, 31st, and 19th in the country. Knowledge creation, knowledge acquisition, and enterprise innovation improved a lot; however, innovation environment ranked last in the country (Table 15.1, Fig. 15.1).

There are only three enterprises from Jilin province in the "China Top 500 Enterprises List 2016"; FAW group plays an important support role in Liaoning's development (Table 15.2).

There are only two enterprises from Jilin province in the "China Top-1000 Innovative Enterprises List 2016." Based on our calculation of invention patent applications from SIPO, the detailed information is showed in Table 15.3.

In general, Jilin's innovation capacity is at the bottom in the country. The momentum of enterprise innovation, aggregation ability, and talent support is relatively weak. Facing with pressure of economy downturn, the contradictions between traditional industry and institutions are exposed. Strong industrial inertia increases the pressure for Jilin in its industry transformation and upgrading with 70% of its economy relying on automobile, petrochemical, and metallurgical.

Faced with the rapid development of the information era, Jilin government is actively changing their ideas in promoting old industrial base with emerging technologies. "The Action Plan of Promoting Big Data" in Jilin Province (2016–2020) was published by the Jilin government to fasten industrial transformation and upgrading, speed up big data construction, and promote innovation development. The State Council approved the establishment of Changchun high-tech district with

X. Liu et al., *Regional Innovation Index of China: 2017*,
https://doi.org/10.1007/978-981-13-1205-2_15

Table 15.1 Jilin innovation capability comprehensive value

Name of indexes	2017 comprehensive value	2017 sub-index rank			
	Index value	Rank	Strength	Efficiency	Potential
Comprehensive value	19	28	21	28	22
1 Knowledge creation	17.24	23	20	15	29
1.1 R&D input	12.09	24	18	15	27
1.2 Patent	15.64	28	20	24	28
1.3 Research paper	30.76	11	18	7	22
2 Knowledge acquisition	13.66	20	19	25	9
2.1 Science and technology cooperation	16.54	27	22	21	28
2.2 Technology transfer	24.49	8	11	12	2
2.3 Foreign-funded enterprise investment	3.37	26	23	19	26
3 Enterprise innovation	17.46	25	21	31	2
3.1 Enterprise R&D input	19.11	28	22	28	23
3.2 Design capability	5.21	31	25	31	22
3.3 Technology improvement	17.98	28	26	30	7
3.4 New product sales income	32.87	11	18	14	1
4 Innovation environment	16.99	31	24	26	28
4.1 Innovation infrastructure	16.98	29	23	18	21
4.2 Market environment	26.83	25	21	18	25
4.3 Employee quality	16.59	28	26	23	28
4.4 Financial environment	7.49	28	25	27	13
4.5 Entrepreneurship	17.05	28	21	21	28
5 Innovation performance	28.75	19	20	17	27
5.1 Macroeconomic	27.7	20	22	12	28
5.2 Industry structure	13.42	26	21	18	28
5.3 Industrial international competitiveness	6.35	27	26	26	22
5.4 Employment	17.54	24	22	11	28
5.5 Sustainable development and environment	78.73	3	6	6	3

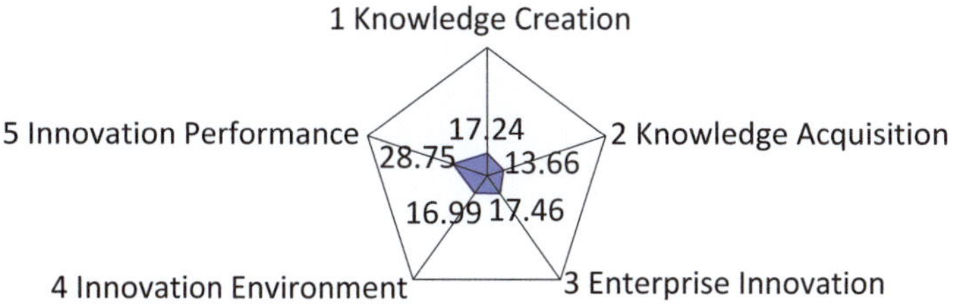

Fig. 15.1 Spider chart of Jilin innovation capability

Table 15.2 Jilin enterprises in the "China Top 500 Enterprises List 2016"

Number	Enterprise	Operation revenue (billion yuan)	Rank
1	China FAW Group	394.99	28
2	Jilin Yatai Group	39.99	329
3	Changchun Ouya Group	36.18	359

Table 15.3 Jilin enterprises in the "China Top-1000 Innovative Enterprises List 2016"

	Enterprises	Ownership	Patent (Nr)	Industry
1	CRRC Changchun Railway Vehicles	SOE	206	Railway vehicles
2	DIRUI Industrial	Listed company	18	R&D, manufacturing, and sales of diagnostic products

the concept of innovation-driven development and planned to build aerospace industry clusters. This will bring new opportunity for Jilin's revitalization, and Jilin should take advantage of this opportunity to add momentums for its development.

Chapter 16
Heilongjiang

In 2017, Heilongjiang province ranked 26th in innovation capability, four places lower than 2016. In terms of different fields, knowledge creation rises 1 place to 15th; the ranking of knowledge acquisition and innovation environment is 17th and 18th, respectively; enterprise innovation and innovation performance rank 31st and 29th, falling 4 places and 1 place, respectively, from the previous year (Table 16.1, Fig. 16.1).

According to the released data from China Enterprise Confederation, it is showed that in "China Top 500 Enterprises List 2016," there are two Heilongjiang enterprises, Heilongjiang Great Northern Wilderness Agribusiness Group Corporation and Harbin Electric Corporation (Table 16.2).

Heilongjiang province has a total of 8 enterprises in the "China Top-1000 Innovative Enterprises List 2016," including 6 state-owned enterprises and 2 private enterprises. From the perspective of industry distribution, the patent innovation of Heilongjiang province is mainly distributed in large equipment manufacturing such as ships and electric machinery (Table 16.3).

In general, Heilongjiang province is still a heavy industry base for China, and other industries develop relatively slowly, while it has improved in knowledge creation. After the 12th 5-year plan, the government is trying to promote the implementation of "Five Major Plans" and the development of the "top ten key industries" in Heilongjiang province, including a batch of emerging industries, such as new materials industry, biological industry, and new energy equipment manufacturing. In addition, after Heilongjiang has participated in the construction of "The Belt and Road," infrastructure construction and the level of opening up have improved. With the continued development, Heilongjiang should be able to attract more foreign investment and venture investment, further promoting the development of science and technology innovation.

X. Liu et al., *Regional Innovation Index of China: 2017*,
https://doi.org/10.1007/978-981-13-1205-2_16

Table 16.1 Heilongjiang innovation capability comprehensive value

	2017 comprehensive value	2017 sub-index rank			
Name of indexes	Index value	Rank	Strength	Efficiency	Potential
Comprehensive value	19.51	26	19	17	30
1 Knowledge creation	21.89	15	18	5	31
1.1 R&D input	10.76	25	17	14	30
1.2 Patent	28.16	11	18	4	26
1.3 Research paper	31.61	10	14	4	31
2 Knowledge acquisition	15.08	17	18	8	30
2.1 Science and technology cooperation	39.30	3	10	1	22
2.2 Technology transfer	8.89	24	26	18	23
2.3 Foreign-funded enterprise investment	1.56	29	25	26	30
3 Enterprise innovation	13.84	31	24	24	30
3.1 Enterprise R&D input	23.43	24	21	15	28
3.2 Design capability	7.90	29	22	25	28
3.3 Technology improvement	15.89	30	25	29	22
3.4 New product sales income	6.31	31	25	27	28
4 Innovation environment	23.41	18	16	21	9
4.1 Innovation infrastructure	21.95	15	11	25	4
4.2 Market environment	30.50	15	17	13	11
4.3 Employee quality	20.72	26	20	16	23
4.4 Financial environment	21.48	12	11	8	6
4.5 Entrepreneurship	22.37	21	15	25	14
5 Innovation performance	23.28	29	26	25	29
5.1 Macroeconomic	20.98	28	21	21	29
5.2 Industry structure	14.15	25	20	19	25
5.3 Industrial international competitiveness	3.92	29	29	29	29
5.4 Employment	7.78	31	31	24	23
5.5 Sustainable development and environment	69.59	11	12	16	1

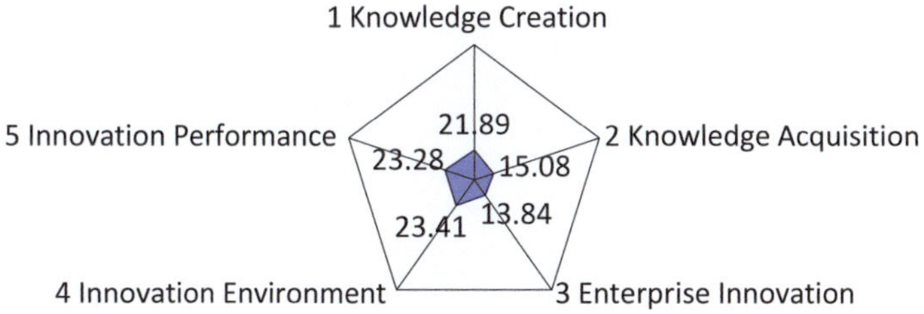

Fig. 16.1 Spider chart of Heilongjiang innovation capability

Table 16.2 Heilongjiang enterprises in the "China Top 500 Enterprises List 2016"

	Enterprise	Operation revenue (billion yuan)	Rank
1	Heilongjiang Great Northern Wilderness Agribusiness Group Corporation	131.34	116
2	Harbin Electric (HRBN)	27.94	453

Table 16.3 Heilongjiang enterprises in the "China Top-1000 Innovative Enterprises List 2016"

	Enterprises	Ownership	Industry
1	AVIC Harbin Aircraft Industry	SOE	Manufacturing of railways, ships, aerospace, and other transport equipment
2	Harbin Motor Factory	SOE	Electrical machinery and equipment manufacturing
3	Harbin Steam Turbine Factory	SOE	General equipment manufacturing
4	Harbin Boiler	SOE	Power generation equipment manufacturing
5	Harbin Dongan Engine	SOE	Transport equipment
6	China Shipbuilding Industry Corporation 703 Research Center	SOE	Ship research
7	Harbin Electric Power Equipment	Private	Unassorted
8	Harbin Antian Technology	Private	Network security products R&D

Chapter 17
Anhui

In 2017, Anhui's innovation capability ranks 10th in China, 1 place lower than the previous year. As to sub-indexes, knowledge creation, knowledge acquisition, enterprise innovation, innovation environment, and innovation performance rank 6th, 24th, 8th, 19th, and 16th, respectively. Among them, knowledge creation and knowledge acquisition are four places higher than last year; the rank of enterprise innovation remains the same; and the rank of innovation environment and innovation performance has declined. Comparing with other regions, Anhui's knowledge acquisition capability is relatively poor (Table 17.1, Fig. 17.1).

From the statistical data, there are 14 enterprises from Anhui in the "China Top 500 Enterprises List 2016"; all of them are state-owned enterprises, and no enterprise ranks in top 100 (Table 17.2).

In the top 1000 enterprise list in innovation capability, there are 81 enterprises from Anhui province, 67 of which are high-tech enterprises, and 21 enterprises have been in the list for the first time, which shows a vitality of patent application in enterprises (Table 17.3).

In general, Anhui province has a relatively complete industrial system with strong project supporting capacity. Among them, automobile and engineering machinery are the largest industrial sectors, and electronic information industry is the most important emerging industry in Anhui. The added value growth rate of the electronic information industry has been maintained at more than 20% for many years. However, knowledge acquisition capability is relatively weak, and Anhui needs to make further efforts in S&T cooperation, technology transformation, and foreign investment in the future and keep using the knowledge spillover effect from the Chinese Academy of Sciences (CAS) and University of Science & Technology of China (USTC) to release enterprise innovation motivation by building a well-developed innovation system.

X. Liu et al., *Regional Innovation Index of China: 2017*,
https://doi.org/10.1007/978-981-13-1205-2_17

Table 17.1 Anhui innovation capability comprehensive value

	2017 comprehensive value	2017 sub-index rank			
Name of indexes	Index value	Rank	Strength	Efficiency	Potential
Comprehensive value	28.36	10	9	13	8
1 Knowledge creation	31.99	6	8	6	2
1.1 R&D input	22.68	10	11	9	6
1.2 Patent	48.11	5	7	3	2
1.3 Research paper	18.35	26	15	24	17
2 Knowledge acquisition	11.6	24	14	31	19
2.1 Science and technology cooperation	22.17	18	7	29	23
2.2 Technology transfer	7.37	29	18	31	20
2.3 Foreign-funded enterprise investment	6.83	21	16	21	14
3 Enterprise innovation	38.82	7	6	7	9
3.1 Enterprise R&D input	44.56	8	8	8	4
3.2 Design capability	40.21	4	4	3	6
3.3 Technology improvement	31.62	12	7	12	21
3.4 New product sales income	35.35	9	7	9	6
4 Innovation environment	22.48	19	13	30	17
4.1 Innovation infrastructure	20.02	22	10	27	10
4.2 Market environment	25.01	29	20	24	29
4.3 Employee quality	26.98	14	11	19	26
4.4 Financial environment	12.01	22	14	24	26
4.5 Entrepreneurship	28.36	14	9	23	7
5 Innovation performance	32.46	16	14	18	10
5.1 Macroeconomic	30.77	18	14	25	9
5.2 Industry structure	21.1	15	16	17	7
5.3 Industrial international competitiveness	19.32	15	14	14	9
5.4 Employment	27.04	9	12	15	6
5.5 Sustainable development and environment	64.1	17	16	22	22

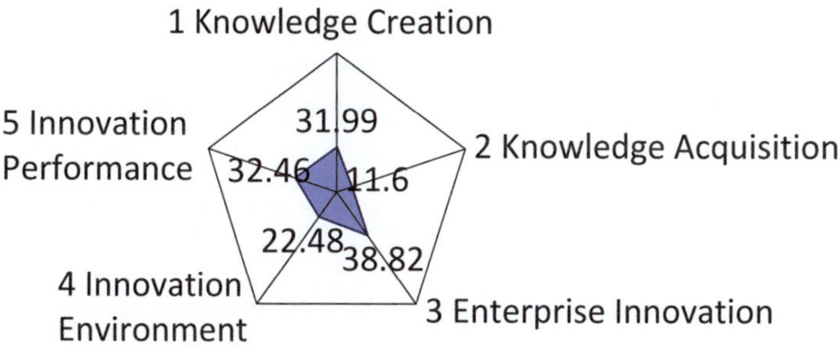

Fig. 17.1 Spider chart of Anhui innovation capability

Table 17.2 Anhui enterprises in the "China Top 500 Enterprises List 2016"

	Enterprise	Operation revenue (billion yuan)	Rank
1	Tongling Nonferrous Metal Group Holding	145.35	107
2	Anhui Conch Group	93.07	157
3	Huanbei Mining	52.24	253
4	Magang (Group) Holding	50.37	264
5	Anhui Jianghuai Automobile Group	47.05	284
6	Huainan Mining Industry	43.36	304
7	Hefei Department Store Group	39.23	331
8	Anhui Construction Engineering Group	36.99	352
9	Anhui International Trade Group	29.15	435
10	Chery Automobile	28.06	450
11	Anhui Transportation Holding	27.02	471
12	Wanbei Coal-Electricity	26.87	474
13	Zhongke Electric Power Equipment Group	25.52	489
14	Huaxin Publishing	24.99	495

Table 17.3 Anhui top 100 enterprises for patents 2016 (top 10)

	Enterprises	Ownership	Industry
1	Chery Automobile	State-owned holding	Automobile production
2	Hefei Midea	Sino-foreign joint venture	Freezer, cold closet
3	Anhui Jianghuai Automobile Group	State-owned holding	Auto and accessories development
4	Hefei Genius New Material	Private	Engineering plastics and other new materials and their products, chemical machinery, automatic control equipment development
5	Magang (Group) Holding	Joint venture	Steel
6	Huainan Mining Industry	SOE	Coal
7	BOE Optoelectronics Technology	SOE	Display, etc.
8	Anhui BBCA Biotech Engineering	SOE	Food production
9	Anhui Huadong Photoelectric Technology Research Institute	SOE	Military special display technology
10	China Tiesiju Civil Engineering Group	SOE	Railway construction

Chapter 18
Fujian

In 2017, Fujian's innovation capability ranks 14th in China, the same as previous year. As to sub-indexes, knowledge creation ranks 19th, 1 place lower than last year; knowledge acquisition and enterprise innovation rank 9th and 13th, the same as previous year; innovation environment and innovation performance rank 13th, 2 places higher than last year (Table 18.1, Fig. 18.1).

According to the statistical data, there are 10 enterprises from Fujian in the "China Top 500 Enterprises List 2016." Industrial Bank is the only enterprise in top 100, ranking 41st (Table 18.2).

In China Top 1000 Enterprises List in innovation capability, there are 20 enterprises from Fujian, mainly from energy, chemical engineering, electric information, and communication industries (Table 18.3).

In recent years, Fujian's innovation capability remains the intermediate level of China. Enterprise innovation is the main part of Fujian's innovation development. R&D expenditure, and R&D personnel are mainly in enterprises. However, Fujian's science and technology foundation is relatively weak, science and technology input still remains low level, and technology transformation needs to be more improved. In the future, with the implementation of the innovation-driven development strategy, it can be expected that Fujian's innovation capacity can be improved greatly.

X. Liu et al., *Regional Innovation Index of China: 2017*,
https://doi.org/10.1007/978-981-13-1205-2_18

Table 18.1 Fujian innovation capability synthetic index

Name of indexes	2017 comprehensive value	2017 sub-index rank			
	Index value	Rank	Strength	Efficiency	Potential
Comprehensive value	25.77	14	11	14	20
1 Knowledge creation	19.57	19	16	26	9
1.1 R&D input	19.65	13	15	11	8
1.2 Patent	22.23	18	17	25	6
1.3 Research paper	14.07	31	19	28	15
2 Knowledge acquisition	21.16	9	8	9	14
2.1 Science and technology cooperation	20.69	23	17	25	9
2.2 Technology transfer	24.62	7	8	5	16
2.3 Foreign-funded enterprise investment	18.91	8	8	8	20
3 Enterprise innovation	24.76	13	11	17	25
3.1 Enterprise R&D input	37.44	12	9	14	22
3.2 Design capability	15.56	18	12	22	10
3.3 Technology improvement	24.92	20	9	21	27
3.4 New product sales income	19.38	16	13	13	24
4 Innovation environment	25.64	13	10	18	24
4.1 Innovation infrastructure	25.61	8	8	8	26
4.2 Market environment	35.14	9	9	10	12
4.3 Employee quality	21.54	25	17	26	20
4.4 Financial environment	24.58	9	7	10	16
4.5 Entrepreneurship	21.34	23	12	28	20
5 Innovation performance	35.31	13	10	11	16
5.1 Macroeconomic	49.33	8	11	7	6
5.2 Industry structure	21.1	14	11	13	21
5.3 Industrial international competitiveness	15.09	18	10	18	16
5.4 Employment	17.01	25	20	9	29
5.5 Sustainable development and environment	74.01	5	10	9	10

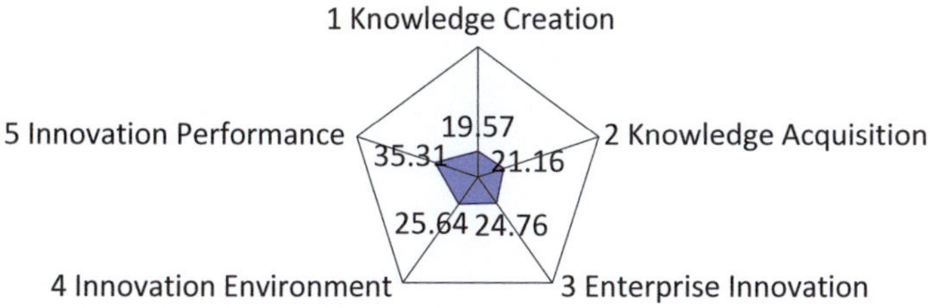

Fig. 18.1 Spider chart of Fujian innovation capability

Table 18.2 Fujian enterprises selected into the "China Top 500 Enterprises List 2016"

	Enterprise	Operation revenue (billion yuan)	Rank
1	Industrial Bank	291.89	41
2	Xiamen C&D Corporation Limited	130.57	117
3	Xiamen Holding Group	102.62	141
4	Zijin Mining Group	74.30	188
5	Xiamen Xiangyu Group	65.61	207
6	Yonghui Superstores	42.15	312
7	Fujian Energy Group	31.39	403
8	China Tobacco Fujian Industrial	28.69	442
9	King Long Motor Group	26.84	475
10	Sansteel MinGuang (Fujian)	24.61	499

Table 18.3 Fujian enterprises in the "China Top-1000 Innovative Enterprises List 2016"

	Enterprises	Patent applications (Nr)
1	Zhangzhou Pien Tze Huang Pharmaceutical	134
2	Xiamen Cheng Shin Tire & Rubber	120
3	Leedarson Group	85
4	Amperex Technology Limited	77
5	Fujian South Highway Machinery	35
6	Kehua Group	23
7	Fujian Star-net Communication	21
8	Motic (Xiamen) Electric Group	20
9	Xiamen Yaxon Network	19
10	Fuzhou Rockchip Electronics	17

Chapter 19
Jiangxi

In 2017, Jiangxi's innovation capacity ranks 19th in China, up 2 places from 2016. In terms of different fields, the innovation environment ranks 14th, up 8 places from the previous year; knowledge acquisition and innovation performance rank 18th, drop 3 and 2 places from the previous year; enterprise innovation and knowledge creation rank 22nd and 29th in China (Table 19.1, Fig. 19.1).

According to the statistics released by the CEC (China Enterprise Confederation), there are seven enterprises from Jiangxi province in the "China Top 500 Enterprises List 2016." Among them, there are three private enterprises, including Zhengbang Group, Twins Group, and Jinko Solar (Table 19.2).

In "China Top-1000 Innovative Enterprises List 2016," 8 enterprises in Jiangxi province are in the list. Among them, the five private enterprises are from the industries of biological pharmaceuticals, electronics manufacturing, and new energy, and the three state-owned enterprises are associated with engineering design industry and traditional manufacturing industries, such as automobile and aerospace (Table 19.3).

In general, in recent years, the vigorous implementation of innovation-driven development strategy in Jiangxi province has achieved initial success. Innovation efficiency has been improved significantly, showing a great innovation potential, and preliminary improvement of innovation environment can be seen in Jiangxi, especially for financial environment. However, knowledge creation and enterprise innovation capability are still relatively weak, so it is necessary to further increase R&D investment and strengthen the integrated development of industry-university-research and establish an industry-leading technology innovation system.

X. Liu et al., *Regional Innovation Index of China: 2017*,
https://doi.org/10.1007/978-981-13-1205-2_19

Table 19.1 Jiangxi innovation capability comprehensive value

Name of indexes	2017 comprehensive value	2017 sub-index rank			
	Index value	Rank	Strength	Efficiency	Potential
Comprehensive value	22.04	19	18	22	7
1 Knowledge creation	13.77	29	23	29	14
1.1 R&D input	12.32	22	21	25	12
1.2 Patent	12.76	29	24	30	15
1.3 Research paper	18.7	24	24	19	10
2 Knowledge acquisition	14.41	18	17	18	18
2.1 Science and technology cooperation	18.01	25	20	20	24
2.2 Technology transfer	16.74	13	14	8	7
2.3 Foreign-funded enterprise investment	9.97	14	10	12	16
3 Enterprise innovation	20.75	22	17	26	3
3.1 Enterprise R&D input	30.09	18	18	26	1
3.2 Design capability	11.95	22	18	23	9
3.3 Technology improvement	25.13	19	17	22	6
3.4 New product sales income	15.59	20	17	22	11
4 Innovation environment	25.48	14	18	11	6
4.1 Innovation infrastructure	19.24	24	19	20	7
4.2 Market environment	25.75	27	25	31	10
4.3 Employee quality	24.39	19	16	18	10
4.4 Financial environment	24.35	10	15	3	8
4.5 Entrepreneurship	33.67	11	16	7	6
5 Innovation performance	31.26	18	15	16	13
5.1 Macroeconomic	29.15	19	18	24	5
5.2 Industry structure	22.42	12	17	10	10
5.3 Industrial international competitiveness	14.28	19	15	17	15
5.4 Employment	26.59	11	14	5	12
5.5 Sustainable development and environment	63.89	18	8	24	25

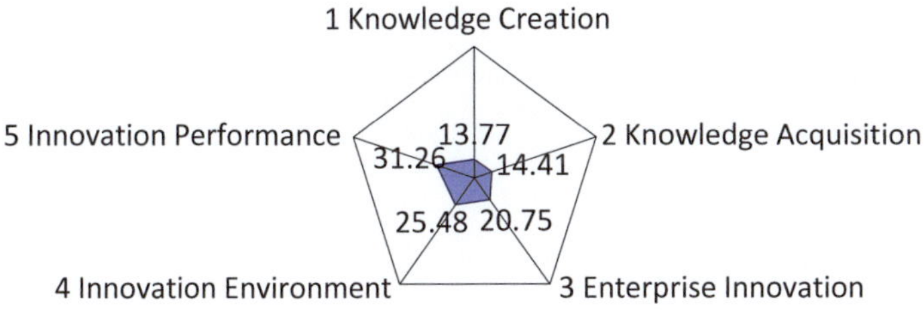

Fig. 19.1 Spider chart of Jiangxi innovation capability

Table 19.2 Jiangxi enterprises in "China Top 500 Enterprises List 2016"

	Enterprise	Operation revenue (billion yuan)	Rank	Ownership	Industry
1	Jiangxi Copper Group	1983.04	75	SOE	Nonferrous metals
2	Jiangling Motors	534.53	249	SOE	Automobile manufacture
3	Jiangxi Construction Industry Group	359.87	362	SOE	Constructional engineering
4	Xinyu Iron and Steel Group	266.6	477	SOE	Metal smelting
5	Zhengbang Group	520.35	254	Private	Agriculture/planting/finance
6	Twins Group	386.18	338	Private	Feed production
7	Jinko Solar	305.61	411	Private	Photovoltaic module manufacturing

Table 19.3 Jiangxi enterprises in the "China Top-1000 Innovative Enterprises List 2016"

Enterprise	Ownership	Industry
Jiangxi Hongdu Aviation Industry Group	SOE	Aerospace and spacecraft manufacturing
Jiangling Motors	SOE	Automobile industry
Jinko Solar	Private	Photovoltaic module manufacturing
NanChang Light Technology Exploitation	Private	Electronic components manufacturing
Huiren Medical	Private	Medicine cultivation/drug development
Jiangxi Tianren Ecological	Private	Biopesticide
Chinese NERIN Engineering	SOE	Engineering survey design/equipment manufacturing integration
Tellhow Polytron Technologies	Private	Electrical machinery and equipment manufacturing

Chapter 20
Shandong

In 2017, the comprehensive index of innovation capability of Shandong province ranks 6th, which remains the same ranking with the previous year. In terms of the sub-index, Shandong province has a strong innovation environment and strong innovation capability, ranking 5th and 6th in China, all the same as last year; knowledge creation, innovation performance, and knowledge acquisition rank 9th, 10th, and 13th, respectively, in the whole country, dropping 1, 3, and 2 respectively, from the previous year (Table 20.1, Fig. 20.1).

In 2016, 49 enterprises in Shandong were listed in "China Top 500 Enterprises List 2016." The majority of top 20 enterprises in Shandong are state-owned enterprises, belonging to energy and heavy industries, but Shandong Weiqiao Pioneering Group, as a private enterprise, ranks first in Shandong province with extremely high operation revenue (Table 20.2).

In "China Top-1000 Innovative Enterprises List 2016," 54 Shandong enterprises were listed. The industry distribution of top 20 enterprises is mainly distributed in electrical machinery and equipment manufacturing, computer, communication, and other electronic equipment manufacturing and biomedical industry (Table 20.3).

In recent years, Shandong province has thoroughly implemented new development idea, accelerating the implementation of innovative development strategy and trying to be an innovative province. Shandong enterprises have strong power, but some traditional industries' innovation capability is slightly insufficient. In the future, Shandong should promote innovation capacity of enterprises by exploring effective economic development mode and drive the industry to transform and upgrade continuously. In order to promote entrepreneurial activity and develop innovative small enterprises, Shandong still needs to exploit potential and encourage innovation and entrepreneurship.

X. Liu et al., *Regional Innovation Index of China: 2017*,
https://doi.org/10.1007/978-981-13-1205-2_20

Table 20.1 Shandong innovation capability comprehensive value

Name of indexes	2017 comprehensive value	2017 sub-index rank			
	Index value	Rank	Strength	Efficiency	Potential
Comprehensive value	33.77	6	4	16	24
1 Knowledge creation	28.93	9	6	24	21
1.1 R&D input	23.34	9	6	12	22
1.2 Patent	39.9	8	5	17	13
1.3 Research paper	18.18	27	9	30	25
2 Knowledge acquisition	18.37	13	6	27	13
2.1 Science and technology cooperation	21.42	20	14	26	14
2.2 Technology transfer	16.82	12	5	28	14
2.3 Foreign-funded enterprise investment	17.25	10	7	11	13
3 Enterprise innovation	39.04	6	4	14	24
3.1 Enterprise R&D input	56.56	4	3	11	21
3.2 Design capability	23.7	9	3	20	13
3.3 Technology improvement	35.47	8	4	27	28
3.4 New product sales income	39.35	7	4	11	22
4 Innovation environment	37.03	5	3	16	21
4.1 Innovation infrastructure	33.37	4	3	15	13
4.2 Market environment	36.14	8	7	17	22
4.3 Employee quality	42.95	4	3	29	13
4.4 Financial environment	32.58	5	4	13	21
4.5 Entrepreneurship	40.12	5	3	12	23
5 Innovation performance	38.26	10	3	12	24
5.1 Macroeconomic	65.49	3	3	10	20
5.2 Industry structure	38.32	5	3	7	19
5.3 Industrial international competitiveness	12.89	20	7	19	23
5.4 Employment	25.16	14	10	8	20
5.5 Sustainable development and environment	49.43	28	31	8	28

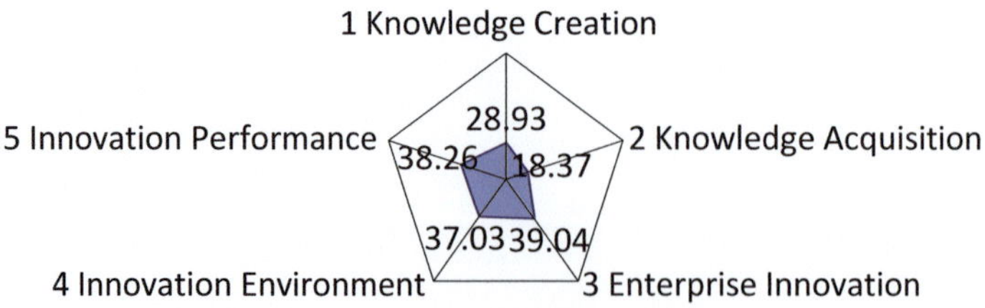

Fig. 20.1 Spider chart of Shandong innovation capability

Table 20.2 Shandong enterprises in "China Top 500 Enterprises List 2016"

	Enterprise	Operation revenue (billion yuan)	Rank
1	Shandong Weiqiao Pioneering Group	333.24	37
2	Haier Group	188.69	84
3	Shandong Energy Group	157.97	103
4	Shandong Great King Group	112.69	126
5	Shandong Iron and Steel Group	102.11	143
6	Yancon Group	101.24	147
7	Hisense Group	99.02	153
8	Shandong Commercial Group	90.70	161
9	Nanshan Group	82.22	167
10	Weichai Holding Group	80.65	171
11	Shandong Gold Group	68.19	198
12	Inspur Group	63.20	218
13	China National Heavy Duty Truck Group	61.66	222
14	Huatai Group	60.24	227
15	Shandong Dahai Group	58.85	230
16	China Wanda Group	55.63	239
17	Shandong Chenming Group	52.73	251
18	Shandong Chambroad Holding	50.19	266
19	Lihuayi Group	50.17	267
20	Shandong Haike Chemical Group	47.90	280

Table 20.3 Shandong enterprises in the "China Top-1000 Innovative Enterprises List 2016"

	Enterprises	Patent (Nr)
1	Hisense Electric	625
2	Hisense Mobile Communications Technology	585
3	Weichai Power	274
4	Inspur Group	195
5	Inspur Common Software	134
6	Valiant	102
7	Inspur Software	100
8	China National Heavy Duty Truck Group	95
9	Wanhua Chemical Group	76
10	Shinva Medical Instrument	61
11	Jinan Shengquan Group	59
12	Shengrui Transmission	53
13	Qingdao Yebio Biological Engineering	43
14	Shandong Electric Power Engineering Consulting	33
15	AUCMA	31
16	Yotani Yoshihikaru Copper	28
17	Disha Pharmaceutical Group	27
18	Qingdao Hisense Network Polytron Technologies	27
19	Shantui Construction Machinery	24
20	Shandong Linglong	24

Chapter 21
Henan

In 2017, Henan ranks 15th nationwide, unchanged from last year. For sub-indexes, knowledge creation ranks 24th, 1 place lower than last year; the rankings of knowledge acquisition, enterprise innovation, and innovation environment are the same as last year, ranking 26th, 20th, and 11th, respectively; innovation performance ranks 7th, dropping 3 places from last year (Table 21.1, Fig. 21.1).

Henan province has been listed for 9 enterprises in "China Top 500 Enterprises List 2016." Among them, Henan Energy and Chemical Industry Group, China Pingmei Shenma Group, and WH Group are listed in the top places, respectively, ranking 101st, 106th, and 111th (Table 21.2).

In "China Top-1000 Innovative Enterprises List 2016," there are 15 enterprises from Henan province, of which top 10 enterprises with the largest number of patent applications in 2016 are shown in the table below, mainly composed of engineering machinery and auto parts enterprises (Table 21.3).

Henan province is a province with large population, and in some gross indexes, Henan ranks relatively high. In recent years, Henan province has established the Zhengluo new national independent innovation demonstration zone, to promote the innovation development of the whole province. Meanwhile, Henan has vigorously implemented three national strategies, speeding up the construction of "One carrier, four systems and six foundations," and the innovation capability has improved significantly, but overall, Henan's innovation capability is still at the middle level of China and needs to be further improved in the areas of core R&D in their main industries and to develop more emerging industries.

X. Liu et al., *Regional Innovation Index of China: 2017*,
https://doi.org/10.1007/978-981-13-1205-2_21

Table 21.1 Henan innovation capability comprehensive value

Name of indexes	2017 comprehensive value	2017 sub-index rank			
	Index value	Rank	Strength	Efficiency	Potential
Comprehensive value	24.23	15	10	26	17
1 Knowledge creation	15.96	24	13	28	28
1.1 R&D input	14.68	19	12	20	23
1.2 Patent	17.45	25	16	27	27
1.3 Research paper	15.56	30	12	25	28
2 Knowledge acquisition	11.24	26	16	28	20
2.1 Science and technology cooperation	19.86	24	12	28	25
2.2 Technology transfer	8.39	26	22	24	17
2.3 Foreign-funded enterprise investment	6.91	20	15	25	11
3 Enterprise innovation	21.38	20	9	30	17
3.1 Enterprise R&D input	33.78	14	5	24	8
3.2 Design capability	9.78	27	14	30	15
3.3 Technology improvement	17.81	29	12	31	25
3.4 New product sales income	23.77	13	8	18	13
4 Innovation environment	26.15	11	7	29	15
4.1 Innovation infrastructure	22.11	14	4	30	12
4.2 Market environment	30.2	17	13	26	7
4.3 Employee quality	30.45	10	4	30	29
4.4 Financial environment	12.76	21	12	25	15
4.5 Entrepreneurship	35.22	8	5	16	11
5 Innovation performance	41.34	7	7	8	6
5.1 Macroeconomic	39.62	12	5	22	13
5.2 Industry structure	31.22	7	6	11	5
5.3 Industrial international competitiveness	44.57	6	5	3	7
5.4 Employment	33.24	5	4	6	8
5.5 Sustainable development and environment	58.04	23	27	20	12

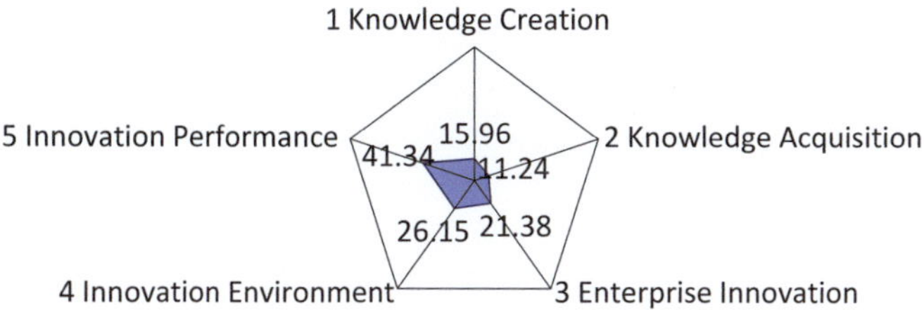

Fig. 21.1 Spider chart of Henan innovation capability

Table 21.2 Henan enterprises in the "China Top 500 Enterprises List 2016"

	Enterprise	Operation revenue (billion yuan)	Rank
1	Henan Energy and Chemical Industry Group	160.25	101
2	China Pingmei Shenma Group	146.80	106
3	WH Group	137.72	111
4	Zhengzhou Yutong Bus	38.20	342
5	An Gang Group	37.14	350
6	Tianrui Group	31.54	400
7	Yuguang Gold & Lead Group	27.72	456
8	Dengfeng Electricity Group	25.50	490
9	Wanji Holding Group	24.93	497

Table 21.3 Henan enterprises in the "China Top-1000 Innovative Enterprises List 2016" (top 10 patent applications)

	Enterprises	Patent applications (Nr)
1	Wuyang Iron & Steel	66
2	Luoyang Sunrui Special Equipment	56
3	XJ-Schindler (Xuchang) Elevator	55
4	Sinosteel Luoyang Institute of Refractories Research	48
5	Zhengzhou Research Institute of Mechanical Engineering	47
6	Yellow River Engineering Consulting	46
7	Henan Senyuan Electric	42
8	Henan Diesel Engine Industry	31
9	Henan Muxiang Veterinary Pharmaceutical	28
10	Zhengzhou Jingyida Auto Parts	25

Chapter 22
Hubei

In 2017, Hubei province's innovation capability ranks 9th in China, improving 3 places compared with 2016, listed in national top 10, which has made a great progress. As to the sub-indexes, knowledge creation, knowledge acquisition, enterprise innovation, innovation environment, and innovation performance index rank 13th, 12th, 11th, 7th, and 8th, respectively. Knowledge creation remains the same ranking with the last year, and other indexes have improved than last year, especially innovation performance, which up 4 places (Table 22.1, Fig. 22.1).

According to the data, there are 11 enterprises in the "China's top 500 enterprises List 2016" from Hubei province. Most of them are state-owned and state-controlled enterprises (Table 22.2).

In "China Top 1000 Enterprises List in Innovation Capability 2016," 25 enterprises from Hubei province were listed. From the State Intellectual Property Office patent search system, top 10 enterprises with the largest number of patent applications in 2016 are shown in the table below. As can be seen, top 10 enterprises are almost state-owned enterprises, mainly belonging to engineering and construction, optical fiber communications, and other industries, all located in the capital city of Wuhan (Table 22.3).

Overall, in the past few years, most of the innovative indexes of Hubei have been improved. In 2016 Hubei province approved to be the national innovation pilot provinces, leading Hubei into a new stage of innovation. And in the 13th 5-year plan for innovation of science and technology, Hubei focuses on the construction of innovative platform to strengthen the basic research and attract talent and so on. Hubei province has also put forward the innovative system of innovation chain, industrial chain, capital chain, talent chain, and policy chain which is called "five chain coordination," and the growth of the number of incubators in the basic indexes also reflected significant improvement of innovation environment in Hubei province.

At present, the advantages of enterprises in Hubei province are still state-owned engineering enterprises. Although the indexes show that the province's patents, papers, scientific, and technological achievements are making progress fastly,

X. Liu et al., *Regional Innovation Index of China: 2017*,
https://doi.org/10.1007/978-981-13-1205-2_22

Table 22.1 Hubei innovation capability comprehensive value

Name of indexes	2017 Comprehensive value	2017 Sub-index rank			
	Index value	Rank	Strength	Efficiency	Potential
Comprehensive value	29.35	9	7	10	9
1 Knowledge creation	23.79	13	10	16	20
1.1 R&D input	17.88	15	9	10	25
1.2 Patent	25.33	12	9	23	14
1.3 Research paper	32.5	9	6	11	20
2 Knowledge acquisition	18.75	12	10	15	11
2.1 Science and technology cooperation	31.25	9	9	10	15
2.2 Technology transfer	18.71	11	10	16	11
2.3 Foreign-funded enterprise investment	9.42	15	13	13	9
3 Enterprise innovation	29.78	11	8	11	10
3.1 Enterprise R&D input	42.99	10	7	10	9
3.2 Design capability	17.63	14	10	19	7
3.3 Technology improvement	26.96	15	11	24	10
3.4 New product sales income	31.03	12	10	10	12
4 Innovation environment	29.85	7	9	9	10
4.1 Innovation infrastructure	24.9	11	14	12	14
4.2 Market environment	33.9	12	10	14	5
4.3 Employee quality	29.82	11	9	22	15
4.4 Financial environment	25.61	8	8	5	27
4.5 Entrepreneurship	35.03	9	8	9	10
5 Innovation performance	40.3	8	8	13	4
5.1 Macroeconomic	42.48	10	8	13	7
5.2 Industry structure	24.71	11	9	14	6
5.3 Industrial international competitiveness	23.58	12	13	11	6
5.4 Employment	40.64	3	7	14	2
5.5 Sustainable development and environment	70.08	9	18	12	13

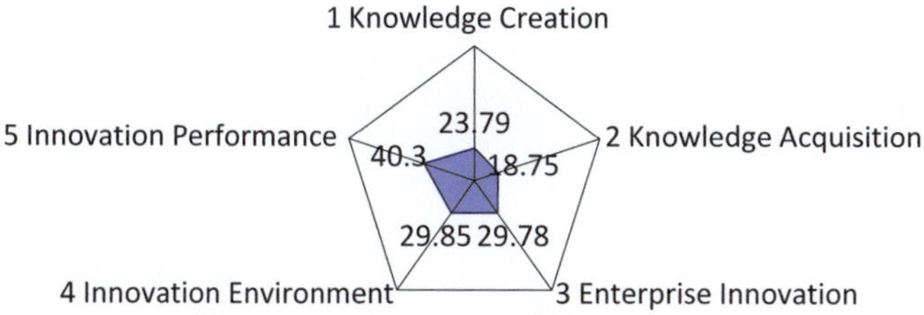

Fig. 22.1 Spider chart of Hubei innovation capability

Table 22.2 Hubei enterprises in the "China Top 500 Enterprises List 2016"

	Enterprise	Operation revenue (billion yuan)	Rank
1	Dongfeng Motor Corporation	520.45	16
2	Wuhan Iron and Steel Company Limited	101.83	144
3	Daye Nonferrous Metals Group Holdings	100.13	152
4	Wuhan Department Store Group	91.32	159
5	Hubei Yihua Chemical Industry	81.04	170
6	Hubei Tobacco Industry	64.05	211
7	Jointown Pharmaceutical Group	49.59	269
8	Hubei Daohuaxiang Group	43.55	302
9	Zall Development Group	31.40	402
10	Shanhe Construction Group	30.49	413
11	Yichang Xingfa Group	28.50	446

Hubei is still relatively lacking the strategic and forward-looking innovation in reality. Hubei's enterprise vitality needs to be further improved, and open innovation, cooperation, and innovation need more attention so that it can take the advantages of science and education to further correct and improve drawbacks of the innovation capacity.

Table 22.3 Hubei enterprises in the "China Top-1000 Innovative Enterprises List 2016" (top 10)

	Enterprise	Ownership	Patent (Nr)	Industry
1	China Railway Siyuan Survey and Design Group	SOE	306	Engineering survey and design
2	Wuhan Marine Machinery	SOE	268	Port machinery, ship supporting products, bridge equipment
3	Fiber Home Telecommunication Technologies	SOE	232	Optical fiber communication and related
4	China Smelter South (Wuhan) Weishi Industrial Furnace	SOE	168	Energy-saving equipment for thermal energy-saving equipment
5	WISDRI Engineering & Research Incorporation Limited	SOE	165	Deep processing of steel products
6	Yangtze River Survey and Planning and Design Research	SOE	133	Survey and design of construction works
7	Yangtze Optical Fibre and Cable Joint Stock	Sino- foreign joint venture	112	Fiber optic preform, fiber, fiber optic cable
8	Wuhan Accelink Technologies	SOE	112	Light and electrical equipment technology and products
9	China Construction Bureau Group	SOE	76	Construction works
10	Wuhan Hongxin Communication Technology	SOE	66	Mobile communication equipment, video surveillance

Chapter 23
Hunan

In 2017, the innovation capacity of Hunan province ranks 12th in the country, ranking 1 position higher than 2016. As to the sub-indexes, knowledge creation and knowledge acquisition rank 16th and 19th, respectively, ranking, respectively, 3 and 4 position higher than last year. Enterprise innovation ranks 9th in the country which is up 1 than last year, while innovative environmental ranks 17th nationwide, down 3 compared with last year. Innovation performance ranked 14th in the country, the same as the previous year (Table 23.1, Fig. 23.1).

Data show that in "China Top 500 Enterprises List 2016," Hunan shortlisted 7 enterprises, of which Sany Group's operating revenue is the highest in Hunan, ranking 185th (Table 23.2).

In the "China Top-1000 Innovative Enterprises List 2016," there are 29 Hunan enterprises, of which the top 10 patent application enterprises are as shown in the table below (Table 23.3).

Hunan has a higher-level innovation capability nationwide, and the ranking is relatively stable in recent years. From the point of view of the geographical city group, the resources of the three cities in Changsha, Zhuzhou, and Xiangtan are the main sources of innovation and development in Hunan province. The construction of the national independent innovation demonstration zones in Changsha, Zhuzhou, and Xiangtan has a stimulating effect on promoting the innovation and development of Hunan province. In addition, the innovation ability of enterprises in Hunan province is relatively good, but high-level, professional-type innovation and entrepreneurship talent is less. In the future it needs to take further efforts to improve the process of knowledge flow with outside.

X. Liu et al., *Regional Innovation Index of China: 2017*,
https://doi.org/10.1007/978-981-13-1205-2_23

Table 23.1 Hunan innovation capability comprehensive value

	2017 comprehensive value	2017 sub-index rank			
Name of indexes	Index value	Rank	Strength	Efficiency	Potential
Comprehensive value	26.63	12	14	11	15
1 Knowledge creation	20.44	16	14	21	17
1.1 R&D input	16.01	17	14	18	13
1.2 Patent	24.84	14	14	22	7
1.3 Research paper	20.48	20	11	14	30
2 Knowledge acquisition	14.04	19	21	13	27
2.1 Science and technology cooperation	27.19	12	15	4	31
2.2 Technology transfer	11.95	17	19	17	10
2.3 Foreign-funded enterprise investment	5.73	23	17	23	17
3 Enterprise innovation	35.3	9	7	9	13
3.1 Enterprise R&D input	43.93	9	10	7	5
3.2 Design capability	20.42	11	9	13	17
3.3 Technology improvement	34.96	9	6	7	24
3.4 New product sales income	45.05	5	6	4	7
4 Innovation environment	23.71	17	14	23	18
4.1 Innovation infrastructure	21.41	17	16	24	3
4.2 Market environment	24.34	31	16	27	30
4.3 Employee quality	28.82	12	8	27	17
4.4 Financial environment	17.34	14	13	14	12
4.5 Entrepreneurship	26.65	18	13	18	16
5 Innovation performance	33.51	14	13	14	9
5.1 Macroeconomic	37.38	13	9	16	12
5.2 Industry structure	22.38	13	10	15	12
5.3 Industrial international competitiveness	17.89	16	19	15	4
5.4 Employment	19.75	20	28	10	11
5.5 Sustainable development and environment	70.13	8	17	13	11

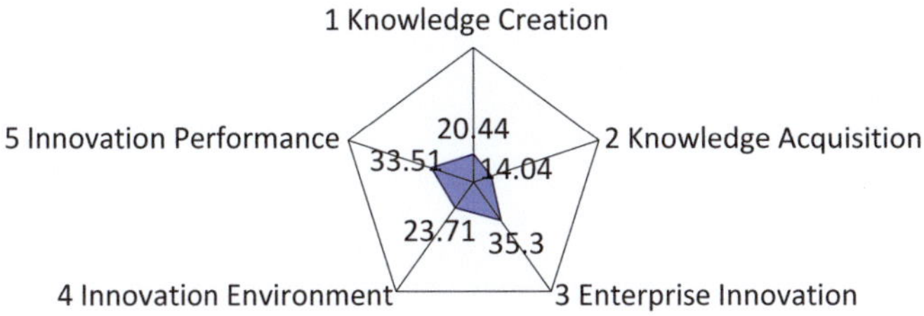

Fig. 23.1 Spider chart of Hunan innovation capability

Table 23.2 Hunan enterprises in the "China Top 500 Enterprises List 2016"

	Enterprise	Operation revenue (billion yuan)	Rank
1	Sany Group	75.05	185
2	Hunan Valin Iron and Steel Group	66.80	204
3	Hunan Construction Engineering Group Corporation	55.70	238
4	Zoomlion Heavy Industry Science and Technology	51.94	255
5	Dahan Holding Group	33.64	384
6	Better Life Commercial Chain Share	31.02	407
7	Hunan Bochang Holding Group	30.51	412

Table 23.3 Hunan enterprises in the "China Top-1000 Innovative Enterprises List 2016" (top 10 patent application)

	Enterprises	Patent application (Nr)
1	Chenzhou Jingui Yinye	165
2	CSR Zhuzhou Electric Locomotive Research Institute	102
3	Weisheng Group	85
4	Zhuzhou Cemented Carbide Group	73
5	Sunward Equipment	58
6	Hunan Er-Kang Pharmaceutical	55
7	Schindler Technology Group	44
8	Changsha Mining and Metallurgy Research Institute	42
9	CSR Zhuzhou Motor	42
10	Zoomlion Heavy Industry Science and Technology	39

Chapter 24
Guangxi

In 2017, the innovation capability of Guangxi Zhuang Autonomous Region ranks 20th in China, 1 place lower than 2016. As to the sub-indexes, knowledge creation ranks 10th; innovation performance ranks 17th; knowledge acquisition, enterprise innovation, and innovation environment rank 31st, 27th, and 28th; innovation performance ranks the same as the previous year; and the rankings of knowledge creation, knowledge acquisition, enterprise innovation, and innovation environment dropped 1, 4, 3, and 4 places, respectively, over the previous year (Table 24.1, Fig. 24.1).

Data show that Guangxi province has 6 companies shortlisted in the "China Top 500 Enterprises List 2016," and all of them are state-owned enterprises, of which the business is involving construction, transportation, machinery manufacturing, and other fields that are still the traditional industries (Table 24.2).

In the "China Top-1000 Innovative Enterprises List 2016," Guangxi Zhuang Autonomous Region has 6 enterprises shortlisted, 4 of which are state-owned enterprises based on chemical and machinery industry. The number of patent applications is in the top, and two private enterprises' performance is also good (Table 24.3).

By comparison, it is found that the economic growth of Guangxi Zhuang Autonomous Region mainly relies on the development of state-owned large enterprises, and state-owned enterprises still occupy the main position of innovation and entrepreneurship activities. With the rise of innovation atmosphere, private enterprises in Guangxi in recent years continue to increase R&D investment and participation in technological innovation cooperation and competition, so that the innovation achievements have also made progress. However, the development of information industry in Guangxi is still slow. Its innovative environment needs to be improved.

X. Liu et al., *Regional Innovation Index of China: 2017*,
https://doi.org/10.1007/978-981-13-1205-2_24

Table 24.1 Guangxi innovation capability comprehensive value

Name of indexes	2017 comprehensive value	2017 sub-index rank			
	Index value	Rank	Strength	Efficiency	Potential
Comprehensive value	21.19	20	20	18	21
1 Knowledge creation	28.59	10	19	3	15
1.1 R&D input	5.91	30	23	28	29
1.2 Patent	56.37	3	10	1	1
1.3 Research paper	18.4	25	22	15	24
2 Knowledge acquisition	9.17	31	24	29	12
2.1 Science and technology cooperation	15.54	28	21	27	18
2.2 Technology transfer	8.37	27	28	29	4
2.3 Foreign-funded enterprise investment	5.01	24	19	20	23
3 Enterprise innovation	16.82	27	19	21	28
3.1 Enterprise R&D input	13.98	29	23	29	27
3.2 Design capability	15.45	19	20	14	20
3.3 Technology improvement	24.51	21	18	9	30
3.4 New product sales income	15.45	21	19	17	20
4 Innovation environment	19.57	28	20	28	14
4.1 Innovation infrastructure	18.13	26	17	26	5
4.2 Market environment	26.89	23	19	15	27
4.3 Employee quality	21.99	24	18	25	14
4.4 Financial environment	9.88	26	24	26	5
4.5 Entrepreneurship	20.96	24	20	20	24
5 Innovation performance	32.12	17	17	20	8
5.1 Macroeconomic	25.94	22	17	26	15
5.2 Industry structure	15.46	23	19	24	15
5.3 Industrial international competitiveness	21.05	14	18	12	5
5.4 Employment	29.83	8	11	17	5
5.5 Sustainable development and environment	68.34	12	9	23	4

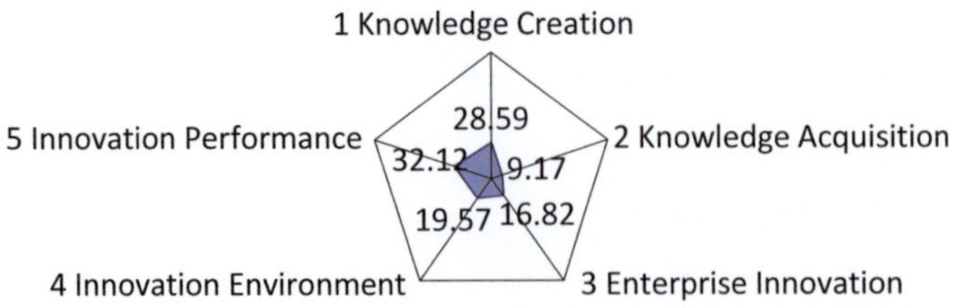

Fig. 24.1 Spider chart of Guangxi innovation capability

Table 24.2 Guangxi enterprises in the "China Top 500 Enterprises List 2016"

	Enterprise	Operation revenue (billion yuan)	Rank
1	Guangxi Investment Group	85.16	166
2	Guangxi Construction Engineering Group	70.08	195
3	Liuzhou Iron & Steel	44.47	296
4	Guangxi Beibu Gulf International Port Group	36.34	357
5	Guangxi Yuchai Machinery	33.92	375
6	Guangxi Shenglong Metallurgy	25.73	486

Table 24.3 Fujian enterprises in the "China Top-1000 Innovative Enterprises List 2016"

Enterprise	Ownership	Patent application (Nr)
Guangxi Pastoral Biochemical	Private	7
Dongfeng Liuzhou Automobile	SOE	33
Liuzhou Liangmianzhen	SOE	13
Guangxi Zhuang Autonomous Region Chemical Research Institute	SOE	6
Liuzhou Ouweimu Machinery	SOE	19
Guilin Woodpecker Medical Equipment.	Private	17

Chapter 25
Hainan

Hainan Province's innovation capacity ranks 16th in 2017, the same as last year, but the indexes vary widely in each area. Knowledge creation ranks 26th, 15 places lower than last year; knowledge acquisition ranks 30th in the country, dropped 16 places from last year; enterprise innovation ranks 14th, dropped 5 places from last year; innovation environment ranks 10th, rising 20 places; innovation performance ranks 20th, 4 places higher than the previous year (Table 25.1, Fig. 25.1).

In "China's top 500 enterprises List 2016," Hainan Province only has the Hainan Airlines Group shortlisted, ranked 85th. In the Chinese top 1000 enterprises list in innovation ability, there is no Hainan enterprise. The enterprise innovation capability is still needed to be further improved.

Overall, the governments of Hainan Province speed up scientific and technological innovation actively, and their work has achieved remarkable results. Their overall rankings in recent years tends to be stable, but as the internal indexes changed greatly, innovation and development stability need to be strengthened. As a tourism resource-rich province, science and technology innovation is the weak spot of Hainan's development. In the future, Hainan can explore how to develop new tourism model, new formats, and new industries to play the core advantages of tourism resources and make further more competitive tourism industry through large data and cloud computing.

X. Liu et al., *Regional Innovation Index of China: 2017*,
https://doi.org/10.1007/978-981-13-1205-2_25

Table 25.1 Hainan innovation capability comprehensive value

Name of indexes	2017 comprehensive value Index value	2017 sub-index rank Rank	Strength	Efficiency	Potential
Comprehensive value	22.49	16	28	9	14
1 Knowledge creation	15.93	26	29	14	24
1.1 R&D input	7.02	29	29	27	26
1.2 Patent	18.89	20	29	13	20
1.3 Research paper	27.86	12	28	9	6
2 Knowledge acquisition	9.54	30	29	23	25
2.1 Science and technology cooperation	17.89	26	30	19	11
2.2 Technology transfer	5.44	30	30	23	24
2.3 Foreign-funded enterprise investment	6.36	22	20	9	25
3 Enterprise innovation	24.14	14	29	10	22
3.1 Enterprise R&D input	31.96	17	29	13	11
3.2 Design capability	17.17	17	29	5	31
3.3 Technology improvement	36.84	7	29	4	2
3.4 New product sales income	10.16	27	29	16	29
4 Innovation environment	28.15	10	28	4	3
4.1 Innovation infrastructure	20.93	21	29	6	6
4.2 Market environment	27.23	21	23	21	21
4.3 Employee quality	14.38	31	28	12	12
4.4 Financial environment	37.24	4	26	2	1
4.5 Entrepreneurship	40.97	4	28	2	8
5 Innovation performance	27.97	20	19	23	22
5.1 Macroeconomic	20.65	29	28	18	23
5.2 Industry structure	15.25	24	28	16	11
5.3 Industrial international competitiveness	7.08	25	27	24	21
5.4 Employment	23.11	15	8	25	26
5.5 Sustainable development and environment	73.77	6	1	14	26

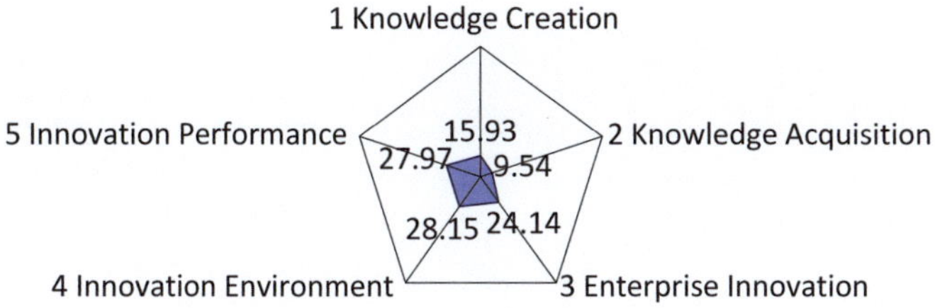

Fig. 25.1 Spider chart of Hainan innovation capability

Chapter 26
Chongqing

In 2017, the innovation capacity of Chongqing ranked No.8 countrywide, which is the same as the last year. Its knowledge creation, knowledge acquisition, corporate innovation, innovation environment, and innovation performance were No.12, No.11, No.10, No.15, and No.6, respectively. Among them, the knowledge creation and corporate innovation have been improved; the rankings of knowledge acquisition and innovation environment have dropped (Table 26.1, Fig. 26.1).

In 2016, there were 12 companies in Chongqing selected into the "China Top 500 Enterprises List 2016." Most of them are from the real estate, energy, and mechanical manufacture industries (Table 26.2).

In the "China Top-1000 Innovative Enterprises List 2016," Chongqing only gets three nominations. Referring to the patent retrieval system by the State Intellectual Property Office, we pick their quantity of patent applications in 2016, as shown in Table 26.3.

In general, the innovation capacity of Chongqing is on the rise, and its regional innovation capacity is in the forefront of the Midwestern China. As one of the youngest municipalities and one of the most important old industrial bases in China, Chongqing has high innovation efficiency and great innovation potential, but its resources of science and education are insufficient, and its companies' innovation is not strong enough as well. In the future, the city needs to further strengthen open innovation, introduce top 100 innovative companies both at home and abroad, improve companies' indigenous innovation, increase endogenous strength for its development of innovation, and establish high-quality innovative ecosystem.

X. Liu et al., *Regional Innovation Index of China: 2017*,
https://doi.org/10.1007/978-981-13-1205-2_26

Table 26.1 Chongqing innovation capability comprehensive value

Name of indexes	2017 comprehensive value	2017 sub-index rank			
	Index value	Rank	Strength	Efficiency	Potential
Comprehensive value	30.05	8	17	6	5
1 Knowledge creation	26.02	12	15	8	5
1.1 R&D input	20.4	12	19	16	5
1.2 Patent	31.71	10	11	6	9
1.3 Research paper	25.9	13	17	10	23
2 Knowledge acquisition	20.15	11	12	5	22
2.1 Science and technology cooperation	24.87	15	18	6	30
2.2 Technology transfer	25.39	6	9	3	29
2.3 Foreign-funded enterprise investment	12.67	11	11	10	7
3 Enterprise innovation	34.27	10	16	8	1
3.1 Enterprise R&D input	42.01	11	16	9	2
3.2 Design capability	25.6	7	13	6	4
3.3 Technology improvement	24.24	22	16	19	17
3.4 New product sales income	45.71	4	11	3	4
4 Innovation environment	23.92	15	19	12	8
4.1 Innovation infrastructure	21.15	20	21	10	17
4.2 Market environment	33.96	10	15	12	4
4.3 Employee quality	18.18	27	22	21	27
4.4 Financial environment	11.5	23	20	21	19
4.5 Entrepreneurship	34.83	10	18	8	1
5 Innovation performance	42.86	6	12	6	5
5.1 Macroeconomic	40.96	11	20	11	2
5.2 Industry structure	33.5	6	15	6	4
5.3 Industrial international competitiveness	48.53	2	4	1	19
5.4 Employment	18.35	22	21	22	10
5.5 Sustainable development and environment	72.93	7	7	7	21

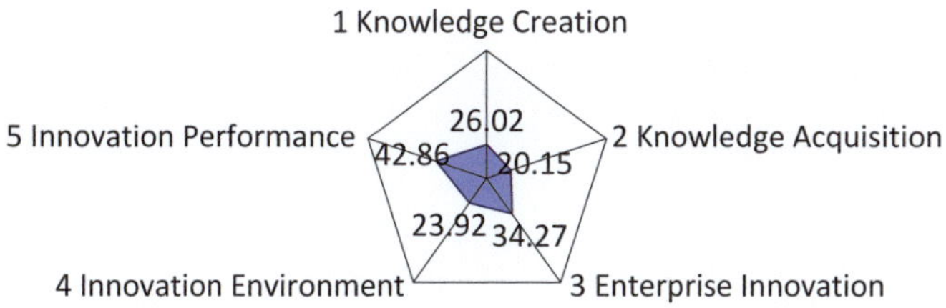

Fig. 26.1 Spider chart of Chongqing innovation capacity

Table 26.2 Chongqing enterprises in the "China Top 500 Enterprises List 2016"

	Enterprise	Operation revenue (billion yuan)	Rank
1	Chongqing General Trading Group	50.50	262
2	Chongqing Construction Investment Holdings	46.25	288
3	Chongqing Longfor Properties	41.26	318
4	Chongqing Rural Commercial Bank	37.00	351
5	Chongqing Jinke Holdings	36.54	354
6	Loncin Holdings	36.24	358
7	Chongqing Energy	36.01	361
8	Chongqing Chemical & Pharmaceutical Group	35.92	363
9	Chongqing Lifan	33.99	374
10	Taiji Group	32.91	390
11	Chongqing Machinery & Electronics Holding Group	27.79	455
12	Chongqing Light Textile Holding	26.57	478

Table 26.3 Chongqing enterprises in the "China Top-1000 Innovative Enterprises List 2016"

	Enterprises	Ownership	Patent (Nr)	Industry
1	CN GPower Gearbox	SOE	20	Gears, shock absorbers, friction plates, etc.
2	HWASDAN	Private	4	Mechanical manufacture
3	Chongqing Longrun Automobile Steering Gears	Private	0	Automobile steering gear

Chapter 27
Sichuan

In 2017, the innovation capacity of Sichuan ranked No.11 countrywide, which is the same as the last year. Its knowledge creation, knowledge acquisition, corporate innovation, innovation environment, and innovation performance were No.8, No.16, No.12, No.8, and No.12, respectively. Among them, its innovation environment has been improved a lot, from No. 13 to No. 8; corporate innovation has been improved; the rankings of knowledge creation and innovation performance have dropped (Table 27.1, Fig. 27.1).

The statistic shows there were 14 companies in Sichuan selected into the "China Top 500 Enterprises List 2016," but none of them entered Top 100 (Table 27.2).

In the "China Top-1000 Innovative Enterprises List 2016," 41 companies are from Sichuan. Referring to the patent retrieval system by the State Intellectual Property Office, we pick top 10 of them with the most quantity of patent applications in 2016. According to the data, these companies are mainly in the fields of electrical equipment, communication equipment, engineering survey, etc. (Table 27.3).

In general, the development of Sichuan is relatively well among other western provinces. It is also the only western province selected into the national innovation pilots. The Sichuan government has always attached great importance to the innovation-driven and talent-driven development and platform support and formed strong innovation capacity. Sichuan has also focused on six industries, including high-tech development, upgrading of dominant traditional industry, modern agricultural science, social development science, ecological environmental protection technology, and S&T services. In the future, Sichuan will further strengthen its foreign cooperation, mitigate geographical disadvantages, and enhance its leading role in the western China.

X. Liu et al., *Regional Innovation Index of China: 2017*,
https://doi.org/10.1007/978-981-13-1205-2_27

Table 27.1 Sichuan innovation capability comprehensive value

Name of indexes	2017 comprehensive value	2017 sub-index rank			
	Index value	Rank	Strength	Efficiency	Potential
Comprehensive value	27.52	11	8	12	16
1 Knowledge creation	30.21	8	7	9	11
1.1 R&D input	25.93	8	7	8	11
1.2 Patent	33	9	8	10	11
1.3 Research paper	33.19	8	8	12	9
2 Knowledge acquisition	15.3	16	11	26	15
2.1 Science and technology cooperation	28.93	11	4	18	12
2.2 Technology transfer	11.07	20	16	21	12
2.3 Foreign-funded enterprise investment	8.24	16	12	15	18
3 Enterprise innovation	24.98	12	13	15	19
3.1 Enterprise R&D input	29.78	19	13	23	15
3.2 Design capability	21.75	10	8	8	25
3.3 Technology improvement	29.61	13	10	18	11
3.4 New product sales income	18.02	19	16	20	15
4 Innovation environment	28.99	8	8	13	13
4.1 Innovation infrastructure	25.57	9	6	22	8
4.2 Market environment	32.03	13	11	11	13
4.3 Employee quality	34.65	7	5	24	16
4.4 Financial environment	16.94	15	10	19	11
4.5 Entrepreneurship	35.77	7	6	5	18
5 Innovation performance	36.02	12	11	9	17
5.1 Macroeconomic	33.6	15	6	23	21
5.2 Industry structure	28.06	9	8	8	8
5.3 Industrial international competitiveness	35.52	7	9	4	27
5.4 Employment	17.66	23	26	12	19
5.5 Sustainable development and environment	65.28	14	21	17	20

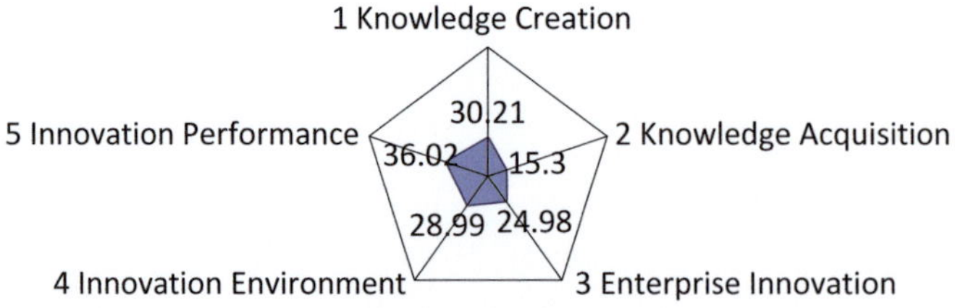

Fig. 27.1 Spider chart of Sichuan innovation capacity

Table 27.2 Sichuan enterprises in the "China Top 500 Enterprises List 2016"

	Enterprise	Operation revenue (billion yuan)	Rank
1	Changhong Holding Group	101.50	146
2	New Hope Group	68.24	197
3	Sichuan Yibin WULIANGYE Group	65.25	208
4	Tongwei Group	53.21	250
5	Sichuan Huashi Group	51.72	256
6	Kechuang Holding Group	46.74	286
7	Sichuan Hongda	40.56	324
8	Sichuan Tranvic Group	40.53	325
9	Sichuan Kelun Group	39.21	332
10	Sichuan CLR Group	33.82	377
11	Sichuan Desheng Group	32.88	391
12	Sichuan Road & Bridge	30.62	410
13	Luzhou Laojiao Group	27.68	459
14	Chengdu Jiangong	25.30	491

Table 27.3 Sichuan enterprises in the "China Top-1000 Innovative Enterprises List 2016" (Top 10)

	Enterprises	Ownership	Patent (Nr)	Industry classification
1	Changhong Holding Group	SOE	674	Electric appliance
2	CNPC Chuanqing Drilling Engineering Company Limited	SOE	273	Exploration of oil, natural gas, and other mineral products
3	Chengdu Engineering	SOE	162	Engineering design and survey
4	Chengdu Qinchuan Technology Development	Private	152	Gas appliances, electronic products, instruments, etc.
5	MAIPU	Private	128	Communication equipment
6	Sichuan Jezetech	Private	120	Radar equipment, communication equipment, electrical appliances, etc.
7	China Railway Eryuan Engineering Group	SOE	98	Survey and design
8	China MCC5 Group	State-holding	97	Real estate development and resource development
9	Sichuan Yuanxing Rubber	Private	84	Tire
10	Dongfang Boiler Group	SOE	68	Boiler

Chapter 28
Guizhou

In 2017, the innovation capacity of Guizhou ranked No.18 countrywide, which fell one rank compared to that of 2016. Its knowledge creation, knowledge acquisition, corporate innovation, innovation environment, and innovation performance were No.18, No.15, No.21, No.26, and No.15, respectively. Among them, only the innovation performance was six ranks higher than last year; others showed a different degree of decline, while the innovation environment drops 19 ranks (Table 28.1, Fig. 28.1).

The data shows that two companies in Guizhou were selected into "China Top 500 Enterprises List 2016." They are China Tobacco Guizhou Industrial Co., Ltd, ranked No.353 and Kweichow Moutai Group ranked No.356 (Table 28.2).

In "China Top-1000 Innovative Enterprises List 2016," there were four companies from Guizhou in 2016. Among them, two private companies are in the field of the pharmaceutical and Internet industry; two state-owned companies are in the field of chemical and aerospace (Table 28.3).

It is found that the traditional dominant industries in Guizhou are still dominated by state-owned companies, and the strength of private companies is still weak. But in the Internet, medicine, and other high-tech fields, some innovative companies are emerging. This is closely related to the development of Guizhou's big-data information technology industry in recent years. In general, the Guizhou's innovation capacity needs to be strengthened, and more efforts should be made in innovation environment construction, so as to further release the potential of innovative development.

X. Liu et al., *Regional Innovation Index of China: 2017*,
https://doi.org/10.1007/978-981-13-1205-2_28

Table 28.1 Guizhou innovation capability comprehensive value

	2017 comprehensive value	2017 sub-index rank			
Name of indexes	Index value	Rank	Strength	Efficiency	Potential
Comprehensive value	19.64	18	26	20	3
1 Knowledge creation	16.91	16	25	18	4
1.1 R&D input	23.83	15	26	31	3
1.2 Patent	16.73	28	23	8	17
1.3 Research paper	15.89	15	26	21	12
2 Knowledge acquisition	15.22	29	27	14	3
2.1 Science and technology cooperation	22.06	9	29	22	27
2.2 Technology transfer	11.77	12	24	4	3
2.3 Foreign-funded enterprise investment	20.88	21	26	27	3
3 Enterprise innovation	23.72	23	23	18	11
3.1 Enterprise R&D input	18.59	12	25	25	6
3.2 Design capability	32.91	10	19	10	18
3.3 Technology improvement	8.03	28	20	6	4
3.4 New product sales income	20.38	26	27	29	19
4 Innovation environment	14.59	30	25	22	7
4.1 Innovation infrastructure	24.69	30	24	28	9
4.2 Market environment	24.17	20	28	25	24
4.3 Employee quality	10.6	24	23	10	2
4.4 Financial environment	27.87	15	22	16	30
4.5 Entrepreneurship	32.7	15	22	19	3
5 Innovation performance	26.32	21	24	22	1
5.1 Macroeconomic	15.93	21	25	29	3
5.2 Industry structure	34.21	8	25	20	9
5.3 Industrial international competitiveness	25.38	12	21	13	1
5.4 Employment	61.65	21	18	16	3
5.5 Sustainable development and environment	54.14	25	19	25	18

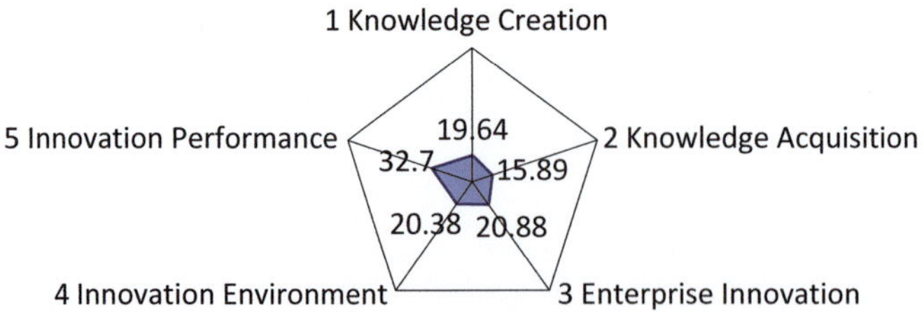

Fig. 28.1 Spider chart of Guizhou innovation capacity

Table 28.2 Guizhou enterprises in the "China Top 500 Enterprises List 2016"

	Enterprise	Operation revenue (billion yuan)	Rank
1	China Tobacco Guizhou Industrial	36.75	353
2	Kweichow Moutai Group	36.36	356

Table 28.3 Guizhou enterprises in the "China Top-1000 Innovative Enterprises List 2016"

Enterprises	Ownership	Industry classification
Guizhou Bailing	Private	Healthcare/health services
Guiyang Longmaster	Private	Internet and medical services/software and information technology services
Wengfu Group	SOE	Chemical/trade/scientific research/engineering
Guizhou Aerospace Electronics	SOE	Aeronautical industry/electronic components

Chapter 29
Yunnan

In 2017, the innovation capacity of Yunnan ranked No.24 countrywide, which increased two ranks compared to that of 2016. Its knowledge creation and corporate innovation were No.17 and No.18, respectively; both were eight ranks higher; the innovation environment was two ranks higher as the No. 21; the innovation performance kept the same as No. 27; its knowledge acquisition dropped three ranks (Table 29.1, Fig. 29.1).

According to the statistics from the China Enterprise Confederation, there were seven companies in Yunnan selected into the "China Top-500 Companies." Most of them are from the tobacco, energy, and chemical industries (Table 29.2).

In the "China Top-1000 Innovative Enterprises List 2016," six companies are from Yunnan, and most of them are state-owned. Based on their industry distributions, medicine, tea, and metal materials industries in Yunnan have stronger patent innovation, which quite have local characteristics (Table 29.3).

In general, the corporate innovation capacity of Yunnan has been improved. Different industries all have made certain progresses in the patents, and the innovation environment has been improved as well. But Yunnan should strengthen its knowledge acquisition capacity, further enhance the level of S&T cooperation with developed areas, improve its technology transfer, and further promote its corporate innovations. In addition, the province also needs to create the innovative and entrepreneurial atmosphere, so as to create the development space for small companies.

X. Liu et al., *Regional Innovation Index of China: 2017*,
https://doi.org/10.1007/978-981-13-1205-2_29

Table 29.1 Yunnan innovation capability comprehensive value

Name of indexes	2017 comprehensive value Index value	2017 sub-index rank Rank	Strength	Efficiency	Potential
Comprehensive value	20.43	24	22	25	6
1 Knowledge creation	20.06	17	21	23	1
1.1 R&D input	20.73	11	20	22	1
1.2 PATENT	18.33	22	22	19	18
1.3 research paper	22.17	17	21	17	7
2 Knowledge acquisition	10.06	28	26	24	23
2.1 Science and technology cooperation	20.92	22	28	17	4
2.2 Technology transfer	10.75	21	25	14	18
2.3 Foreign-funded enterprise investment	1.40	30	23	29	31
3 Enterprise innovation	21.46	18	25	20	5
3.1 Enterprise R&D input	32.10	16	24	16	3
3.2 Design capability	17.31	16	21	17	2
3.3 Technology improvement	22.22	26	23	20	16
3.4 New product sales income	10.95	24	24	26	17
4 Innovation environment	22.33	21	22	14	12
4.1 Innovation infrastructure	21.40	19	22	9	16
4.2 Market environment	27.06	22	22	23	18
4.3 Employee quality	27.16	13	15	13	3
4.4 Financial environment	5.24	29	28	30	18
4.5 Entrepreneurship	30.80	13	23	4	12
5 Innovation performance	24.81	27	25	28	11
5.1 Macroeconomic	22.55	25	23	30	10
5.2 Industry structure	12.38	28	24	27	16
5.3 Industrial international competitiveness	9.00	23	22	21	25
5.4 Employment	14.39	29	29	27	9
5.5 Sustainable development and environment	65.73	13	14	26	5

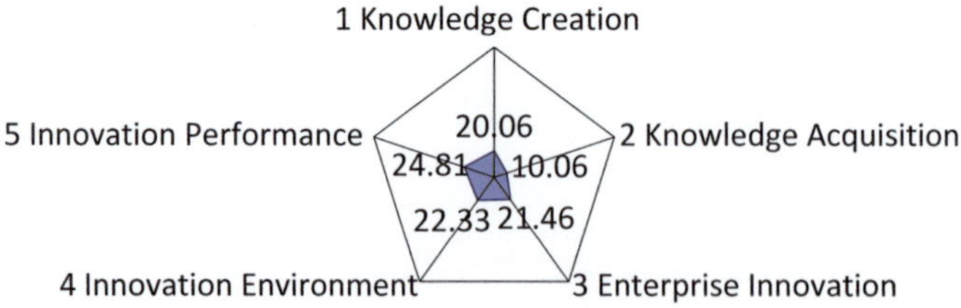

Fig. 29.1 Spider chart of Yunnan innovation capacity

Table 29.2 Yunnan enterprises in the "China Top 500 Enterprises List 2016"

	Enterprise	Operation revenue (billion yuan)	Rank
1	Hongta Group	103.39	139
2	Hongyunhonghe Group	92.88	158
3	Yuntianhua Group	58.56	233
4	Yunnan Construction and Investment Holding Group	44.24	298
5	Yunnan Provincial Energy Investment Group	41.89	313
6	Yunnan Tin Group	41.88	314
7	KISC	35.68	366

Table 29.3 Yunnan enterprises in the "China Top-1000 Innovative Enterprises List 2016"

	Enterprises
1	Yunnan Baiyao
2	Yunnan Dahongshan Pipeline Co., Ltd.
3	Yunnan Chihong Zn&Ge
4	Kunming Metallurgical Research Institute
5	Yunnan TASLY Deepure Biological Tea
6	Institute of Medical Biology Chinese Academy of Medical Sciences

Chapter 30
Tibet

In 2017, the innovation capacity of Tibet ranked No.31 countrywide. Its knowledge creation, knowledge acquisition, corporate innovation, innovation environment, and innovation performance were No.30, No.22, No.26, No.29, and No.26, respectively (Table 30.1, Fig. 30.1).

None of the companies in Tibet were selected into the "China Top 500 Enterprises List 2016" and "China Top-1000 Innovative Companies" in 2016.

In general, the innovation capacity of Tibet has made a great progress. In recent years, the Tibet government has actively implemented the innovation-driven strategies and conducted various factors to the innovation in companies, especially in the aspect of technology transfer. However, due to the weak basis and its geographical location, there are still many deficiencies in its economy, infrastructure construction, talent and education resources, etc. In the future, Tibet should further enhance the knowledge acquisition, promote the technology transfer, and improve the overall capacity through technology applications.

X. Liu et al., *Regional Innovation Index of China: 2017*,
https://doi.org/10.1007/978-981-13-1205-2_30

Table 30.1 Tibet innovation capability comprehensive value

	2017 comprehensive value	2017 sub-index rank			
Name of indexes	Index value	Rank	Strength	Efficiency	Potential
Comprehensive value	17.70	31	30	30	1
1 Knowledge creation	13.02	30	31	30	6
1.1 R&D input	15.6	18	31	26	4
1.2 Patent	7.06	31	31	28	31
1.3 Research paper	19.8	21	31	27	1
2 Knowledge acquisition	11.84	22	30	10	1
2.1 Science and technology cooperation	39.27	4	25	3	1
2.2 Technology transfer	11.43	19	31	10	6
2.3 Foreign-funded enterprise investment	20.65	7	31	22	1
3 Enterprise innovation	16.84	26	31	28	4
3.1 Enterprise R&D input	1.29	31	31	31	31
3.2 Design capability	31.55	5	31	11	1
3.3 Technology improvement	11.28	31	31	23	31
3.4 New product sales income	23.66	14	31	28	2
4 Innovation environment	19.13	29	31	25	2
4.1 Innovation infrastructure	5.71	31	31	31	22
4.2 Market environment	30.96	14	31	22	1
4.3 Employee quality	40.01	5	31	2	1
4.4 Financial environment	4.21	30	31	29	29
4.5 Entrepreneurship	14.77	31	31	30	15
5 Innovation performance	24.89	26	22	24	26
5.1 Macroeconomic	22.85	24	31	28	1
5.2 Industry structure	9.87	30	31	22	26
5.3 Industrial international competitiveness	0	31	31	31	31
5.4 Employment	21.97	16	9	31	15
5.5 Sustainable development and environment	69.74	10	3	3	31

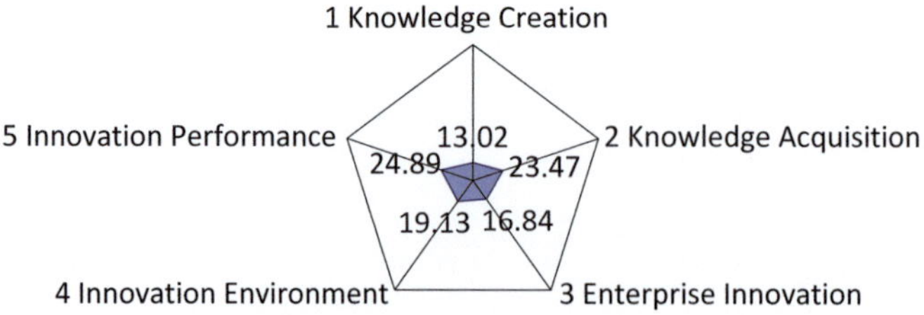

Fig. 30.1 Spider chart of Tibet innovation capacity

Chapter 31
Shaanxi

In 2017, the innovation capacity of Shaanxi ranked No.13 countrywide, which is three ranks decline compared to that of the last year. Its knowledge creation ranked No.7 countrywide, dropped one rank; the knowledge acquisition and corporate innovation ranked No.14 and No.19, respectively; both fell four ranks; innovation environment ranked No.12, the same as the last year; the innovation performance was No.11, dropped two ranks (Table 31.1, Fig. 31.1).

There were seven companies in Shaanxi listed into the "China Top 500 Enterprises List 2016," as shown below. Two of them entered the top 100: Shaanxi Yanchang Petroleum (group) Co., Ltd. ranked No.61 and Shaanxi Coal and Chemical Industry Group Co., Ltd. ranked No.82 (Table 31.2).

In the "China Top-1000 Innovative Enterprises List 2016," 18 companies are from Shaanxi. Table below shows the top 10 of them with the most quantity of patent applications in 2016. Among them, three companies applied for more than 100 patents: China National Heavy Machinery Research Institute with 162 applications, Shaanxi Sunger Road Bio-science Co., Ltd. with 153 applications, and Xi'an Thermal Power Research Institute Co., Ltd. with 116 applications (Table 31.3).

In general, innovation in Shaanxi is more science-driven, with great advantages of its rich S&T resources, universities, and institutes. In recent years, the local government has actively promoted the innovation-driven development strategy and carried out a batch of technology projects, including the overall Shaanxi S&T innovation plan, major funds for tech innovation, Shaanxi S&T research and development project, etc., so as to build a regional innovation system with Shaanxi characteristics. But meanwhile, large companies with strong innovation capacity are few in Shaanxi. In the future, Shaanxi should strengthen the university-industry-research cooperation, as well as the construction on the ecological environment for innovation and entrepreneurship.

X. Liu et al., *Regional Innovation Index of China: 2017*,
https://doi.org/10.1007/978-981-13-1205-2_31

Table 31.1 Shaanxi innovation capability comprehensive value

Name of indexes	2017 comprehensive value	2017 sub-index rank			
	Index value	Rank	Strength	Efficiency	Potential
Comprehensive value	26.05	13	15	8	23
1 Knowledge creation	30.41	7	9	4	27
1.1 R&D input	28.58	7	8	4	15
1.2 Patent	22.97	16	12	11	30
1.3 Research paper	48.95	2	5	2	21
2 Knowledge acquisition	16.19	14	13	20	8
2.1 Science and technology cooperation	32.75	7	6	13	8
2.2 Technology transfer	11.94	18	15	22	8
2.3 Foreign-funded enterprise investment	6.95	19	18	14	12
3 Enterprise innovation	21.41	19	18	16	23
3.1 Enterprise R&D input	35.62	13	17	12	18
3.2 Design capability	11.9	23	17	18	27
3.3 Technology improvement	25.4	18	19	15	14
3.4 New product sales income	10.39	26	20	25	21
4 Innovation environment	25.83	12	17	8	19
4.1 Innovation infrastructure	21.4	18	18	13	15
4.2 Market environment	33.92	11	12	7	14
4.3 Employee quality	31.31	8	14	9	6
4.4 Financial environment	15.32	17	18	12	28
4.5 Entrepreneurship	27.21	16	19	10	19
5 Innovation performance	36.25	11	18	7	20
5.1 Macroeconomic	32.09	17	15	14	22
5.2 Industry structure	17.02	19	18	21	13
5.3 Industrial international competitiveness	45.68	4	12	2	8
5.4 Employment	21.67	17	15	13	18
5.5 Sustainable development and environment	64.81	16	13	15	29

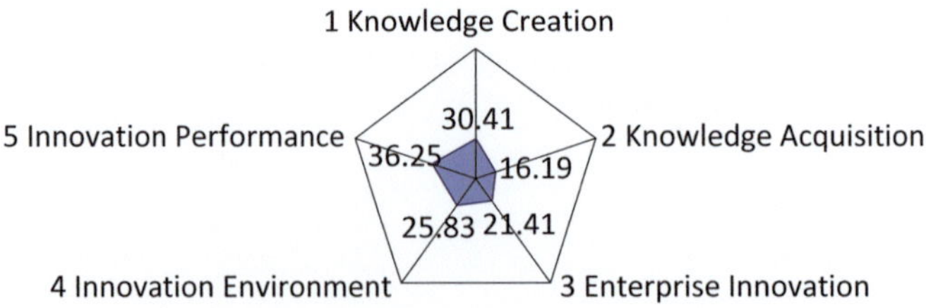

Fig. 31.1 Spider chart of Shaanxi innovation capacity

Table 31.2 Shaanxi enterprises in the "China Top 500 Enterprises List 2016"

	Enterprise	Operation revenue (billion yuan)	Rank
1	Shaanxi Yanchang Petroleum	220.89	61
2	Shaanxi Coal and Chemical Industry Group	190.61	82
3	Shaanxi Youser	116.18	124
4	Shaanxi Dongling	76.11	182
5	Xi'an Maike	71.00	192
6	Shaanxi Construction Engineering Group	68.11	199
7	Shaanxi Auto Holding Group	30.02	422

Table 31.3 Shaanxi enterprises in the "China Top-1000 Innovative Companies List 2016"

	Enterprise	Patent applications (Nr)
1	China National Heavy Machinery Research Institute	162
2	Shaanxi Sunger Road Bio-science	153
3	Xi'an Thermal Power Research Institute	116
4	CNPC BAOJI Oilfield Machinery	69
5	Northwest Institute For Nonferrous Metal Research	63
6	JINDUICHENG Molybdenum	58
7	China Railway First Survey & Design institute Group	57
8	Xi'an Focuslight	52
9	Shaanxi Meibang	51
10	Xi'an Shaangu Power	45

Chapter 32
Gansu

In 2017, the innovation capacity of Gansu ranked No. 21 countrywide, which is one rank lower compared to that of the last year. Its corporate innovation ranked No. 16, innovation performance ranked No. 22, and knowledge creation, knowledge acquisition, and innovation environment were No. 20, No. 27, and No. 24, respectively. In details, the ranking of the corporate innovation is the same as last year; innovation performance increased by four ranks; knowledge creation, knowledge acquisition, and innovation environment dropped by three, nine, and three places, respectively (Table 32.1, Fig. 32.1).

From the data, there were four companies in Gansu listed into "China Top 500 Enterprises List 2016:" Jinchuan Group ranked No. 63, JISCO ranked No. 149, Baiyin Nonferrous Group Co., Ltd. ranked No. 245, and Gansu Construction & Investment Group ranked No. 308 (Table 32.2).

In the "China Top-1000 Innovative Enterprises List 2016," two companies are from Gansu – Northwest Research Institute of Mining and Metallurgy and Tianhua Institute of Chemical Machinery & Automation Co., Ltd. – both state-owned. Their patent applications in 2016 were 55 and 14, respectively (Table 32.3).

In general, the basis of Gansu's economic development is weak, and the state-owned companies are still the main force for the economic development. The position of private companies still needs to be improved. Under the guidance of the government's support policies of innovation and entrepreneurship, the quantity of incubators in Gansu has increased significantly in recent years, which also stimulated the activities of local innovative companies. But Gansu still needs to strengthen the knowledge accumulation, innovation environment, and corporate innovation capacity, which cannot be separated from more precise policy guidance and effective industry-university-research cooperation.

X. Liu et al., *Regional Innovation Index of China: 2017*,
https://doi.org/10.1007/978-981-13-1205-2_32

Table 32.1 Gansu innovation capability comprehensive value

Name of indexes	2017 comprehensive value Index value	2017 subindex rank Rank	Strength	Efficiency	Potential
Comprehensive value	20.82	21	25	19	11
1 Knowledge creation	19.47	20	24	10	22
1.1 R&D input	12.17	23	24	17	18
1.2 Patent	17.13	27	25	20	22
1.3 Research paper	38.75	4	20	3	14
2 Knowledge acquisition	10.19	27	25	21	29
2.1 Science and technology cooperation	22.08	19	24	12	21
2.2 Technology transfer	10.2	23	21	19	13
2.3 Foreign-funded enterprise investment	1.28	31	29	31	27
3 Enterprise innovation	22.06	16	26	13	14
3.1 Enterprise R&D input	27.42	21	26	20	13
3.2 Design capability	10.35	26	26	27	8
3.3 Technology improvement	43.08	5	21	2	8
3.4 New product sales income	10.59	25	23	21	27
4 Innovation environment	21.73	24	27	24	4
4.1 Innovation infrastructure	24.11	12	26	29	1
4.2 Market environment	25.54	28	30	19	16
4.3 Employee quality	24.92	18	27	7	4
4.4 Financial environment	13.18	19	21	17	10
4.5 Entrepreneurship	20.91	25	26	26	9
5 Innovation performance	27.09	22	21	27	12
5.1 Macroeconomic	15.9	30	27	31	16
5.2 industry structure	12.98	27	27	25	14
5.3 Industrial international competitiveness	16.7	17	23	16	3
5.4 Employment	26.86	10	5	29	16
5.5 Sustainable development and environment	62.99	19	11	27	15

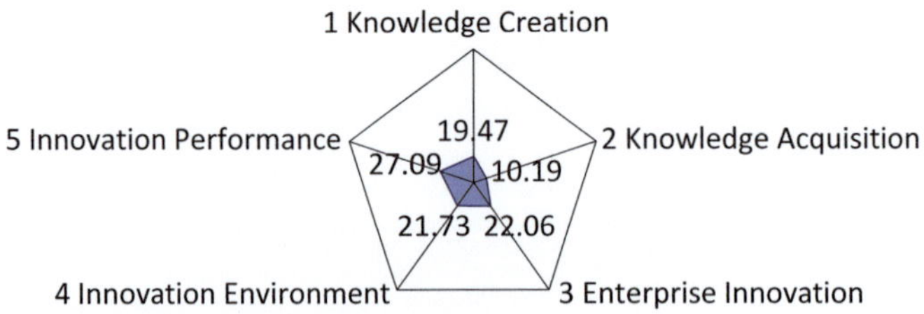

Fig. 32.1 Spider chart of Gansu innovation capacity

Table 32.2 Gansu enterprises in the "China Top 500 Enterprises List 2016"

	Enterprise	Operation revenue (billion yuan)	Rank
1	Jinchuan Group	220.41	63
2	JISCO	100.86	149
3	Baiyin Nonferrous Group Co., Ltd.	54.90	245
4	Gansu Construction & Investment Group	42.51	308

Table 32.3 Gansu enterprises in the "China Top-1000 Innovative Enterprises List 2016"

Enterprises	Ownership	Patent applications (Nr)
Northwest Research Institute of Mining and Metallurgy	SOE	55
Tianhua Institute of Chemical Machinery & Automation	SOE	14

Chapter 33
Qinghai

In 2017, the innovation capacity of Qinghai ranked No. 30 countrywide, increased one rank compared to that of the last year. Its innovation performance was relatively strong as No. 23 countrywide, seven ranks higher than the last year; the knowledge acquisition and innovation environment were both No. 25, and knowledge creation was No. 28, the same as the last year; the corporate innovation ranked No. 29, one rank higher (Table 33.1, Fig. 33.1).

According to the statistics, only the Western Mining Group was selected into the "China Top 500 Enterprises List 2016."

In recent years, Qinghai has promoted the innovation-driven strategy, implemented key technological support projects focused on "123" and "1020," and overcome key technical problems that restricted economic and social development. Among them, the "123" technology support project targets the innovation chain around the industrial chain. It focuses on ten characteristic industries, such as new energy, new materials, equipment manufacturing, and so on. The "1020" technology supports project of ecological agriculture, and animal husbandry aims at the implementation of provincial technology planning around ten characteristic industries of agricultural and animal husbandry industry. It is expected to solve a number of key technology and common technical problems. Qinghai will further promote the construction of the rural informatization. In general, the economic strength of Qinghai is weak, so does its innovation capacity. In the future, Qinghai should combine with regional advantages to further enhance the innovation capacity.

X. Liu et al., *Regional Innovation Index of China: 2017*,
https://doi.org/10.1007/978-981-13-1205-2_33

Table 33.1 Qinghai innovation capability comprehensive value

Name of indexes	2017 comprehensive value Index Value	2017 subindex rank Rank	Strength	Efficiency	Potential
Comprehensive value	18.13	30	31	24	10
1 Knowledge creation	13.95	28	30	20	26
1.1 R&D input	1.32	31	30	29	31
1.2 Patent	22.39	17	30	14	5
1.3 Research paper	22.33	16	30	13	5
2 Knowledge acquisition	11.29	25	31	22	5
2.1 Science and technology cooperation	23.47	16	31	9	6
2.2 Technology transfer	4.72	31	29	27	28
2.3 Foreign-funded Enterprise investment	7.08	18	30	28	5
3 Enterprise innovation	15.85	29	30	29	7
3.1 Enterprise R&D Input	8.68	30	30	30	30
3.2 Design capability	12.82	21	30	16	21
3.3 Technology improvement	28.32	14	30	11	1
3.4 New product sales income	18.68	17	30	31	3
4 Innovation environment	20.77	25	30	6	22
4.1 Innovation infrastructure	31.32	5	30	1	27
4.2 Market environment	27.6	20	29	9	26
4.3 Employee quality	14.99	30	30	6	24
4.4 Financial environment	2.86	31	30	28	31
4.5 Entrepreneurship	27.07	17	29	17	4
5 Innovation performance	25.92	23	29	29	3
5.1 Macroeconomic	21.14	27	30	17	14
5.2 Industry structure	15.6	22	30	29	3
5.3 Industrial international competitiveness	22.14	13	30	20	2
5.4 Employment	20.66	19	16	23	7
5.5 Sustainable development and environment	50.08	27	15	29	27

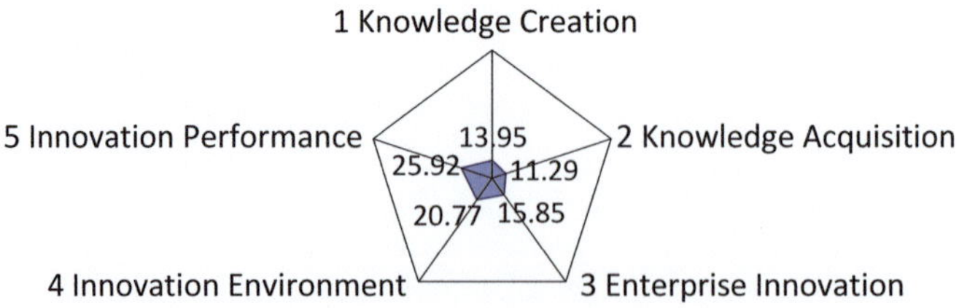

Fig. 33.1 Spider chart of Qinghai innovation capacity

Chapter 34
Ningxia

In 2017, the innovation capacity of Ningxia ranked No. 22 countrywide, dropped by two ranks. Its innovation environment declined from No. 26 to No. 30; innovation performance kept the same at No. 31; other indicators all increased, especially the knowledge acquisition from No. 19 in the last year to this year's No. 5 (Table 34.1, Fig. 34.1).

According to the statistics from the China Enterprise Confederation, there were two companies in Ningxia selected into the "China Top 500 Enterprises List 2016." Most of them are from the petrochemical and metal resources industries. There were no companies from Ningxia in the "China Top 500 Enterprises List 2016" (Table 34.2).

In general, Ningxia has few innovative companies, so there is large space to improve its innovation capacity. In recent years, with the implementation of the "The Belt and Road Initiative" and the Internet planning, Ningxia has progressed a lot in the field of information technology popularization and relevant applications, especially creating a well policy environment for the development of Internet hospital. In the future, Ningxia should further strengthen the application capacity of technology innovation and bring new driving force to the economic and social development.

X. Liu et al., *Regional Innovation Index of China: 2017*,
https://doi.org/10.1007/978-981-13-1205-2_34

Table 34.1 Ningxia innovation capability comprehensive value

Name of indexes	2017 comprehensive value Index value	2017 subindex rank Rank	Strength	Efficiency	Potential
Comprehensive value	20.68	22	29	23	2
1 Knowledge creation	15.96	24	28	25	8
1.1 R&D input	10.35	26	28	21	16
1.2 Patent	21.32	19	28	16	4
1.3 Research paper	16.44	29	29	23	4
2 Knowledge acquisition	26.10	5	15	4	2
2.1 Science and technology cooperation	20.99	21	19	24	5
2.2 Technology transfer	52.47	1	6	1	1
2.3 Foreign-funded enterprise investment	10.15	13	28	17	4
3 Enterprise innovation	22.68	15	28	12	8
3.1 Enterprise R&D input	26.47	22	28	19	16
3.2 Design capability	17.52	15	28	12	5
3.3 Technology improvement	32.76	11	28	5	3
3.4 New product sales income	14.67	23	28	15	15
4 Innovation environment	18.40	30	29	31	1
4.1 Innovation infrastructure	19.87	23	28	23	2
4.2 Market environment	30.29	16	26	28	2
4.3 Employee quality	16.33	29	29	8	9
4.4 Financial environment	8.82	27	29	31	3
4.5 Entrepreneurship	16.67	29	30	31	2
5 Innovation performance	20.52	31	31	31	7
5.1 Macroeconomic	22.18	26	29	15	17
5.2 Industry structure	17.15	18	29	26	2
5.3 Industrial international competitiveness	7.16	24	28	22	30
5.4 Employment	14.69	27	30	30	4
5.5 Sustainable development and environment	41.44	30	20	31	19

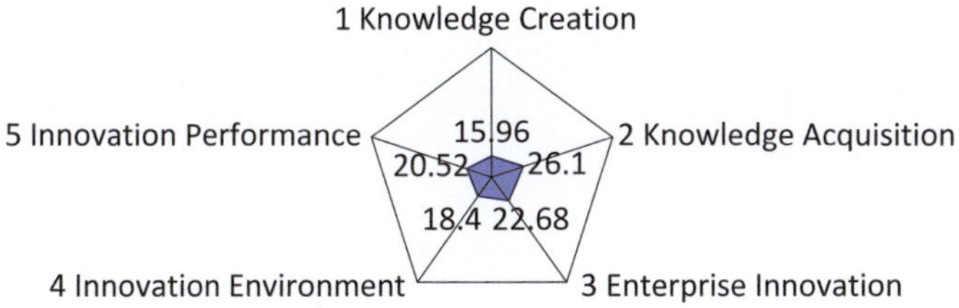

Fig. 34.1 Spider chart of Ningxia innovation capacity

Table 34.2 Ningxia enterprises in the "China Top 500 Enterprises List 2016"

	Enterprise	Operation revenue (billion yuan)	Rank
1	Baota Petrochemical Group	40.93	322
2	Ningxia Tianyuan Manganese Limited Company	37.28	349

Chapter 35
Xinjiang

In 2017, the innovation capacity of Xinjiang ranked No. 25 countrywide, the same as the last year. Its knowledge creation, knowledge acquisition, and corporate innovation dropped by two ranks, one rank, and one rank, respectively. The innovation environment and innovation performance both increased by four ranks (Table 35.1, Fig. 35.1).

According to the statistics from the China Enterprise Confederation, there were four companies in Xinjiang selected into the "China Top 500 Enterprises List 2016," including TBEA SunOasis and Goldwind S&T Co., Ltd. The Goldwind S&T was also selected into the list of "2016 China Top-1000 Innovative Companies" (Table 35.2).

In general, the innovation capacity of Xinjiang has been improved, with great potentials. In recent years, Xinjiang has invested heavily in key technologies such as large radio telescope, new energy, and UHV transmission and transformation and has achieved some outcomes. With the implementation of the "Belt and Road Initiative," Xinjiang will have better development opportunities. The government could increase the investments in technology transfer and promote the transformation and upgrading of local industries.

X. Liu et al., *Regional Innovation Index of China: 2017*,
https://doi.org/10.1007/978-981-13-1205-2_35

Table 35.1 Xinjiang innovation capability comprehensive value

Name of indexes	2017 comprehensive value	2017 subindex rank			
	Index value	Rank	Strength	Efficiency	Potential
Comprehensive value	20.04	25	27	27	4
1 Knowledge creation	17.91	22	26	17	12
1.1 R&D input	8.33	28	27	30	20
1.2 Patent	18.79	21	26	21	12
1.3 Research paper	35.30	6	25	5	3
2 Knowledge acquisition	13.09	21	28	16	7
2.1 Science and technology cooperation	26.27	13	26	11	2
2.2 Technology transfer	13.51	15	27	9	5
2.3 Foreign-funded enterprise investment	2.88	28	27	30	22
3 Enterprise innovation	18.93	23	27	23	6
3.1 Enterprise R&D Input	21.35	27	27	27	14
3.2 Design capability	14.78	20	27	15	11
3.3 Technology improvement	25.61	17	27	17	5
3.4 New product sales income	14.84	22	26	23	5
4 Innovation environment	22.09	23	26	15	5
4.1 Innovation infrastructure	18.62	25	27	11	20
4.2 Market environment	26.77	26	27	20	15
4.3 Employee quality	25.80	17	25	5	11
4.4 Financial environment	16.29	16	27	18	2
4.5 Entrepreneurship	22.95	20	27	22	5
5 Innovation performance	25.66	25	30	30	2
5.1 Macroeconomic	25.90	23	26	20	8
5.2 Industry structure	19.88	16	26	30	1
5.3 Industrial international competitiveness	3.71	30	24	30	28
5.4 Employment	37.93	4	13	28	1
5.5 Sustainable development and environment	40.88	31	24	30	30

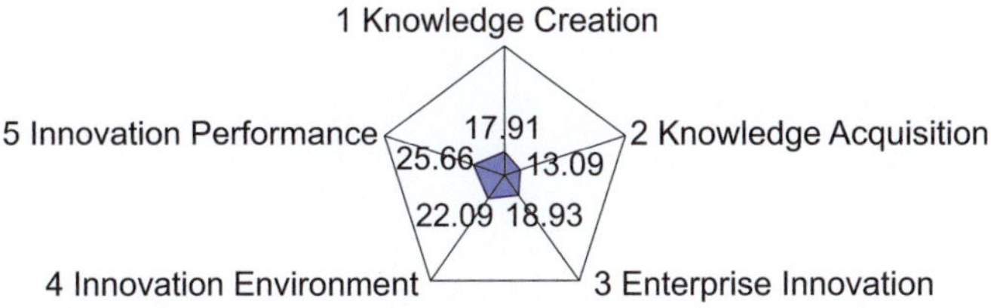

Fig. 35.1 Spider chart of Xinjiang innovation capacity

Table 35.2 Xinjiang enterprises in the "China Top 500 Enterprises List 2016"

	Enterprise	Operation revenue (billion yuan)	Rank
1	Xinjiang Guanghui Industry Investment Group	105.04	135
2	TBEA SunOasis	48.26	277
3	Xinjiang Goldwind S&T	30.06	420
4	The Xinjiang Production and Construction Corps	27.13	468

Appendices

Appendix 1: Definition of the Evaluation Index of Regional Innovation Capacity and Data Sources

Knowledge creation

Category	Code	Indicator	Definition	Data source
Knowledge creation	11101	Full-time R&D personnel (man-year)	Measuring the S&T human investment in a region	China Statistical Yearbook
	11102	Average full-time R&D personnel per 10,000 people (man-year /10,000 people)	Proportion of full-time R&D personnel to the residents	Data analysis
	11103	Growth rate of full-time R&D personnel(%)	The growth compared with last year	Data analysis
	11201	Government R&D investment(100 million Yuan)	Measuring the R&D investment of local government	China Statistical Yearbook
	11202	Ratio of government R&D investment to regional GDP (%)	Ratio of government R&D investment to regional GDP	Data analysis
	11203	Growth rate of government R&D investment(%)	The growth compared with last year	Data analysis
	12101	Quantity of patent application (excluding companies' applications)	Measure the R&D level of universities and institutes in a region	China Statistical Yearbook
	12102	Patent application per 10,000 R& D personnel	Ratio of patent application to R&D personnel	Data analysis
	12103	Growth rate of patent applications(excluding companies' applications)(%)	The growth compared with last year	Data analysis

(continued)

X. Liu et al., *Regional Innovation Index of China: 2017*,
https://doi.org/10.1007/978-981-13-1205-2

Category	Code	Indicator	Definition	Data source
	12104	Patent application per 100 million Yuan R&D internal expenditure	Ratio of patent applications to R&D investment	Data analysis
	12201	Quantity of granted patents	Measure the technology level of universities and institutes in a region	China Statistical Yearbook
	12202	Granted patents per 10,000 R&D personnel	Ratio of granted patents to R&D personnel	Data analysis
	12203	Growth rate of granted patents(%)	The growth compared with last year	Data analysis
	12204	Granted patents per100 million Yuan R&D internal expenditure	Ratio of granted patents to R&D investment	Data analysis
	13101	Amount of domestic papers	Measure the level of regional publication in domestic journals	Chinese S&T Papers Statistics and Analysis
	13102	Average amount of domestic papers per 100,000 R&D personnel	Ratio of domestic papers to R&D personnel	Data analysis
	13103	Growth rate of domestic papers(%)	The growth compared with last year	Data analysis
	13201	Amount of international papers	Measure the level of regional publication in international journals	Chinese S&T Papers Statistics and Analysis
	13202	Average amount of international papers per 100,000 R&D personnel	Ratio of international papers to R&D personnel	Data analysis
	13203	Growth rate of international papers(%)	The growth compared with last year	Data analysis

Knowledge acquisition

Category	Code	Indicator	Definition	Data source
Knowledge acquisition	21111	Coauthored S&T papers amount by different units in the same province	Measure the knowledge cooperation level among different units in the region	Chinese S&T Papers Statistics and Analysis
	21112	Average coauthored amount of S&T papers by different units in the same province per 100,000 R&D personnel	Ratio of S&T papers by different units in the same province to R&D personnel	Data analysis
	21113	Growth rate of coauthored S&T papers by different units in the same province (%)	The growth compared with last year	Data analysis
	21121	Coauthored S&T papers amount by cross-region cooperation	Measure the cross-region knowledge cooperation level	Chinese S&T Papers Statistics and Analysis
	21122	Average amount of coauthored S&T papers by cross-region cooperation per 100,000 R&D personnel	Ratio of S&T papers by cross-province units to R & D personnel	Data analysis
	21123	Growth rate of coauthored S&T papers by cross-region cooperation(%)	The growth compared with last year	Data analysis
	21131	Coauthored S&T papers amount by cross-nation cooperation	Measure the knowledge cooperation level with international institutions	Chinese S&T Papers Statistics and Analysis
	21132	Average amount of coauthored S&T papers by cross-nation cooperation per 1,000,000 R&D personnel	Ratio of S&T papers by cross-nation units to R&D personnel	Data analysis
	21133	Growth rate of coauthored S&T papers by cross-nation cooperation(%)	The growth compared with last year	Data analysis
	21201	Corporate funds for the internal R&D expenditure of universities and institutes (10,000 Yuan)	Measure the cooperation between enterprises and universities /institutes	China Statistical Yearbook
	21202	Ratio of corporate funds for the internal R&D expenditure of universities and institutes(%)	Ratio of corporate funds for the internal R&D expenditure of universities and institutes	Data analysis
	21203	Growth rate of corporate funds for the internal R&D expenditure of universities and institutes(%)	The growth compared with last year	Data analysis

(continued)

Category	Code	Indicator	Definition	Data source
	22101	Transaction amount in technology market(flow direction)(10,000 Yuan)	Measure the flow of technology	China Statistical Yearbook
	22102	Average transaction amount of technology market(flow direction)(10,000 Yuan per project)	Ratio of the technical transaction amount to the number of the technical transaction items	Data analysis
	22103	Growth rate of the transaction amount in technology market(flow direction)(%)	The growth compared with last year	Data analysis
	22201	Domestic technology acquisition expenditure of above-scale industrial companies (10,000 Yuan)	Measure the domestic technology acquisition of regional companies	China Statistical Yearbook
	22202	Average domestic technology acquisition expenditure of above-scale industrial companies(10,000 Yuan per project)	Ratio of domestic technology acquisition expenditure to the number of the transaction items	Data analysis
	22203	Growth rate of domestic technology acquisition expenditure of above-scale industrial companies(%)	The growth compared with last year	Data analysis
	22301	Foreign technology acquisition expenditure of above-scale industrial companies (10,000 Yuan)	Measure the foreign technology acquisition of regional companies	China Statistical Yearbook
	22302	Average foreign technology acquisition expenditure of above-scale industrial companies(10,000 Yuan per project)	Ratio of contract amount for introducing foreign technology to the number of introduced projects	Data analysis
	22303	Growth rate of foreign technology acquisition expenditure of above-scale industrial companies(%)	The growth compared with last year	Data analysis
	23001	Registered foreign capital of foreign direct investment (100 million USD)	Measuring the use of foreign capital	China Statistical Yearbook
	23002	Average registered foreign capital of foreign direct investment (10,000 USD)	Ratio of foreign capitals to the quantity of residents	Data analysis
	23003	Growth rate of registered foreign capital of foreign direct investment (%)	The growth compared with last year	Data analysis

Corporate innovation

Category	Code	Indicator	Definition	Data source
Corporate innovation	31101	R&D personnel of large industrial companies(10,000 people)	Measuring the R&D human investment capacity	China Statistical Yearbook
	31102	Ratio of R&D personnel to the total personnel in the large industrial companies(%)	Ratio of R&D personnel to the total personnel in the corporation	Data analysis
	31103	Growth rate of R&D personnel of large industrial companies(%)	The growth compared with last year	Data analysis
	31201	Internal R&D expenditure of large industrial companies (100 million Yuan)	Measuring the R&D fund investment capacity	China Statistical Yearbook
	31202	Ratio of internal R&D expenditure to the sales revenue of large industrial companies(%)	Ratio of R&D investment to the sales revenue	Data analysis
	31203	Growth rate of the internal R&D expenditure of large industrial companies(%)	The growth compared with last year	Data analysis
	31301	Amount of large industrial companies owned R&D institutions	Measuring the R&D infrastructure of a company	China Statistical Yearbook
	31302	Ratio of large industrial companies owned R&D institutions to the total number of companies(%)	Ratio of companies owned R& D departments to the total number of companies	Data analysis
	31303	Growth rate of the amount of large industrial companies owned R&D institutions(%)	The growth compared with last year	Data analysis
	32101	Patent application amount of large industrial companies	Measuring the R&D output of a company	China Statistical Yearbook
	32102	Average patent application amount of large industrial companies per 10,000 R&D personnel	Ratio of patent applications to R&D personnel	Data analysis
	32103	Growth rate of patent application amount of large industrial companies(%)	The growth compared with last year	Data analysis
	32201	Granted patent amount of large industrial companies	Measuring the core technology level of a company	China Statistical Yearbook
	32202	Average granted patent amount of large industrial companies per 100,000 people	Ratio of the quantity of authorized patents to the total number of companies	Data analysis

(continued)

Category	Code	Indicator	Definition	Data source
	32203	Growth rate of granted patent amount of large industrial companies(%)	The growth compared with last year	Data analysis
	33101	External R&D expenditure of large industrial companies (100 million Yuan)	Measure R&D cooperation between company and external units	China Statistical Yearbook
	33102	Average external R&D expenditure of large industrial companies(10,000Yuan for each)	Ratio of R&D external expenditure of a company to the total number of companies	Data analysis
	33103	Growth rate of external R&D expenditure of large industrial companies(%)	The growth compared with last year	Data analysis
	33201	Technology renovation expenditure of large industrial companies(10,000Yuan)	Measure the technology upgrading capacity of the company	China Statistical Yearbook
	33202	Average technology renovation expenditure of large industrial companies (10,000Yuan for each)	Ratio of technical transformation expenditure of a company to the total number of companies	Data analysis
	33203	Growth rate of technology renovation expenditure of large industrial companies(%)	The growth compared with last year	Data analysis
	33301	Companies with E-commerce	Measure the modern information technology capacity of the company	China Statistical Yearbook
	33302	Ratio of companies with E-commerce to the total number of companies(%)	Ratio of companies with E-commerce to the total number of companies	Data analysis
	33303	Growth rate of companies with E-commerce(%)	The growth compared with last year	Data analysis
	34001	Sales income of new product in the large industrial companies(100 million Yuan)	Measure the new product R&D capacity of the company	China Statistical Yearbook
	34002	Ratio of the sales income of new product to the total sales income(%)	Ratio of the sales income of new product to the total sales income	Data analysis
	34003	Growth rate of sales income of new product in the large industrial companies(%)	The growth compared with last year	Data analysis

Innovation environment

Category	Code	Indicator	Definition	Data source
Innovation environment	41111	Number of mobile phone users(10,000 people)	Measuring the communication infrastructure	China Statistical Yearbook
	41112	Mobile phone penetration rate (per 100 people)	Measuring the communication infrastructure	China Statistical Yearbook
	41113	Growth rate of mobile phone users(%)	The growth compared with last year	Data analysis
	41121	The Internet users(10,000 people)	Measuring the information infrastructure	China Statistical Yearbook
	41122	The Internet penetration rate (%)	Measuring the information infrastructure	China Statistical Yearbook
	41123	Growth rate of the Internet users(%)	The growth compared with last year	Data analysis
	41211	Number of Technology Business Incubators	Measuring the entrepreneurial incubator infrastructure in a region	China Torch Statistical Yearbook
	41212	Average entrepreneurial mentors quantity of every technology business incubator	Ratio of entrepreneurial mentors quantity to the incubator quantity	Data analysis
	41213	Growth rate of technology business incubators(%)	The growth compared with last year	Data analysis
	42101	Total export-import volume divided by destination and sources(100 million USD)	Measure the market openness of a region	China Statistical Yearbook
	42102	Ratio of total import and export to regional GDP(%)	Ratio of total import and export to regional GDP	Data analysis
	42103	Growth rate of the total export-import volume divided by destination and sources(%)	The growth compared with last year	Data analysis
	42201	Number of personnel of S&T services industry(10,000 people)	Measure the level of regional technology service	China Statistical Yearbook
	42202	Ratio of personnel of S&T services industry to the personnel in the tertiary industry (%)	Ratio of personnel of S&T services industry to the personnel in the tertiary industry	Data analysis
	42203	Growth rate of personnel amount of S&T services industry(%)	The growth compared with last year	Data analysis
	42301	Household consumption level (Yuan)	Measuring the regional market environment	China Statistical Yearbook

(continued)

Category	Code	Indicator	Definition	Data source
	42303	Growth rate of household consumption level(%)	The growth compared with last year	Data analysis
	43101	Educational expenditure (100 million Yuan)	Measuring the awareness to the talent training in a region	China Statistical Yearbook
	43102	Ratio of educational expenditure to the regional GDP(%)	Ratio of educational expenditure to the regional GDP	Data analysis
	43103	Growth rate of educational expenditure(%)	The growth compared with last year	Data analysis
	43201	College and above educated population(samples)	Measuring the quality of workers in a region	China Statistical Yearbook
	43202	Ratio of college and above educated population to the sample population(%)	Ratio of college and above educated population to the sample population	Data analysis
	43203	Growth rate of college and above educated population(%)	The growth compared with last year	Data analysis
	44111	Financial loans in the internal R&D expenditure of above-scale industrial companies (10,000 Yuan)	Measuring the financial environment of corporate R&D	China Statistical Yearbook
	44112	Average financial loans in the internal R&D expenditure of above-scale industrial companies(10,000 Yuan for each)	Ratio of R&D loans to the quantity of companies	Data analysis
	44113	Growth rate of financial loans in the internal R&D expenditure of above-scale industrial companies(%)	The growth compared with last year	Data analysis
	44211	The venture capital amount of technology business incubators(10,000 Yuan)	Measuring the financial ability of regional technology business incubator	China Torch Statistical Yearbook
	44212	Venture capital intensity of technology business incubators(10,000 Yuan per project)	Ratio of venture capitals to the quantity of investment projects	Data analysis
	44213	Growth rate of venture capital amount of technology business incubators(%)	The growth compared with last year	Data analysis
	44221	Total incubator funds for technology business incubators(10,000 Yuan)	Measuring the financial ability of regional technology business incubator	China Torch Statistical Yearbook

(continued)

Category	Code	Indicator	Definition	Data source
	44222	Average incubator funds for technology business incubators(10,000 Yuan for each)	Ratio of total incubator funds to the quantity of technology business incubators	Data analysis
	44223	Growth rate of incubator funds for technology business incubators(%)	The growth compared with last year	Data analysis
	45101	Number of high-tech companies	Measuring the level of regional high-tech industrial innovation	China Statistical Yearbook
	45102	Ratio of high-tech companies to the total companies(%)	Ratio of high-tech companies to the total companies	Data analysis
	45103	Growth rate of high-tech companies(%)	The growth compared with last year	Data analysis
	45201	Amount of companies graduated from technology business incubators	Measuring the hatching ability of regional technology business incubator	China Torch Statistical Yearbook
	45202	Average amount of companies graduated from technology business incubators (per every incubator)	Ratio of graduate companies to the total incubators	Data analysis
	45203	Growth rate of companies graduated from technology business incubators(%)	The growth compared with last year	Data analysis

Innovation performance

Category	Code	Indicator	Definition	Data source
Innovation performance	51001	Regional GDP(100 million Yuan)	Measuring the level of regional economic development	China Statistical Yearbook
	51002	Per capita GDP	Measuring the level of regional economic development	Data analysis
	51003	Growth rate of Regional GDP(%)	The growth compared with last year	Data analysis
	52101	Added value of the tertiary industry(100 million Yuan)	Measuring the regional industrial structure	China Statistical Yearbook
	52102	Ratio of added value of the tertiary industry to the regional GDP(%)	Ratio of added value of the tertiary industry to the regional GDP	Data analysis
	52103	Growth rate of added value of the tertiary industry(%)	The growth compared with last year	Data analysis
	52201	Main business income of the high-tech industry(100 million Yuan)	Measuring the development of regional high-tech industry	China Statistical Yearbook of High-tech industry
	52202	Ratio of main business income of the high-tech industry to the regional GDP (%)	Ratio of main business income of the high-tech industry to the regional GDP	Data analysis
	52203	Growth rate of main business income of the high-tech industry(%)	The growth compared with last year	Data analysis
	53001	High-tech product export volume(100 million USD)	Measuring the international competiveness of regional high-tech industry	China Statistical Yearbook
	53002	Ratio of high-tech product exports to the total exports (%)	Ratio of high-tech product exports to the total exports	Data analysis
	53003	Growth rate of high-tech product exports(%)	The growth compared with last year	Data analysis
	54101	The registered urban unemployment rate(%)	Measuring the level of regional employment	China Statistical Yearbook
	54103	Growth rate of registered urban unemployment rate (%)	The growth compared with last year	Data analysis
	54201	High-tech employments (10,000 people)	Measuring the job creation ability of regional high-tech industry	China Statistical Yearbook of High-tech industry

(continued)

Category	Code	Indicator	Definition	Data source
	54202	Ratio of high-tech employments to the total employments(%)	Ratio of high-tech employments to the total employments	Data analysis
	54203	Growth rate of high-tech employments(%)	The growth compared with last year	Data analysis
	55101	Regional energy consumption(value equivalent)(tons of standard coal/10,000 Yuan)	Measuring regional energy consumption and sustainable capacity	China Statistical Yearbook
	55103	Decrease rate of regional energy consumption(value equivalent)(%)	The growth compared with last year	Data analysis
	55201	Total power consumption (100 million kWh)	Measuring regional power consumption and sustainable capacity	China Statistical Yearbook
	55202	Total power consumption per 10,000 Yuan GDP (kwh/10,000 Yuan)	Ratio of total power consumption to the regional GDP	Data analysis
	55203	Growth rate of total power consumption(%)	The growth compared with last year	Data analysis
	55301	Total industrial sewage discharge(10,000 tons)	Measuring regional industrial sewage discharge and sustainable capacity	China Statistical Yearbook
	55302	Industrial sewage discharge per 10,000 Yuan GDP (ton/10,000 Yuan)	Ratio of industrial sewage discharge to the regional GDP	Data analysis
	55303	Growth rate of the total industrial sewage discharge (%)	The growth compared with last year	Data analysis
	55401	Major pollutant emissions in exhaust gas(10,000 tons)	Measuring regional pollutant emissions in exhaust gas and sustainable capacity	China Statistical Yearbook
	55402	Major pollutant emissions in exhaust gas per 100 million Yuan GDP(ton/100 million Yuan)	The ratio of major pollutant emissions in exhaust gas to the regional GDP	Data analysis
	55403	Growth rate of major pollutant emissions in exhaust gas(%)	The growth compared with last year	Data analysis

Appendix 2 Basic Indicators of Regional Innovation Capacity by Regions

11101 R&D of full time personnel equivalent/person year

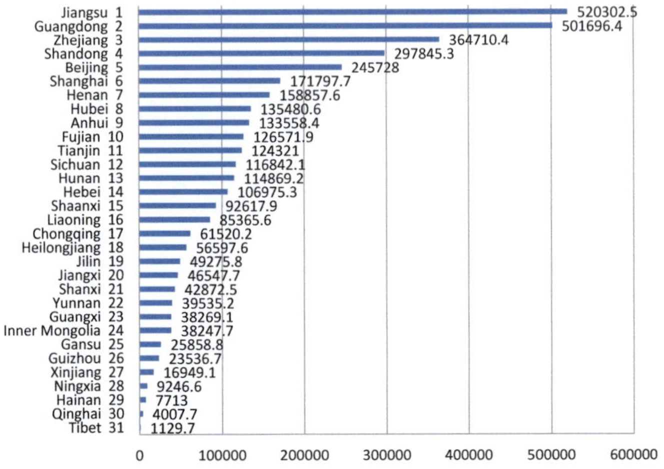

11102 Average R&D personnel per 10,000 people equivalent

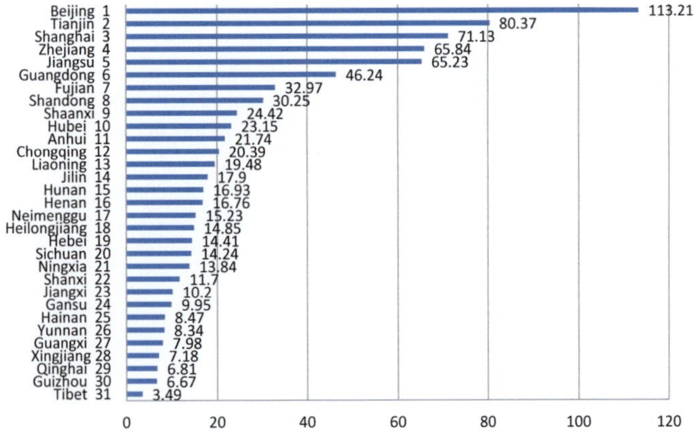

11103 The growth rate of R&D persernnel (%)

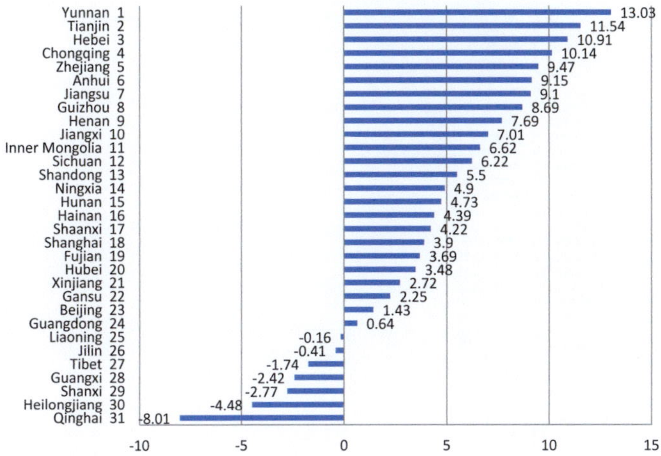

11201 Government R&D investment (100 million yuan)

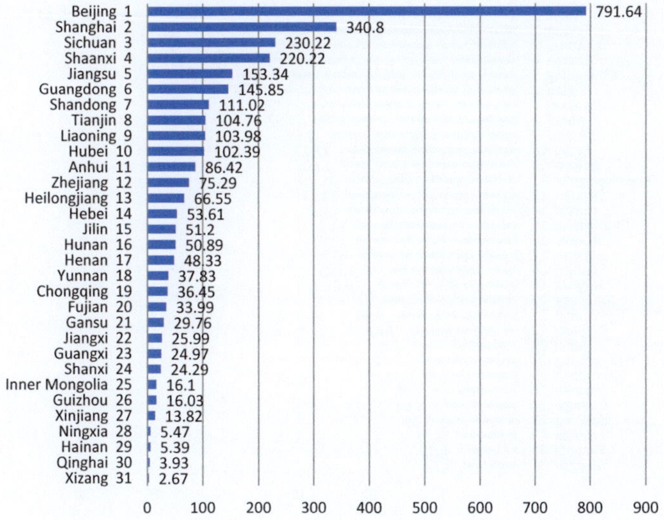

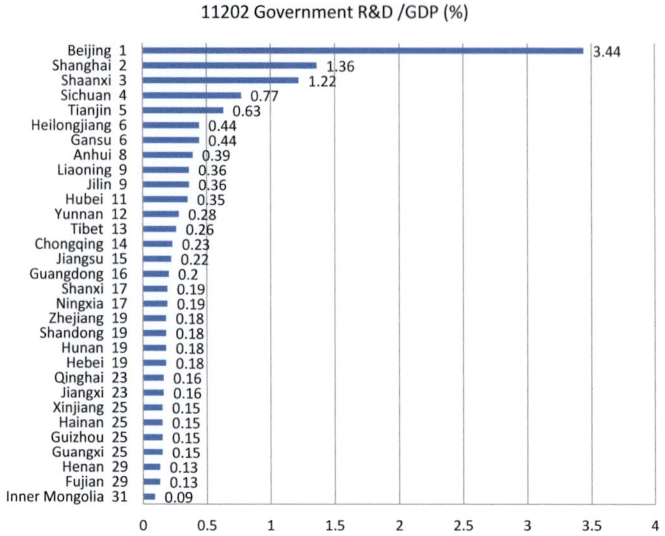
11202 Government R&D /GDP (%)
Beijing 1
3.44
Shanghai 2
1.36
Shaanxi 3
1.22
Sichuan 4
0.77
Tianjin 5
0.63
Heilongjiang 6
0.44
Gansu 6
0.44
Anhui 8
0.39
Liaoning 9
0.36
Jilin 9
0.36
Hubei 11
0.35
Yunnan 12
0.28
Tibet 13
0.26
Chongqing 14
0.23
Jiangsu 15
0.22
Guangdong 16
0.2
Shanxi 17
0.19
Ningxia 17
0.19
Zhejiang 19
0.18
Shandong 19
0.18
Hunan 19
0.18
Hebei 19
0.18
Qinghai 23
0.16
Jiangxi 23
0.16
Xinjiang 25
0.15
Hainan 25
0.15
Guizhou 25
0.15
Guangxi 25
0.15
Henan 29
0.13
Fujian 29
0.13
Inner Mongolia 31
0.09
0
0.5
1
1.5
2
2.5
3
3.5
4

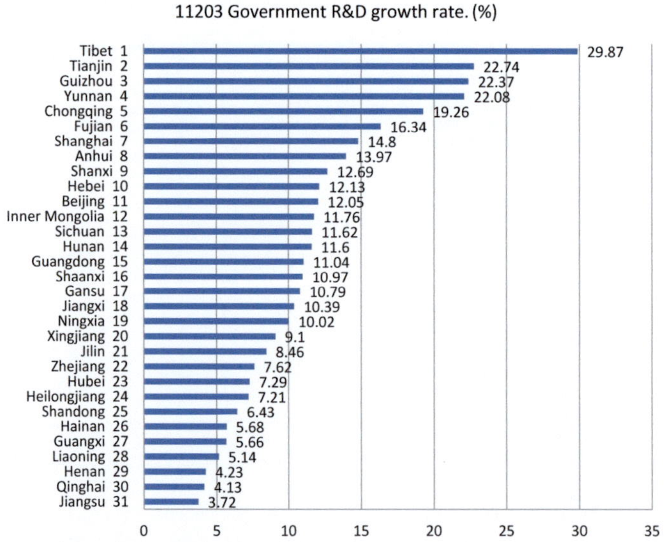
11203 Government R&D growth rate. (%)
Tibet 1
29.87
Tianjin 2
22.74
Guizhou 3
22.37
Yunnan 4
22.08
Chongqing 5
19.26
Fujian 6
16.34
Shanghai 7
14.8
Anhui 8
13.97
Shanxi 9
12.69
Hebei 10
12.13
Beijing 11
12.05
Inner Mongolia 12
11.76
Sichuan 13
11.62
Hunan 14
11.6
Guangdong 15
11.04
Shaanxi 16
10.97
Gansu 17
10.79
Jiangxi 18
10.39
Ningxia 19
10.02
Xingjiang 20
9.1
Jilin 21
8.46
Zhejiang 22
7.62
Hubei 23
7.29
Heilongjiang 24
7.21
Shandong 25
6.43
Hainan 26
5.68
Guangxi 27
5.66
Liaoning 28
5.14
Henan 29
4.23
Qinghai 30
4.13
Jiangsu 31
3.72
0
5
10
15
20
25
30
35

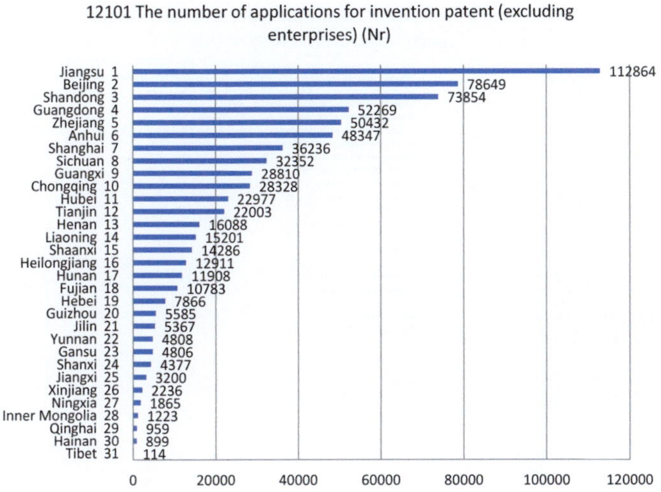
12101 The number of applications for invention patent (excluding enterprises) (Nr)
Jiangsu 1
Beijing 2
Shandong 3
Guangdong 4
Zhejiang 5
Anhui 6
Shanghai 7
Sichuan 8
Guangxi 9
Chongqing 10
Hubei 11
Tianjin 12
Henan 13
Liaoning 14
Shaanxi 15
Heilongjiang 16
Hunan 17
Fujian 18
Hebei 19
Guizhou 20
Jilin 21
Yunnan 22
Gansu 23
Shanxi 24
Jiangxi 25
Xinjiang 26
Ningxia 27
Inner Mongolia 28
Qinghai 29
Hainan 30
Tibet 31
112864
78649
73854
52269
50432
48347
36236
32352
28810
28328
22977
22003
16088
15201
14286
12911
11908
10783
7866
5585
5367
4808
4806
4377
3200
2236
1865
1223
959
899
114
0
20000
40000
60000
80000
100000
120000

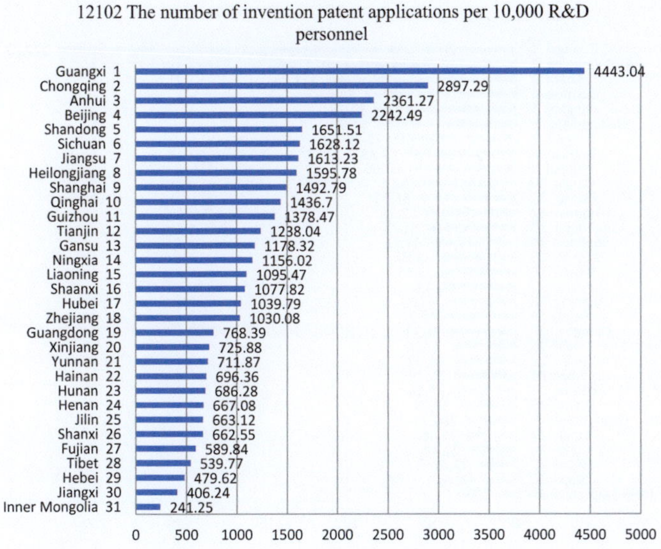
12102 The number of invention patent applications per 10,000 R&D personnel
Guangxi 1
Chongqing 2
Anhui 3
Beijing 4
Shandong 5
Sichuan 6
Jiangsu 7
Heilongjiang 8
Shanghai 9
Qinghai 10
Guizhou 11
Tianjin 12
Gansu 13
Ningxia 14
Liaoning 15
Shaanxi 16
Hubei 17
Zhejiang 18
Guangdong 19
Xinjiang 20
Yunnan 21
Hainan 22
Hunan 23
Henan 24
Jilin 25
Shanxi 26
Fujian 27
Tibet 28
Hebei 29
Jiangxi 30
Inner Mongolia 31
4443.04
2897.29
2361.27
2242.49
1651.51
1628.12
1613.23
1595.78
1492.79
1436.7
1378.47
1238.04
1178.32
1156.02
1095.47
1077.82
1039.79
1030.08
768.39
725.88
711.87
696.36
686.28
667.08
663.12
662.55
589.84
539.77
479.62
406.24
241.25
0
500
1000
1500
2000
2500
3000
3500
4000
4500
5000

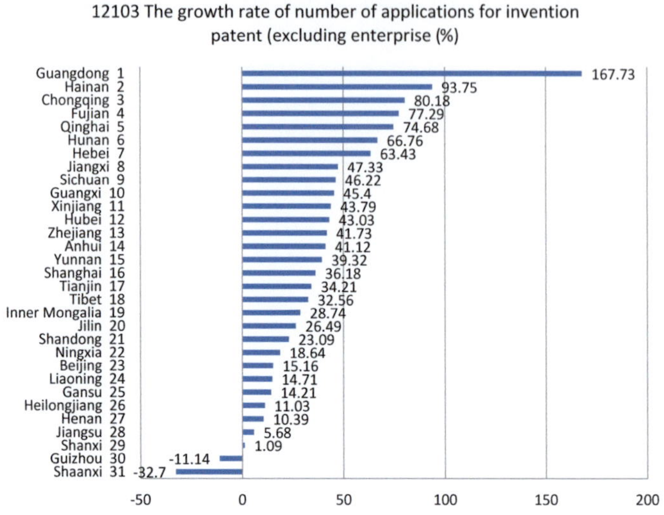
12103 The growth rate of number of applications for invention patent (excluding enterprise (%)
Guangdong 1
Hainan 2
Chongqing 3
Fujian 4
Qinghai 5
Hunan 6
Hebei 7
Jiangxi 8
Sichuan 9
Guangxi 10
Xinjiang 11
Hubei 12
Zhejiang 13
Anhui 14
Yunnan 15
Shanghai 16
Tianjin 17
Tibet 18
Inner Mongalia 19
Jilin 20
Shandong 21
Ningxia 22
Beijing 23
Liaoning 24
Gansu 25
Heilongjiang 26
Henan 27
Jiangsu 28
Shanxi 29
Guizhou 30
Shaanxi 31
167.73
93.75
80.18
77.29
74.68
66.76
63.43
47.33
46.22
45.4
43.79
43.03
41.73
41.12
39.32
36.18
34.21
32.56
28.74
26.49
23.09
18.64
15.16
14.71
14.21
11.03
10.39
5.68
1.09
-11.14
-32.7
-50
0
50
100
150
200

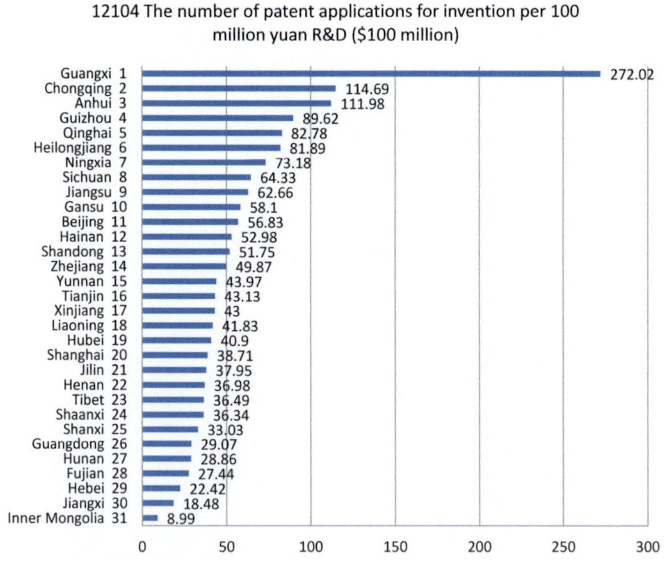
12104 The number of patent applications for invention per 100 million yuan R&D ($100 million)
Guangxi 1
Chongqing 2
Anhui 3
Guizhou 4
Qinghai 5
Heilongjiang 6
Ningxia 7
Sichuan 8
Jiangsu 9
Gansu 10
Beijing 11
Hainan 12
Shandong 13
Zhejiang 14
Yunnan 15
Tianjin 16
Xinjiang 17
Liaoning 18
Hubei 19
Shanghai 20
Jilin 21
Henan 22
Tibet 23
Shaanxi 24
Shanxi 25
Guangdong 26
Hunan 27
Fujian 28
Hebei 29
Jiangxi 30
Inner Mongolia 31
272.02
114.69
111.98
89.62
82.78
81.89
73.18
64.33
62.66
58.1
56.83
52.98
51.75
49.87
43.97
43.13
43
41.83
40.9
38.71
37.95
36.98
36.49
36.34
33.03
29.07
28.86
27.44
22.42
18.48
8.99
0
50
100
150
200
250
300

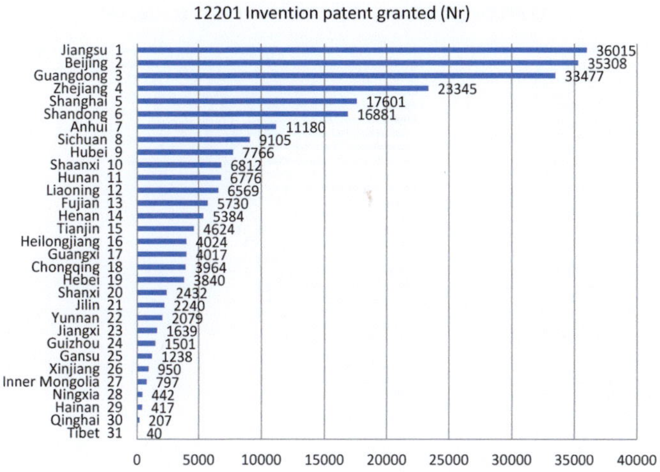
12201 Invention patent granted (Nr)
Jiangsu 1
Beijing 2
Guangdong 3
Zhejiang 4
Shanghai 5
Shandong 6
Anhui 7
Sichuan 8
Hubei 9
Shaanxi 10
Hunan 11
Liaoning 12
Fujian 13
Henan 14
Tianjin 15
Heilongjiang 16
Guangxi 17
Chongqing 18
Hebei 19
Shanxi 20
Jilin 21
Yunnan 22
Jiangxi 23
Guizhou 24
Gansu 25
Xinjiang 26
Inner Mongolia 27
Ningxia 28
Hainan 29
Qinghai 30
Tibet 31
36015
35308
33477
23345
17601
16881
11180
9105
7766
6812
6776
6569
5730
5384
4624
4024
4017
3964
3840
2432
2240
2079
1639
1501
1238
950
797
442
417
207
40
0
5000
10000
15000
20000
25000
30000
35000
40000

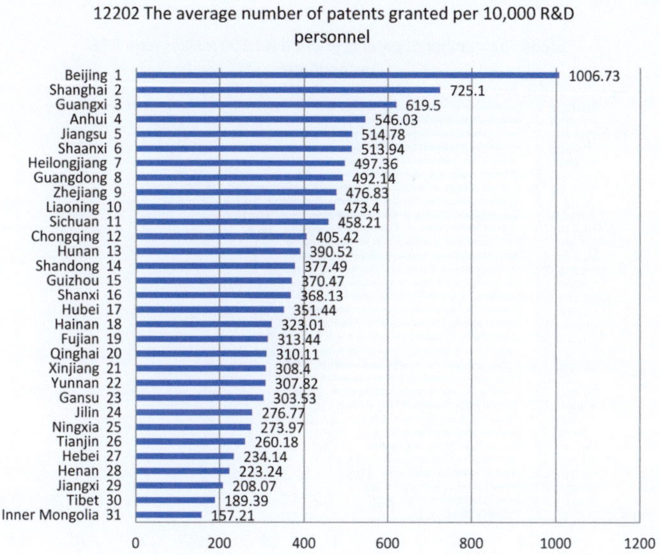
12202 The average number of patents granted per 10,000 R&D personnel
Beijing 1
Shanghai 2
Guangxi 3
Anhui 4
Jiangsu 5
Shaanxi 6
Heilongjiang 7
Guangdong 8
Zhejiang 9
Liaoning 10
Sichuan 11
Chongqing 12
Hunan 13
Shandong 14
Guizhou 15
Shanxi 16
Hubei 17
Hainan 18
Fujian 19
Qinghai 20
Xinjiang 21
Yunnan 22
Gansu 23
Jilin 24
Ningxia 25
Tianjin 26
Hebei 27
Henan 28
Jiangxi 29
Tibet 30
Inner Mongolia 31
1006.73
725.1
619.5
546.03
514.78
513.94
497.36
492.14
476.83
473.4
458.21
405.42
390.52
377.49
370.47
368.13
351.44
323.01
313.44
310.11
308.4
307.82
303.53
276.77
273.97
260.18
234.14
223.24
208.07
189.39
157.21
0
200
400
600
800
1000
1200

12203 The growth rate of the invention patent granted (%)

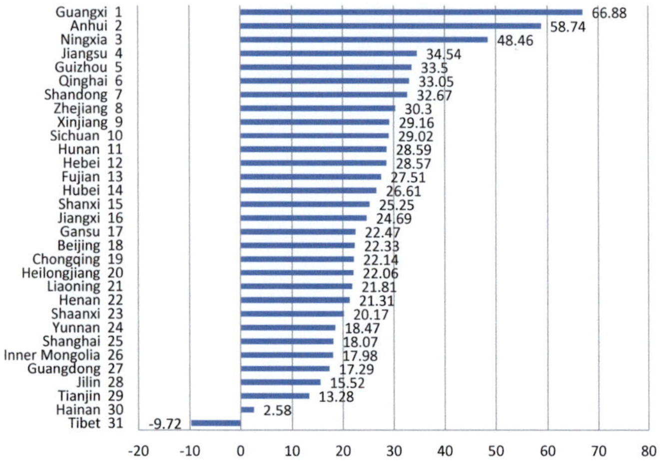

12204 The number of patents granted for 100 million yuan R&D expenditure

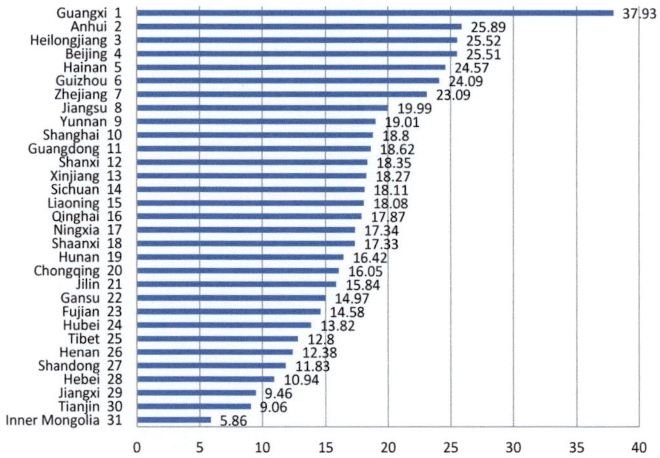

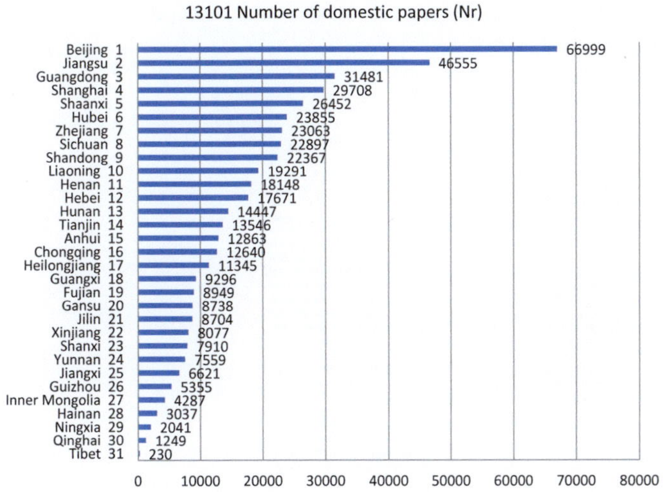
13101 Number of domestic papers (Nr)
Beijing 1
Jiangsu 2
Guangdong 3
Shanghai 4
Shaanxi 5
Hubei 6
Zhejiang 7
Sichuan 8
Shandong 9
Liaoning 10
Henan 11
Hebei 12
Hunan 13
Tianjin 14
Anhui 15
Chongqing 16
Heilongjiang 17
Guangxi 18
Fujian 19
Gansu 20
Jilin 21
Xinjiang 22
Shanxi 23
Yunnan 24
Jiangxi 25
Guizhou 26
Inner Mongolia 27
Hainan 28
Ningxia 29
Qinghai 30
Tibet 31
66999
46555
31481
29708
26452
23855
23063
22897
22367
19291
18148
17671
14447
13546
12863
12640
11345
9296
8949
8738
8704
8077
7910
7559
6621
5355
4287
3037
2041
1249
230
0
10000
20000
30000
40000
50000
60000
70000
80000

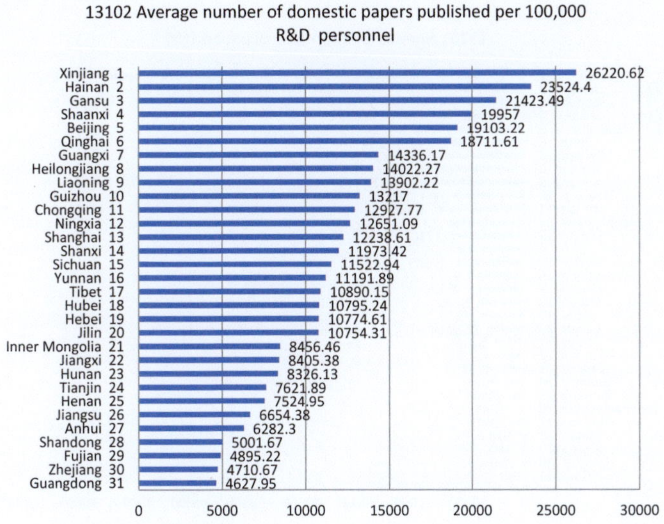
13102 Average number of domestic papers published per 100,000 R&D personnel
Xinjiang 1
Hainan 2
Gansu 3
Shaanxi 4
Beijing 5
Qinghai 6
Guangxi 7
Heilongjiang 8
Liaoning 9
Guizhou 10
Chongqing 11
Ningxia 12
Shanghai 13
Shanxi 14
Sichuan 15
Yunnan 16
Tibet 17
Hubei 18
Hebei 19
Jilin 20
Inner Mongolia 21
Jiangxi 22
Hunan 23
Tianjin 24
Henan 25
Jiangsu 26
Anhui 27
Shandong 28
Fujian 29
Zhejiang 30
Guangdong 31
26220.62
23524.4
21423.49
19957
19103.22
18711.61
14336.17
14022.27
13902.22
13217
12927.77
12651.09
12238.61
11973.42
11522.94
11191.89
10890.15
10795.24
10774.61
10754.31
8456.46
8405.38
8326.13
7621.89
7524.95
6654.38
6282.3
5001.67
4895.22
4710.67
4627.95
0
5000
10000
15000
20000
25000
30000

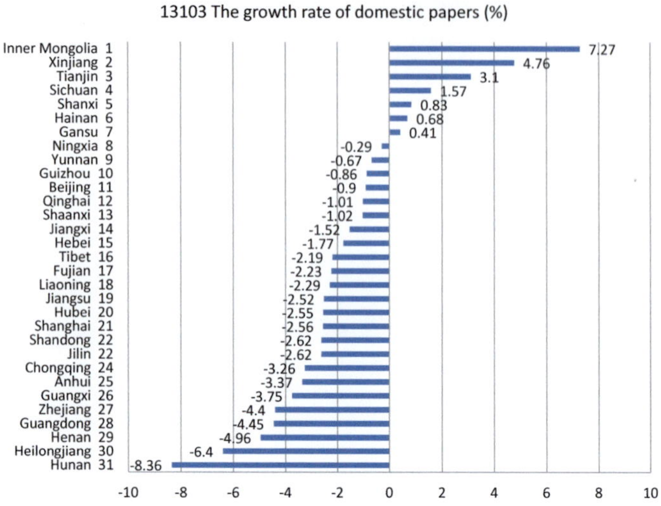
13103 The growth rate of domestic papers (%)
Inner Mongolia 1
Xinjiang 2
Tianjin 3
Sichuan 4
Shanxi 5
Hainan 6
Gansu 7
Ningxia 8
Yunnan 9
Guizhou 10
Beijing 11
Qinghai 12
Shaanxi 13
Jiangxi 14
Hebei 15
Tibet 16
Fujian 17
Liaoning 18
Jiangsu 19
Hubei 20
Shanghai 21
Shandong 22
Jilin 22
Chongqing 24
Anhui 25
Guangxi 26
Zhejiang 27
Guangdong 28
Henan 29
Heilongjiang 30
Hunan 31
7.27
4.76
3.1
1.57
0.83
0.68
0.41
-0.29
-0.67
-0.86
-0.9
-1.01
-1.02
-1.52
-1.77
-2.19
-2.23
-2.29
-2.52
-2.55
-2.56
-2.62
-2.62
-3.26
-3.37
-3.75
-4.4
-4.45
-4.96
-6.4
-8.36
-10 -8 -6 -4 -2 0 2 4 6 8 10

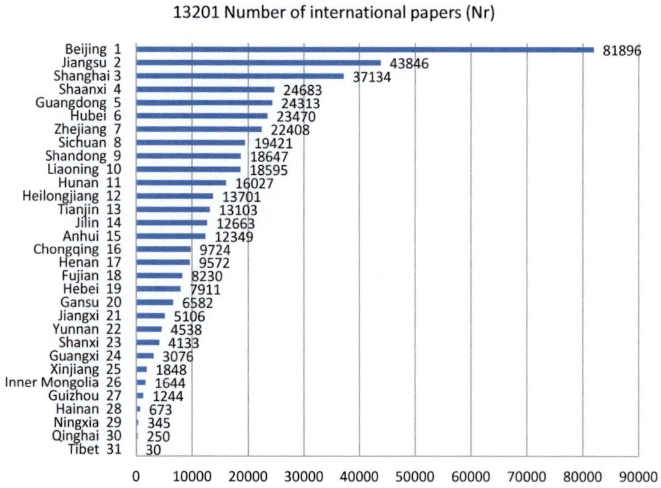
13201 Number of international papers (Nr)
Beijing 1
Jiangsu 2
Shanghai 3
Shaanxi 4
Guangdong 5
Hubei 6
Zhejiang 7
Sichuan 8
Shandong 9
Liaoning 10
Hunan 11
Heilongjiang 12
Tianjin 13
Jilin 14
Anhui 15
Chongqing 16
Henan 17
Fujian 18
Hebei 19
Gansu 20
Jiangxi 21
Yunnan 22
Shanxi 23
Guangxi 24
Xinjiang 25
Inner Mongolia 26
Guizhou 27
Hainan 28
Ningxia 29
Qinghai 30
Tibet 31
81896
43846
37134
24683
24313
23470
22408
19421
18647
18595
16027
13701
13103
12663
12349
9724
9572
8230
7911
6582
5106
4538
4133
3076
1848
1644
1244
673
345
250
30
0 10000 20000 30000 40000 50000 60000 70000 80000 90000

13202 The number of international papers published per 100,000 R&D personnel

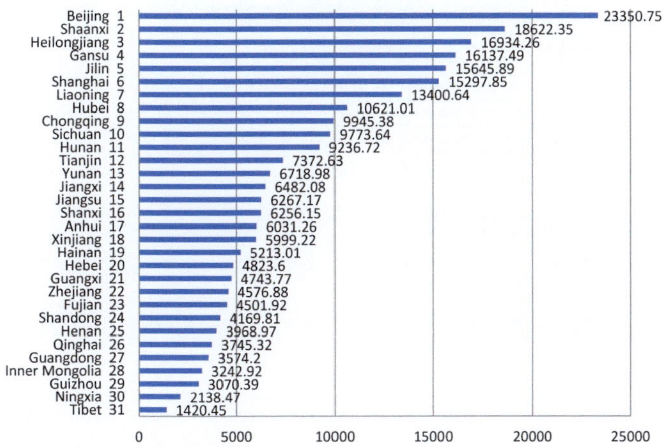

13203 The growth rate of international papers (%)

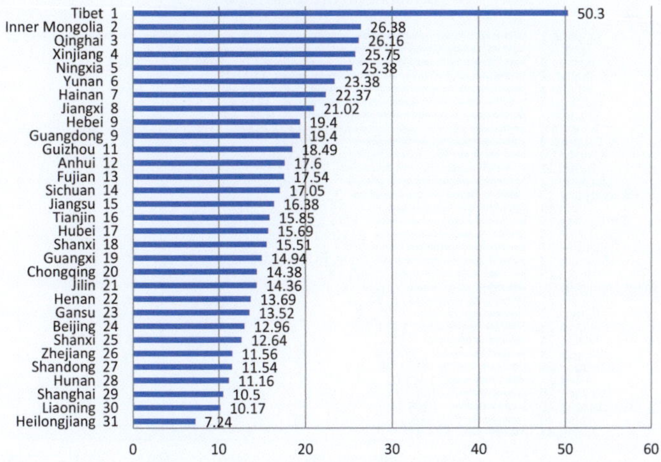

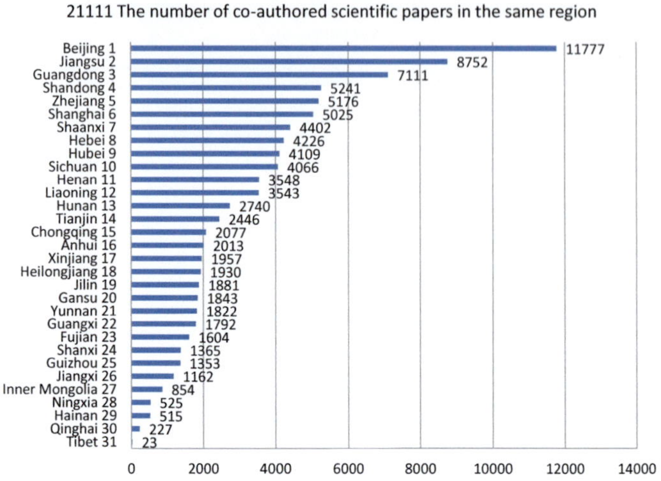
21111 The number of co-authored scientific papers in the same region
Beijing 1
Jiangsu 2
Guangdong 3
Shandong 4
Zhejiang 5
Shanghai 6
Shaanxi 7
Hebei 8
Hubei 9
Sichuan 10
Henan 11
Liaoning 12
Hunan 13
Tianjin 14
Chongqing 15
Anhui 16
Xinjiang 17
Heilongjiang 18
Jilin 19
Gansu 20
Yunnan 21
Guangxi 22
Fujian 23
Shanxi 24
Guizhou 25
Jiangxi 26
Inner Mongolia 27
Ningxia 28
Hainan 29
Qinghai 30
Tibet 31
11777
8752
7111
5241
5176
5025
4402
4226
4109
4066
3548
3543
2740
2446
2077
2013
1957
1930
1881
1843
1822
1792
1604
1365
1353
1162
854
525
515
227
23
0
2000
4000
6000
8000
10000
12000
14000

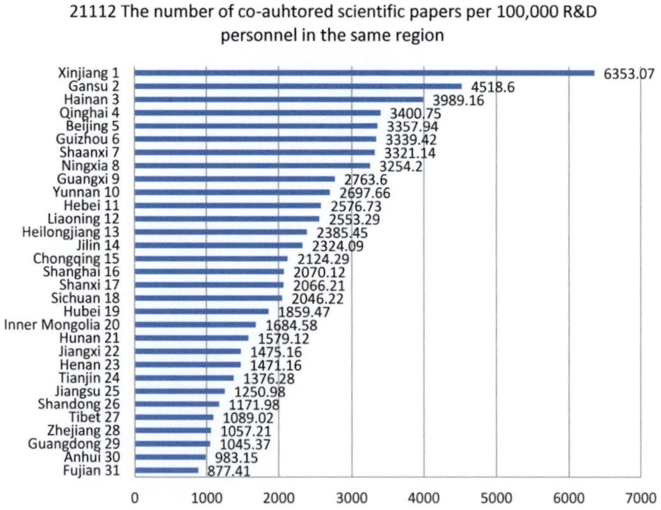
21112 The number of co-auhtored scientific papers per 100,000 R&D personnel in the same region
Xinjiang 1
Gansu 2
Hainan 3
Qinghai 4
Beijing 5
Guizhou 6
Shaanxi 7
Ningxia 8
Guangxi 9
Yunnan 10
Hebei 11
Liaoning 12
Heilongjiang 13
Jilin 14
Chongqing 15
Shanghai 16
Shanxi 17
Sichuan 18
Hubei 19
Inner Mongolia 20
Hunan 21
Jiangxi 22
Henan 23
Tianjin 24
Jiangsu 25
Shandong 26
Tibet 27
Zhejiang 28
Guangdong 29
Anhui 30
Fujian 31
6353.07
4518.6
3989.16
3400.75
3357.94
3339.42
3321.14
3254.2
2763.6
2697.66
2576.73
2553.29
2385.45
2324.09
2124.29
2070.12
2066.21
2046.22
1859.47
1684.58
1579.12
1475.16
1471.16
1376.28
1250.98
1171.98
1089.02
1057.21
1045.37
983.15
877.41
0
1000
2000
3000
4000
5000
6000
7000

21113 The growth rate of number of co-auhtored scientific papers in the same region (%)

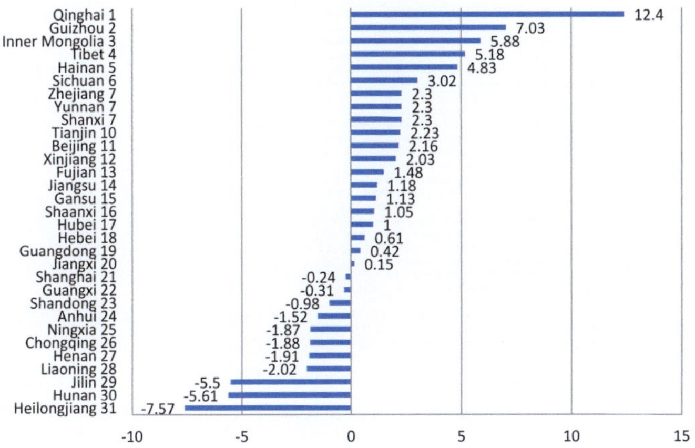

21121 The cross-region co-authored scientific papers

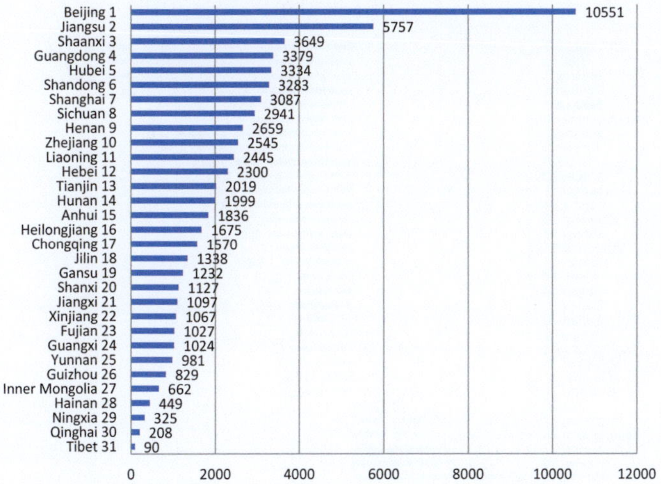

21122 The number of cross-region co-authored papers among regions per 100,000 R&D personnel

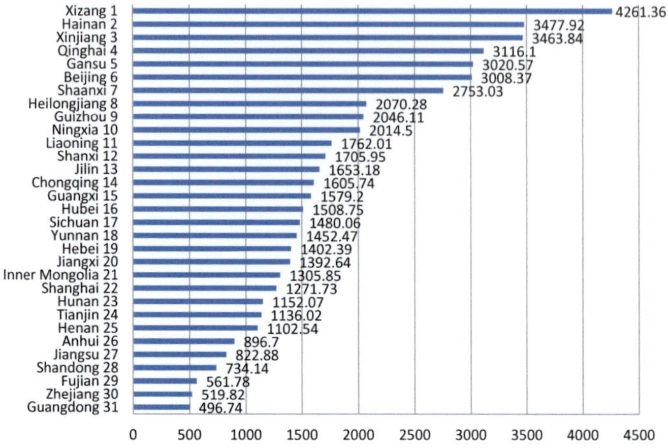

21123 The growth rate of the cross region co-author's scientific papers (%)

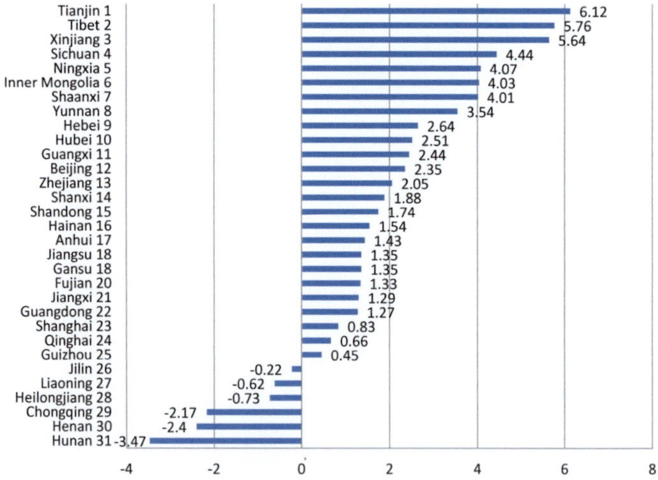

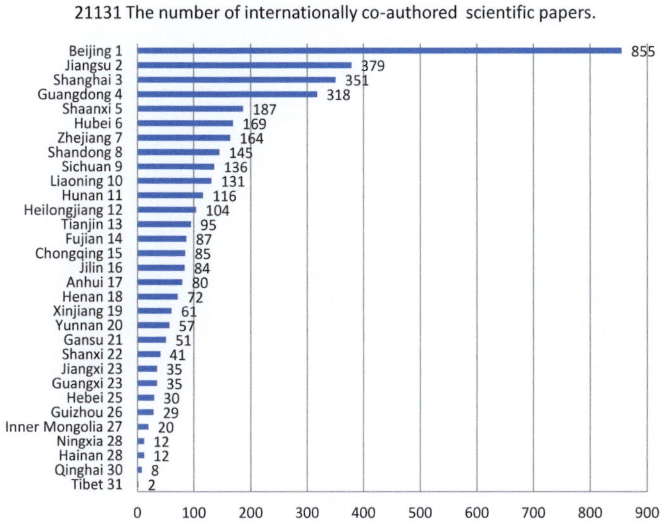
21131 The number of internationally co-authored scientific papers.
Beijing 1
Jiangsu 2
Shanghai 3
Guangdong 4
Shaanxi 5
Hubei 6
Zhejiang 7
Shandong 8
Sichuan 9
Liaoning 10
Hunan 11
Heilongjiang 12
Tianjin 13
Fujian 14
Chongqing 15
Jilin 16
Anhui 17
Henan 18
Xinjiang 19
Yunnan 20
Gansu 21
Shanxi 22
Jiangxi 23
Guangxi 23
Hebei 25
Guizhou 26
Inner Mongolia 27
Ningxia 28
Hainan 28
Qinghai 30
Tibet 31
855
379
351
318
187
169
164
145
136
131
116
104
95
87
85
84
80
72
61
57
51
41
35
35
30
29
20
12
12
8
2
0
100
200
300
400
500
600
700
800
900

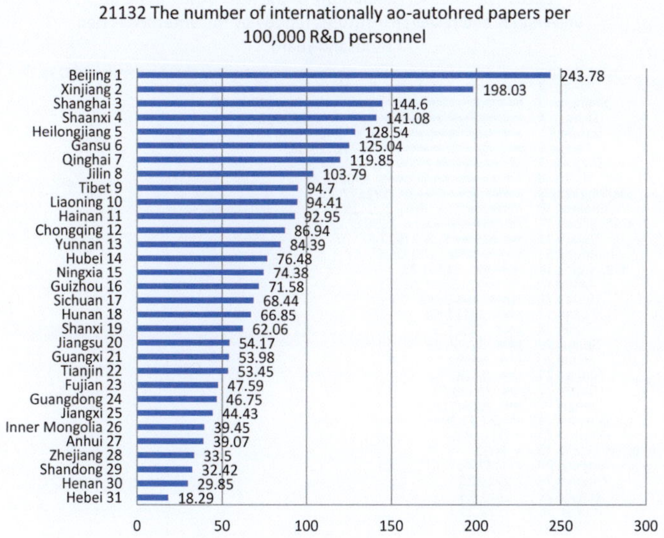
21132 The number of internationally ao-autohred papers per 100,000 R&D personnel
Beijing 1
Xinjiang 2
Shanghai 3
Shaanxi 4
Heilongjiang 5
Gansu 6
Qinghai 7
Jilin 8
Tibet 9
Liaoning 10
Hainan 11
Chongqing 12
Yunnan 13
Hubei 14
Ningxia 15
Guizhou 16
Sichuan 17
Hunan 18
Shanxi 19
Jiangsu 20
Guangxi 21
Tianjin 22
Fujian 23
Guangdong 24
Jiangxi 25
Inner Mongolia 26
Anhui 27
Zhejiang 28
Shandong 29
Henan 30
Hebei 31
243.78
198.03
144.6
141.08
128.54
125.04
119.85
103.79
94.7
94.41
92.95
86.94
84.39
76.48
74.38
71.58
68.44
66.85
62.06
54.17
53.98
53.45
47.59
46.75
44.43
39.45
39.07
33.5
32.42
29.85
18.29
0
50
100
150
200
250
300

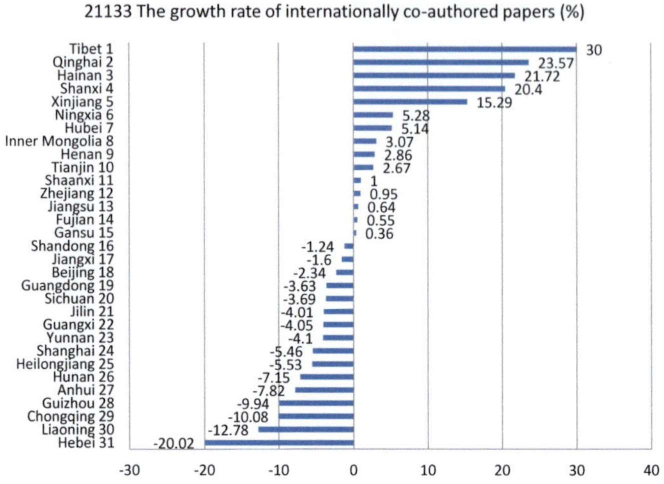
21133 The growth rate of internationally co-authored papers (%)
Tibet 1 30
Qinghai 2 23.57
Hainan 3 21.72
Shanxi 4 20.4
Xinjiang 5 15.29
Ningxia 6 5.28
Hubei 7 5.14
Inner Mongolia 8 3.07
Henan 9 2.86
Tianjin 10 2.67
Shaanxi 11 1
Zhejiang 12 0.95
Jiangsu 13 0.64
Fujian 14 0.55
Gansu 15 0.36
Shandong 16 -1.24
Jiangxi 17 -1.6
Beijing 18 -2.34
Guangdong 19 -3.63
Sichuan 20 -3.69
Jilin 21 -4.01
Guangxi 22 -4.05
Yunnan 23 -4.1
Shanghai 24 -5.46
Heilongjiang 25 -5.53
Hunan 26 -7.15
Anhui 27 -7.82
Guizhou 28 -9.94
Chongqing 29 -10.08
Liaoning 30 -12.78
Hebei 31 -20.02
-30 -20 -10 0 10 20 30 40

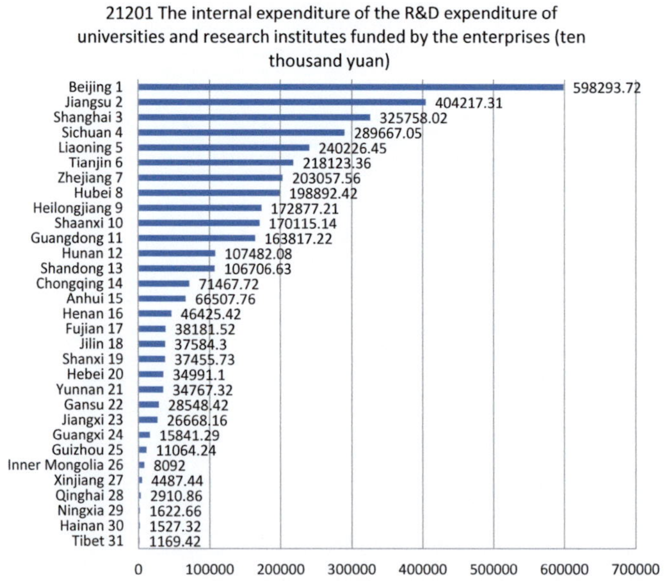
21201 The internal expenditure of the R&D expenditure of universities and research institutes funded by the enterprises (ten thousand yuan)
Beijing 1 598293.72
Jiangsu 2 404217.31
Shanghai 3 325758.02
Sichuan 4 289667.05
Liaoning 5 240226.45
Tianjin 6 218123.36
Zhejiang 7 203057.56
Hubei 8 198892.42
Heilongjiang 9 172877.21
Shaanxi 10 170115.14
Guangdong 11 163817.22
Hunan 12 107482.08
Shandong 13 106706.63
Chongqing 14 71467.72
Anhui 15 66507.76
Henan 16 46425.42
Fujian 17 38181.52
Jilin 18 37584.3
Shanxi 19 37455.73
Hebei 20 34991.1
Yunnan 21 34767.32
Gansu 22 28548.42
Jiangxi 23 26668.16
Guangxi 24 15841.29
Guizhou 25 11064.24
Inner Mongolia 26 8092
Xinjiang 27 4487.44
Qinghai 28 2910.86
Ningxia 29 1622.66
Hainan 30 1527.32
Tibet 31 1169.42
0 100000 200000 300000 400000 500000 600000 700000

21202 The share of internal expenditure of the R&D expenditure of universities and research institutes funded by the enterprises (10,000 yuan)

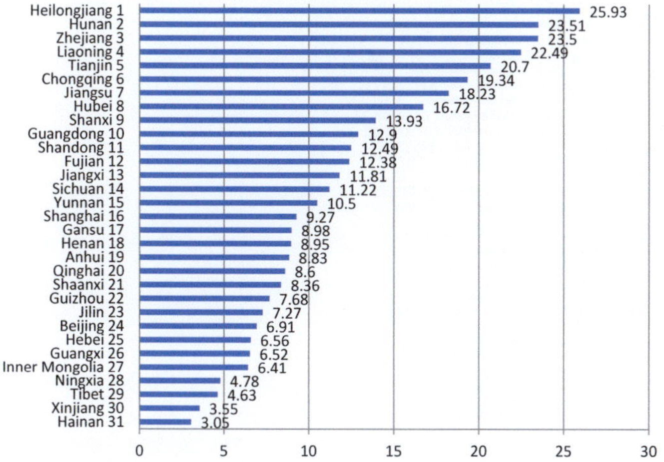

21203 The growth rate of R&D expenditure of universities and research institutes comes from enterprise (%)

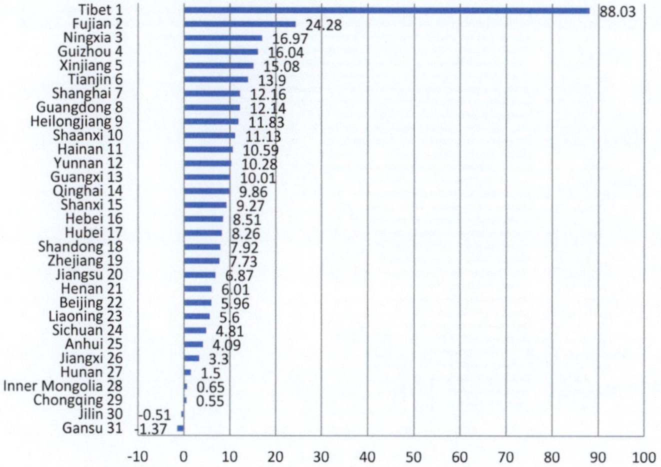

22101 Value of technical market transaction (by flow) (10,000 yuan)

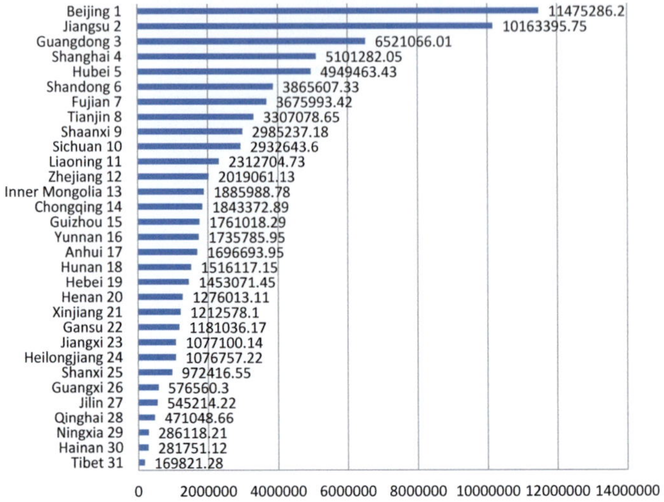

22102 Average value of technology market from enterprises (by flow)(10,000 yuan/item)

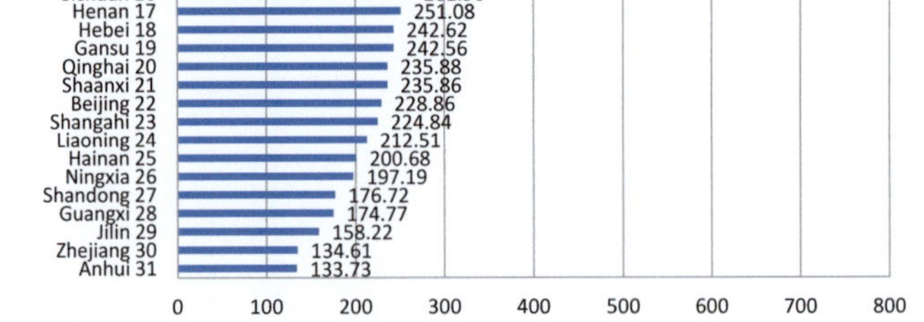

22103 Growth rate of value of technology market transactions (by flow)(%)

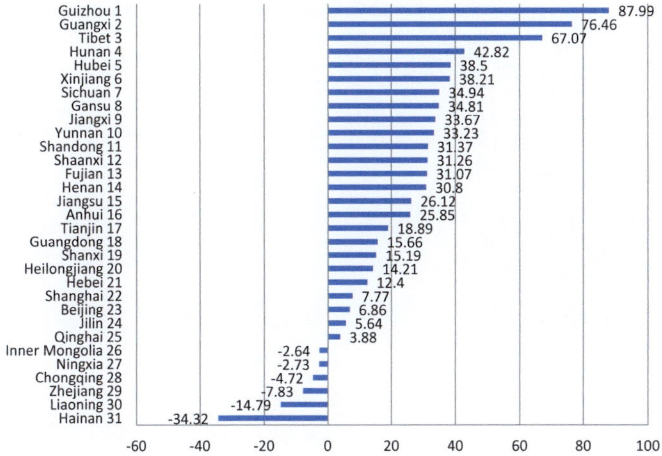

22201 Domestic technical expenses (10,000 yuan) of large industrial enterprises

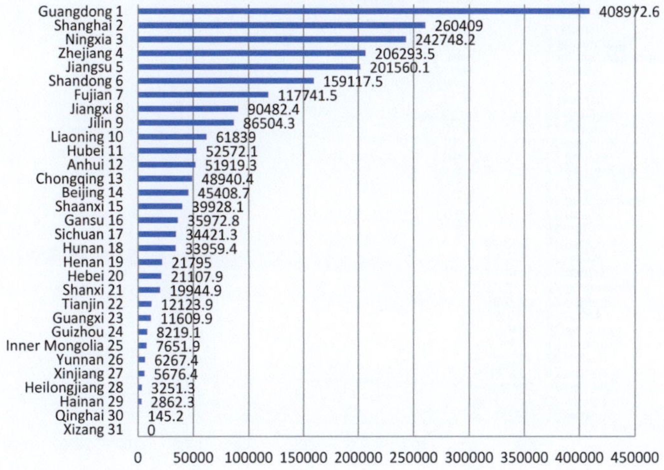

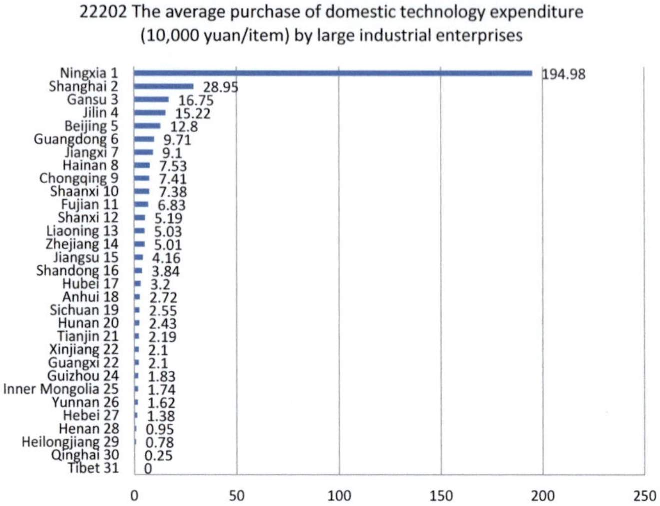
22202 The average purchase of domestic technology expenditure (10,000 yuan/item) by large industrial enterprises
Ningxia 1 194.98
Shanghai 2 28.95
Gansu 3 16.75
Jilin 4 15.22
Beijing 5 12.8
Guangdong 6 9.71
Jiangxi 7 9.1
Hainan 8 7.53
Chongqing 9 7.41
Shaanxi 10 7.38
Fujian 11 6.83
Shanxi 12 5.19
Liaoning 13 5.03
Zhejiang 14 5.01
Jiangsu 15 4.16
Shandong 16 3.84
Hubei 17 3.2
Anhui 18 2.72
Sichuan 19 2.55
Hunan 20 2.43
Tianjin 21 2.19
Xinjiang 22 2.1
Guangxi 22 2.1
Guizhou 24 1.83
Inner Mongolia 25 1.74
Yunnan 26 1.62
Hebei 27 1.38
Henan 28 0.95
Heilongjiang 29 0.78
Qinghai 30 0.25
Tibet 31 0
0 50 100 150 200 250

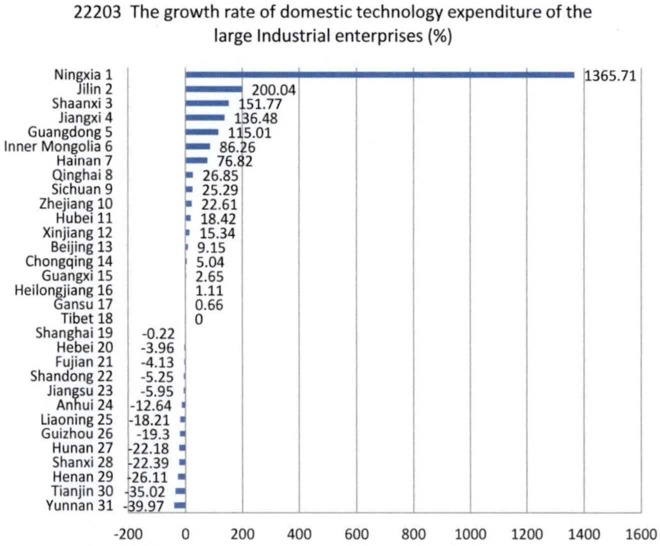
22203 The growth rate of domestic technology expenditure of the large Industrial enterprises (%)
Ningxia 1 1365.71
Jilin 2 200.04
Shaanxi 3 151.77
Jiangxi 4 136.48
Guangdong 5 115.01
Inner Mongolia 6 86.26
Hainan 7 76.82
Qinghai 8 26.85
Sichuan 9 25.29
Zhejiang 10 22.61
Hubei 11 18.42
Xinjiang 12 15.34
Beijing 13 9.15
Chongqing 14 5.04
Guangxi 15 2.65
Heilongjiang 16 1.11
Gansu 17 0.66
Tibet 18 0
Shanghai 19 -0.22
Hebei 20 -3.96
Fujian 21 -4.13
Shandong 22 -5.25
Jiangsu 23 -5.95
Anhui 24 -12.64
Liaoning 25 -18.21
Guizhou 26 -19.3
Hunan 27 -22.18
Shanxi 28 -22.39
Henan 29 -26.11
Tianjin 30 -35.02
Yunnan 31 -39.97
-200 0 200 400 600 800 1000 1200 1400 1600

22301 Expenditure on imported technology by large Industrial enterprises (10,000 yuan)

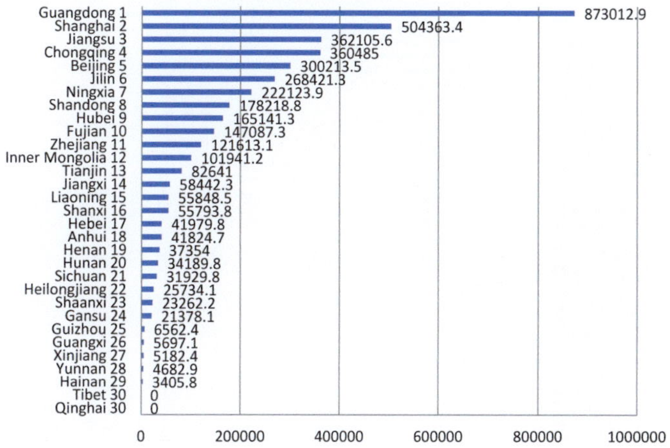

22302 The average expenditure on imported technology by large Industrial enterprises (10,000 yuan)

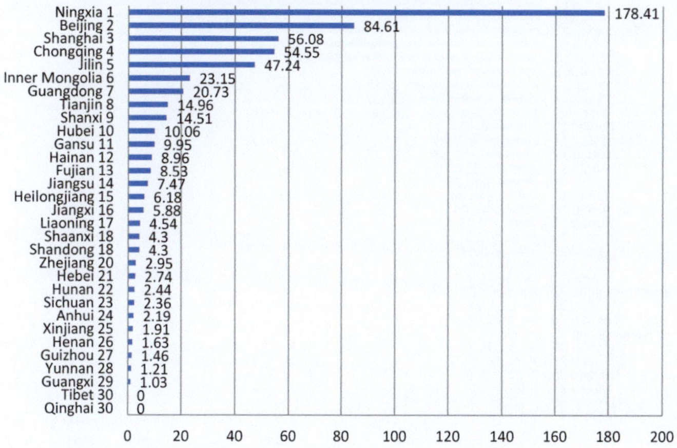

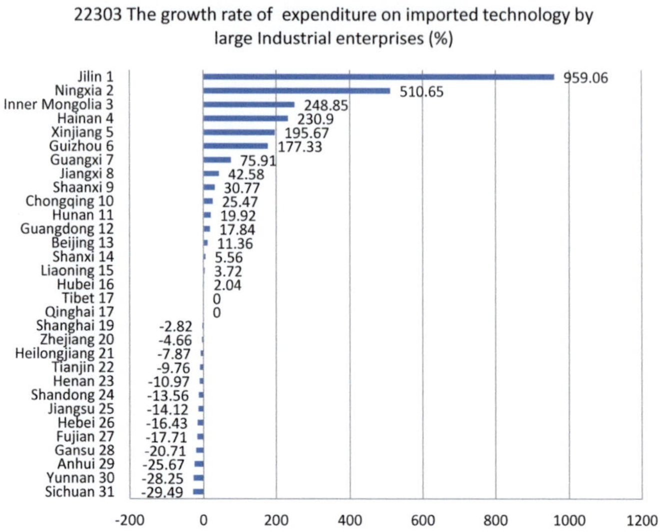
22303 The growth rate of expenditure on imported technology by large Industrial enterprises (%)
Jilin 1
Ningxia 2
Inner Mongolia 3
Hainan 4
Xinjiang 5
Guizhou 6
Guangxi 7
Jiangxi 8
Shaanxi 9
Chongqing 10
Hunan 11
Guangdong 12
Beijing 13
Shanxi 14
Liaoning 15
Hubei 16
Tibet 17
Qinghai 17
Shanghai 19
Zhejiang 20
Heilongjiang 21
Tianjin 22
Henan 23
Shandong 24
Jiangsu 25
Hebei 26
Fujian 27
Gansu 28
Anhui 29
Yunnan 30
Sichuan 31
959.06
510.65
248.85
230.9
195.67
177.33
75.91
42.58
30.77
25.47
19.92
17.84
11.36
5.56
3.72
2.04
0
0
-2.82
-4.66
-7.87
-9.76
-10.97
-13.56
-14.12
-16.43
-17.71
-20.71
-25.67
-28.25
-29.49
-200
0
200
400
600
800
1000
1200

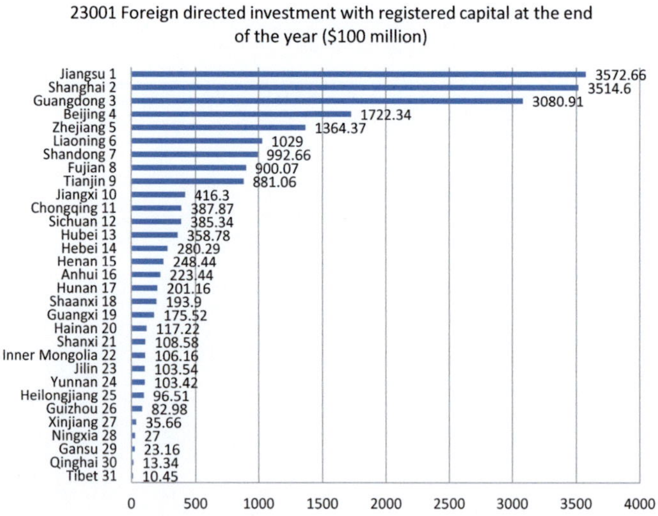
23001 Foreign directed investment with registered capital at the end of the year ($100 million)
Jiangsu 1
Shanghai 2
Guangdong 3
Beijing 4
Zhejiang 5
Liaoning 6
Shandong 7
Fujian 8
Tianjin 9
Jiangxi 10
Chongqing 11
Sichuan 12
Hubei 13
Hebei 14
Henan 15
Anhui 16
Hunan 17
Shaanxi 18
Guangxi 19
Hainan 20
Shanxi 21
Inner Mongolia 22
Jilin 23
Yunnan 24
Heilongjiang 25
Guizhou 26
Xinjiang 27
Ningxia 28
Gansu 29
Qinghai 30
Tibet 31
3572.66
3514.6
3080.91
1722.34
1364.37
1029
992.66
900.07
881.06
416.3
387.87
385.34
358.78
280.29
248.44
223.44
201.16
193.9
175.52
117.22
108.58
106.16
103.54
103.42
96.51
82.98
35.66
27
23.16
13.34
10.45
0
500
1000
1500
2000
2500
3000
3500
4000

23002 Foreign direct investment per capita (usd/person)

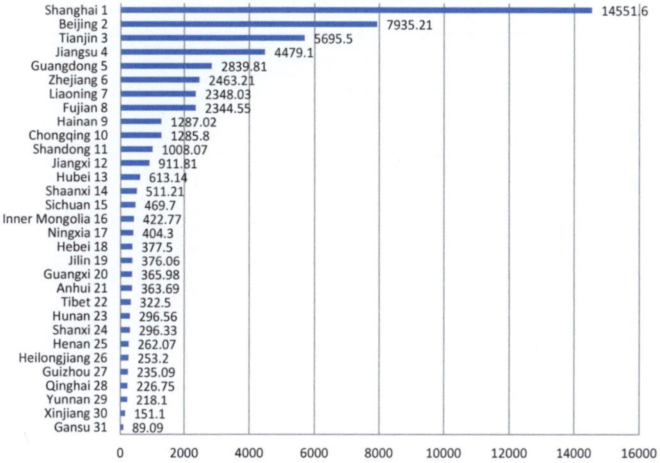

23003 The growth rate of FDI(registered) capital (%)

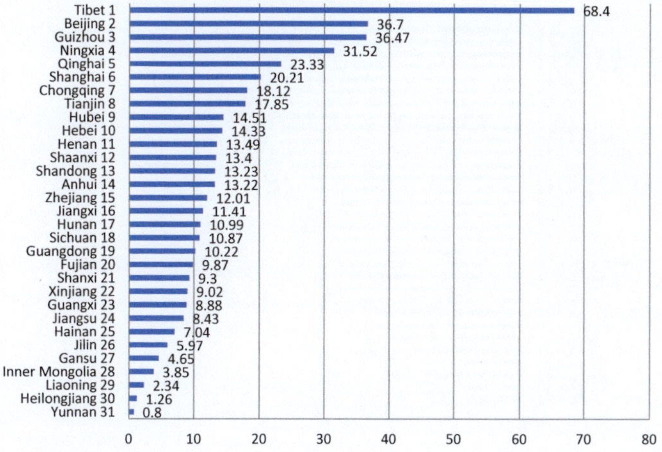

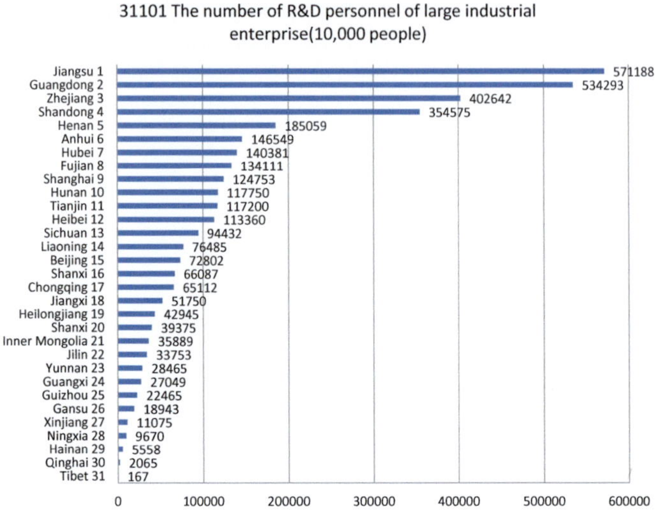
31101 The number of R&D personnel of large industrial enterprise(10,000 people)
Jiangsu 1
Guangdong 2
Zhejiang 3
Shandong 4
Henan 5
Anhui 6
Hubei 7
Fujian 8
Shanghai 9
Hunan 10
Tianjin 11
Heibei 12
Sichuan 13
Liaoning 14
Beijing 15
Shanxi 16
Chongqing 17
Jiangxi 18
Heilongjiang 19
Shanxi 20
Inner Mongolia 21
Jilin 22
Yunnan 23
Guangxi 24
Guizhou 25
Gansu 26
Xinjiang 27
Ningxia 28
Hainan 29
Qinghai 30
Tibet 31
571188
534293
402642
354575
185059
146549
140381
134111
124753
117750
117200
113360
94432
76485
72802
66087
65112
51750
42945
39375
35889
33753
28465
27049
22465
18943
11075
9670
5558
2065
167
0
100000
200000
300000
400000
500000
600000

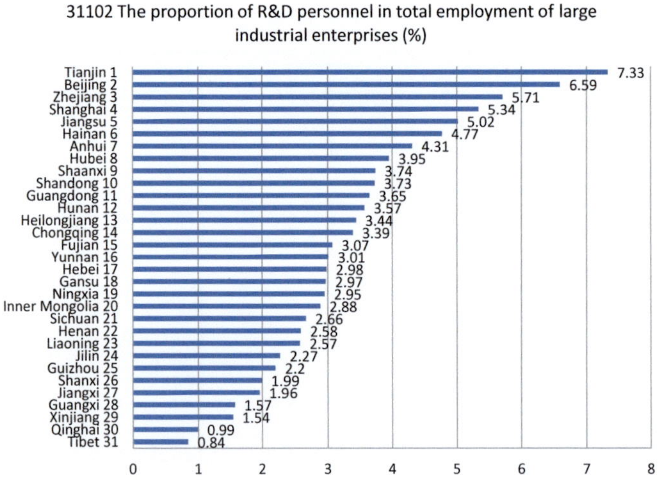
31102 The proportion of R&D personnel in total employment of large industrial enterprises (%)
Tianjin 1
Beijing 2
Zhejiang 3
Shanghai 4
Jiangsu 5
Hainan 6
Anhui 7
Hubei 8
Shaanxi 9
Shandong 10
Guangdong 11
Hunan 12
Heilongjiang 13
Chongqing 14
Fujian 15
Yunnan 16
Hebei 17
Gansu 18
Ningxia 19
Inner Mongolia 20
Sichuan 21
Henan 22
Liaoning 23
Jilin 24
Guizhou 25
Shanxi 26
Jiangxi 27
Guangxi 28
Xinjiang 29
Qinghai 30
Tibet 31
7.33
6.59
5.71
5.34
5.02
4.77
4.31
3.95
3.74
3.73
3.65
3.57
3.44
3.39
3.07
3.01
2.98
2.97
2.95
2.88
2.66
2.58
2.57
2.27
2.2
1.99
1.96
1.57
1.54
0.99
0.84
0
1
2
3
4
5
6
7
8

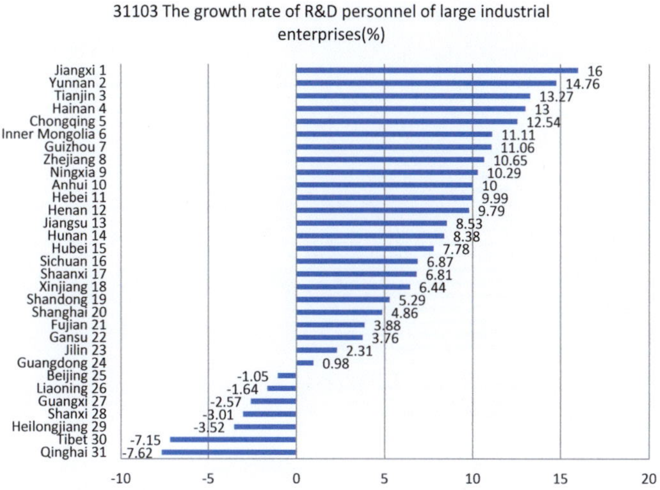
31103 The growth rate of R&D personnel of large industrial enterprises(%)
Jiangxi 1 16
Yunnan 2 14.76
Tianjin 3 13.27
Hainan 4 13
Chongqing 5 12.54
Inner Mongolia 6 11.11
Guizhou 7 11.06
Zhejiang 8 10.65
Ningxia 9 10.29
Anhui 10 10
Hebei 11 9.99
Henan 12 9.79
Jiangsu 13 8.53
Hunan 14 8.38
Hubei 15 7.78
Sichuan 16 6.87
Shaanxi 17 6.81
Xinjiang 18 6.44
Shandong 19 5.29
Shanghai 20 4.86
Fujian 21 3.88
Gansu 22 3.76
Jilin 23 2.31
Guangdong 24 0.98
Beijing 25 -1.05
Liaoning 26 -1.64
Guangxi 27 -2.57
Shanxi 28 -3.01
Heilongjiang 29 -3.52
Tibet 30 -7.15
Qinghai 31 -7.62
-10 -5 0 5 10 15 20

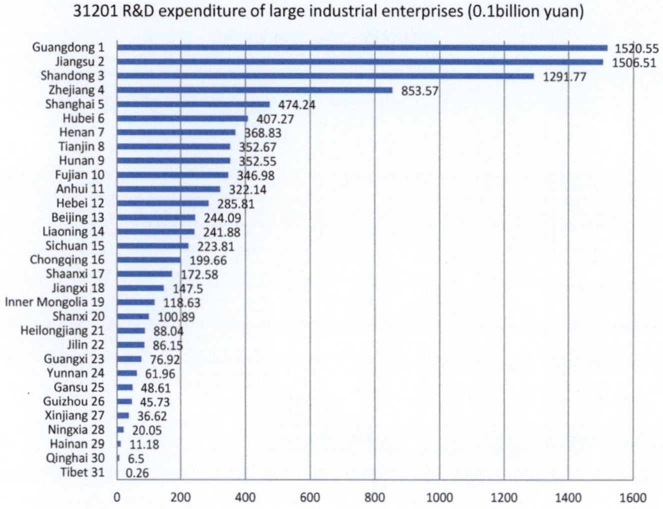
31201 R&D expenditure of large industrial enterprises (0.1billion yuan)
Guangdong 1 1520.55
Jiangsu 2 1506.51
Shandong 3 1291.77
Zhejiang 4 853.57
Shanghai 5 474.24
Hubei 6 407.27
Henan 7 368.83
Tianjin 8 352.67
Hunan 9 352.55
Fujian 10 346.98
Anhui 11 322.14
Hebei 12 285.81
Beijing 13 244.09
Liaoning 14 241.88
Sichuan 15 223.81
Chongqing 16 199.66
Shaanxi 17 172.58
Jiangxi 18 147.5
Inner Mongolia 19 118.63
Shanxi 20 100.89
Heilongjiang 21 88.04
Jilin 22 86.15
Guangxi 23 76.92
Yunnan 24 61.96
Gansu 25 48.61
Guizhou 26 45.73
Xinjiang 27 36.62
Ningxia 28 20.05
Hainan 29 11.18
Qinghai 30 6.5
Tibet 31 0.26
0 200 400 600 800 1000 1200 1400 1600

31202 The R&D expenditure of large industrial enterprises/sales revenue. (%)

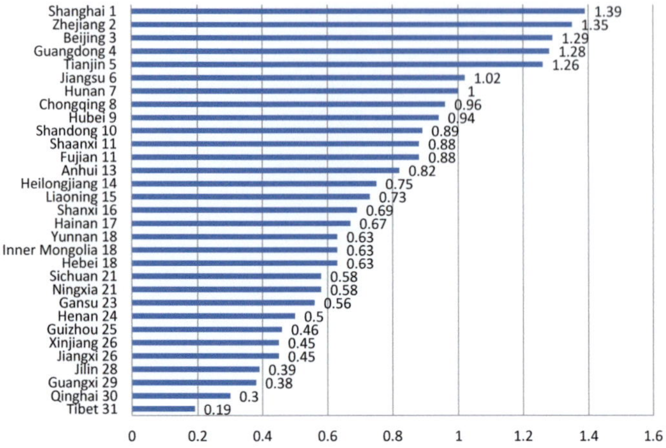

31203 The growth rate of R&D expenditure of large industrial enterprises/sales revenue (%)

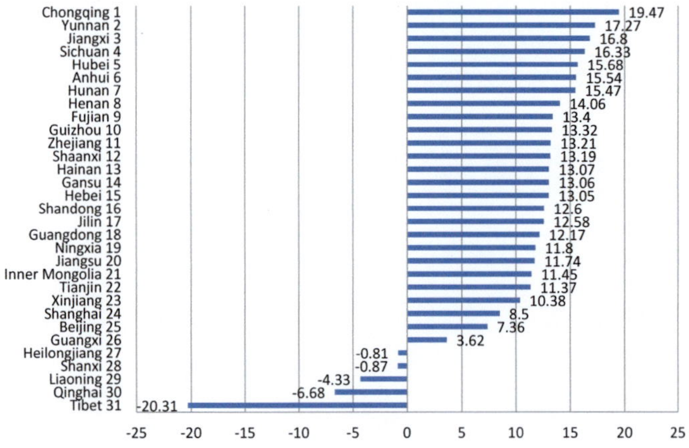

31301 The number of large industrial enterprises with R&D institutions

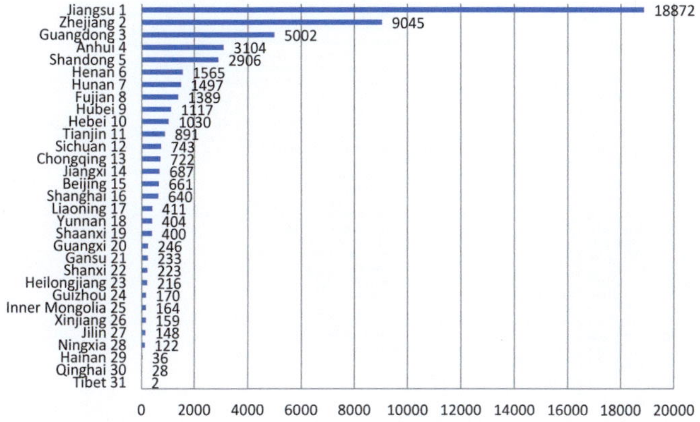

31302 The proportion of large industrial enterprises with R&D institutions in total enterprises (%)

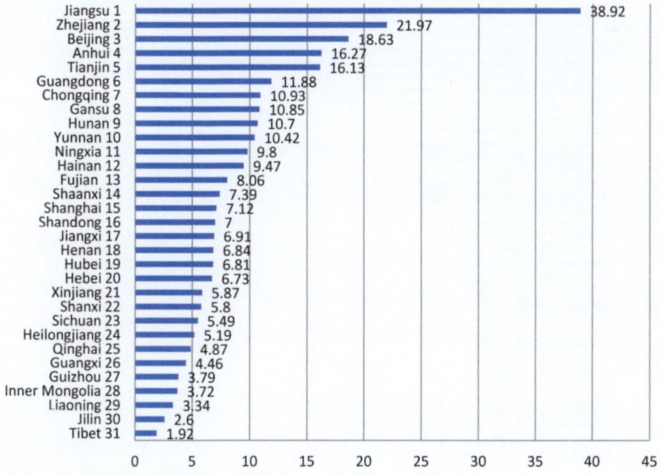

31303The growth rate of the number of large industrial enterprises with R&D institutions (%)

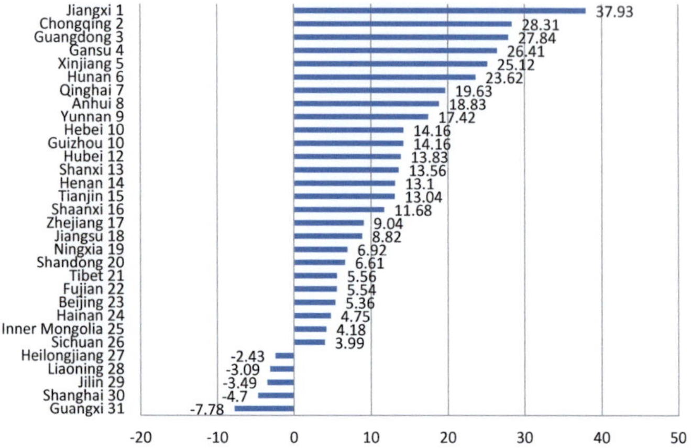

32101 The number of invention patent applications in large industrial enterprises (Nr)

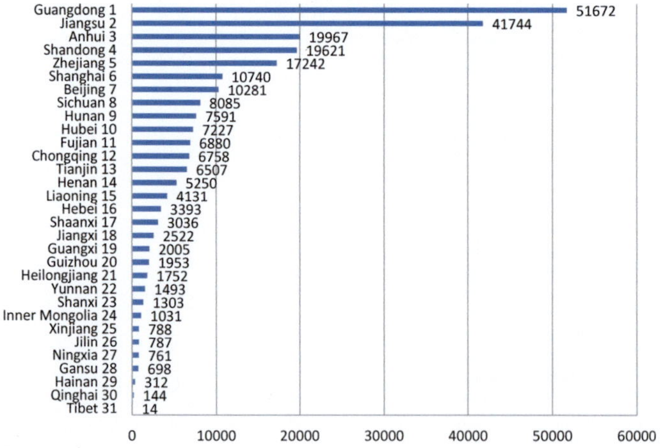

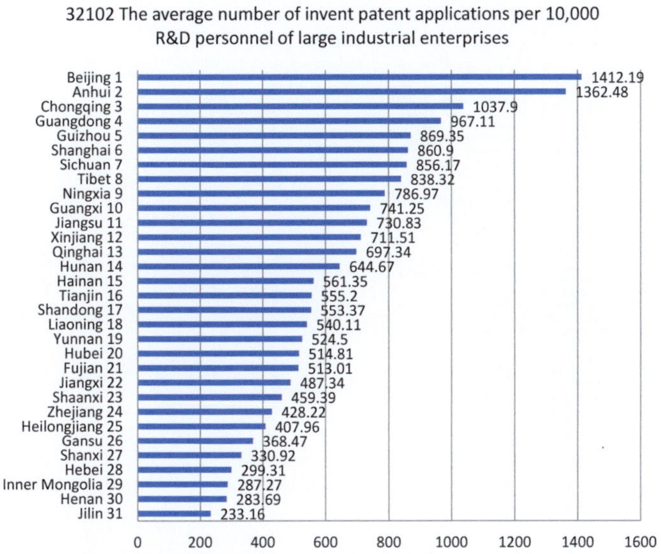
32102 The average number of invent patent applications per 10,000 R&D personnel of large industrial enterprises
Beijing 1 1412.19
Anhui 2 1362.48
Chongqing 3 1037.9
Guangdong 4 967.11
Guizhou 5 869.35
Shanghai 6 860.9
Sichuan 7 856.17
Tibet 8 838.32
Ningxia 9 786.97
Guangxi 10 741.25
Jiangsu 11 730.83
Xinjiang 12 711.51
Qinghai 13 697.34
Hunan 14 644.67
Hainan 15 561.35
Tianjin 16 555.2
Shandong 17 553.37
Liaoning 18 540.11
Yunnan 19 524.5
Hubei 20 514.81
Fujian 21 513.01
Jiangxi 22 487.34
Shaanxi 23 459.39
Zhejiang 24 428.22
Heilongjiang 25 407.96
Gansu 26 368.47
Shanxi 27 330.92
Hebei 28 299.31
Inner Mongolia 29 287.27
Henan 30 283.69
Jilin 31 233.16
0 200 400 600 800 1000 1200 1400 1600

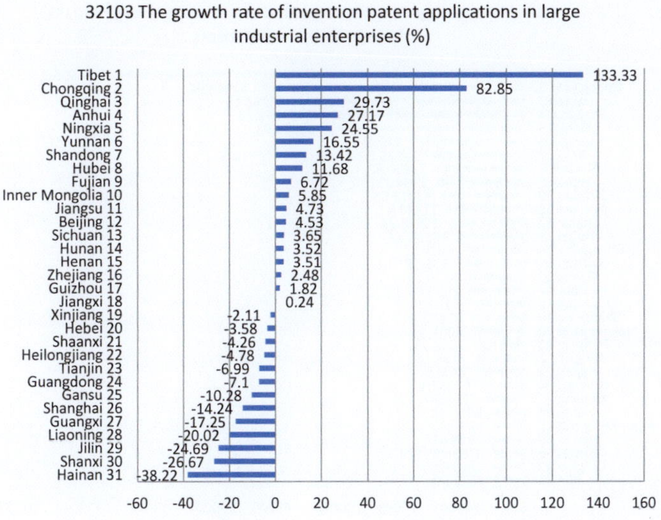
32103 The growth rate of invention patent applications in large industrial enterprises (%)
Tibet 1 133.33
Chongqing 2 82.85
Qinghai 3 29.73
Anhui 4 27.17
Ningxia 5 24.55
Yunnan 6 16.55
Shandong 7 13.42
Hubei 8 11.68
Fujian 9 6.72
Inner Mongolia 10 5.85
Jiangsu 11 4.73
Beijing 12 4.53
Sichuan 13 3.65
Hunan 14 3.52
Henan 15 3.51
Zhejiang 16 2.48
Guizhou 17 1.82
Jiangxi 18 0.24
Xinjiang 19 -2.11
Hebei 20 -3.58
Shaanxi 21 -4.26
Heilongjiang 22 -4.78
Tianjin 23 -6.99
Guangdong 24 -7.1
Gansu 25 -10.28
Shanghai 26 -14.24
Guangxi 27 -17.25
Liaoning 28 -20.02
Jilin 29 -24.69
Shanxi 30 -26.67
Hainan 31 -38.22
-60 -40 -20 0 20 40 60 80 100 120 140 160

32201 The number of granted invention patents in large industrial enterprises (Nr)

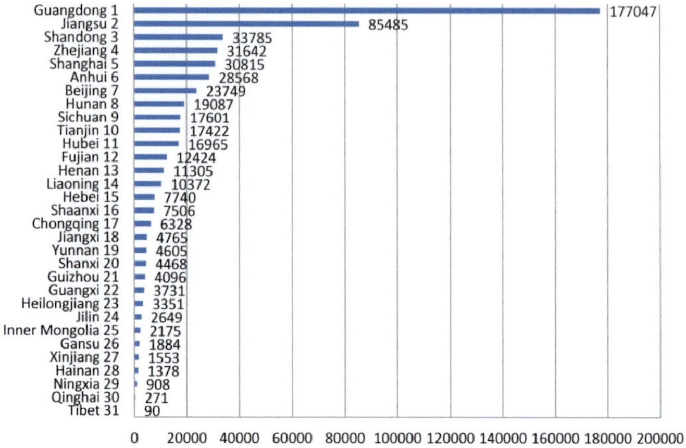

32202 The average number of granted invention patents per 10,000 large industrial enterprises

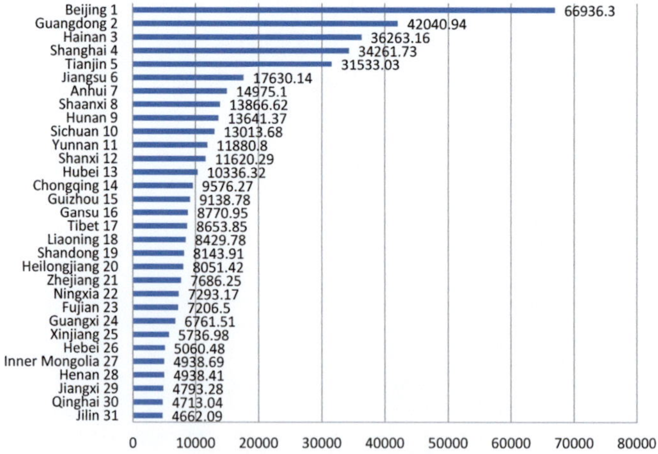

32203 The growth rate of the number of granted invention patents in large industrial enterprises (%)

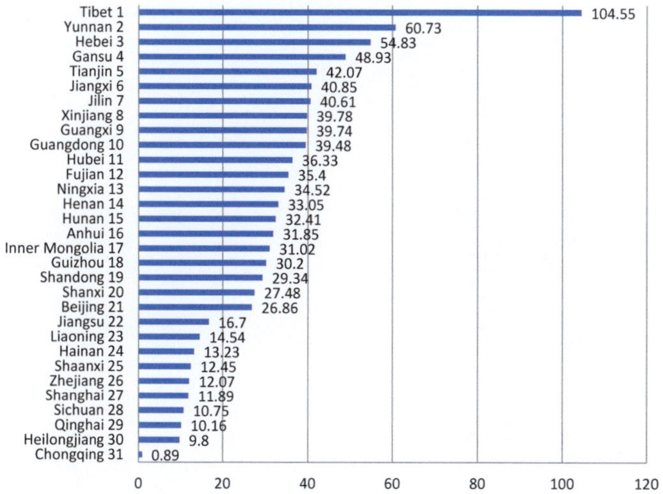

33101 External R&D expenditure of large industrial enterprises (10,000 yuan)

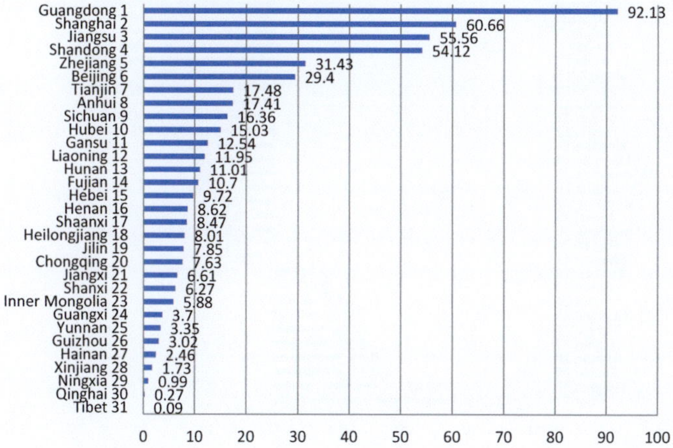

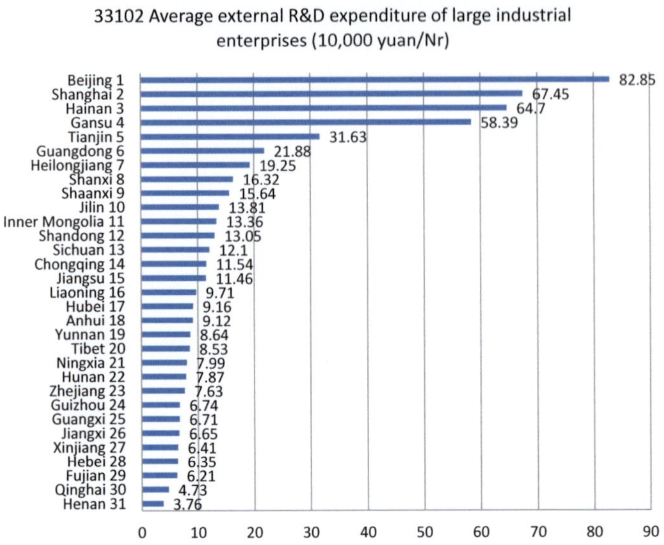
33102 Average external R&D expenditure of large industrial enterprises (10,000 yuan/Nr)
Beijing 1
Shanghai 2
Hainan 3
Gansu 4
Tianjin 5
Guangdong 6
Heilongjiang 7
Shanxi 8
Shaanxi 9
Jilin 10
Inner Mongolia 11
Shandong 12
Sichuan 13
Chongqing 14
Jiangsu 15
Liaoning 16
Hubei 17
Anhui 18
Yunnan 19
Tibet 20
Ningxia 21
Hunan 22
Zhejiang 23
Guizhou 24
Guangxi 25
Jiangxi 26
Xinjiang 27
Hebei 28
Fujian 29
Qinghai 30
Henan 31
82.85
67.45
64.7
58.39
31.63
21.88
19.25
16.32
15.64
13.81
13.36
13.05
12.1
11.54
11.46
9.71
9.16
9.12
8.64
8.53
7.99
7.87
7.63
6.74
6.71
6.65
6.41
6.35
6.21
4.73
3.76
0
10
20
30
40
50
60
70
80
90

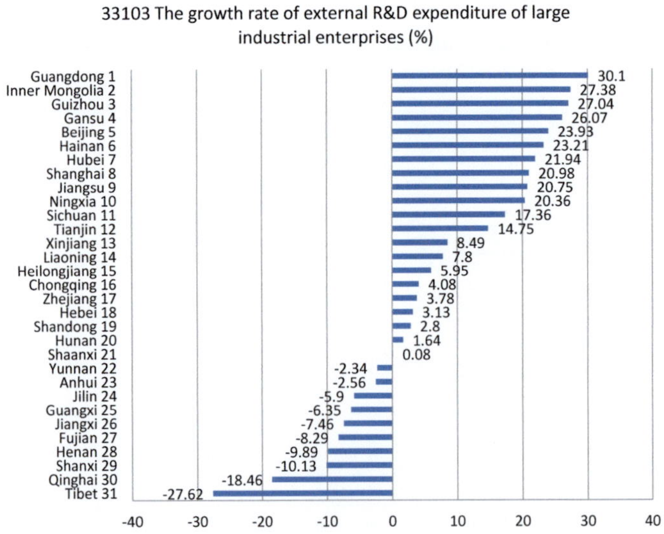
33103 The growth rate of external R&D expenditure of large industrial enterprises (%)
Guangdong 1
Inner Mongolia 2
Guizhou 3
Gansu 4
Beijing 5
Hainan 6
Hubei 7
Shanghai 8
Jiangsu 9
Ningxia 10
Sichuan 11
Tianjin 12
Xinjiang 13
Liaoning 14
Heilongjiang 15
Chongqing 16
Zhejiang 17
Hebei 18
Shandong 19
Hunan 20
Shaanxi 21
Yunnan 22
Anhui 23
Jilin 24
Guangxi 25
Jiangxi 26
Fujian 27
Henan 28
Shanxi 29
Qinghai 30
Tibet 31
30.1
27.38
27.04
26.07
23.93
23.21
21.94
20.98
20.75
20.36
17.36
14.75
8.49
7.8
5.95
4.08
3.78
3.13
2.8
1.64
0.08
-2.34
-2.56
-5.9
-6.35
-7.46
-8.29
-9.89
-10.13
-18.46
-27.62
-40
-30
-20
-10
0
10
20
30
40

33201 Technical renovation expenditure of large industrial enterprises (10,000 yuan)

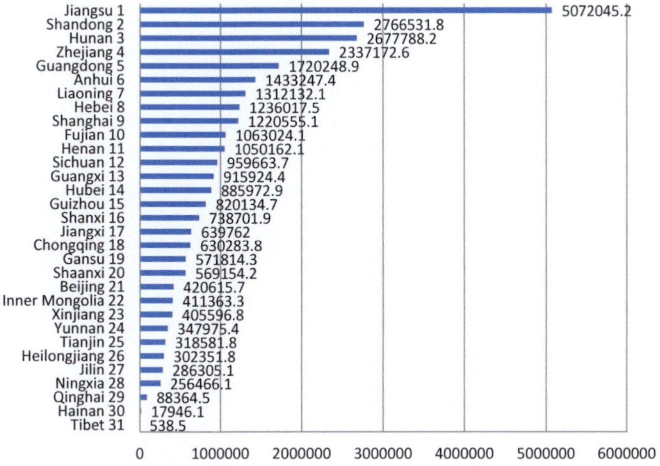

33202 AverageTechnical renovation expenditure of large industrial enterprises (10,000 yuan）

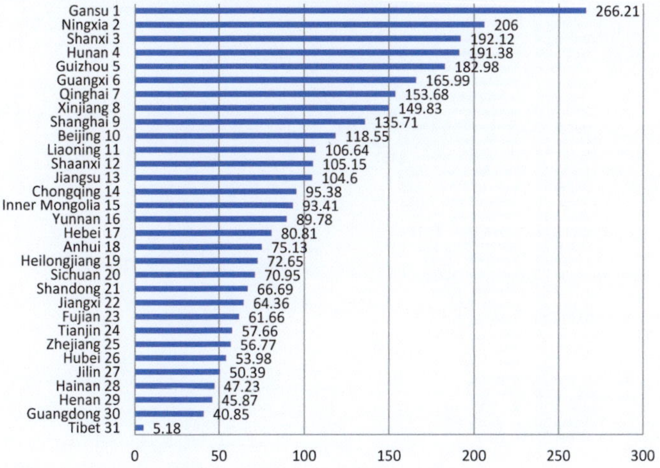

33203 The growth rate of technical renovation expenditure of large industrial enterprises (%)

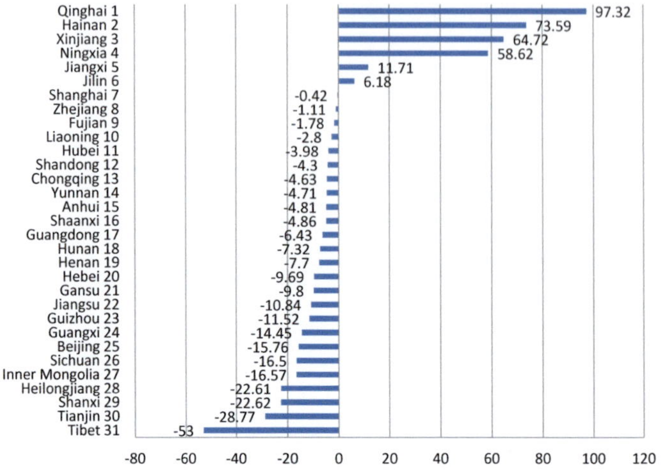

33301 The number of enterprises with e-commerce transactions

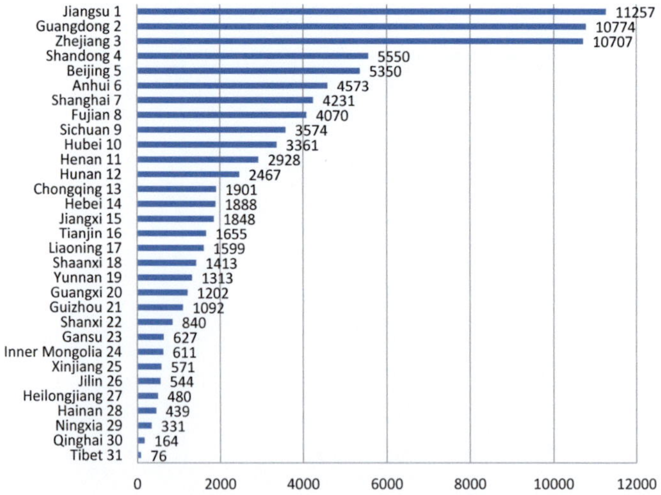

33302 The proportion of enterprises with e-commerce transactions in the total enterprises (%).

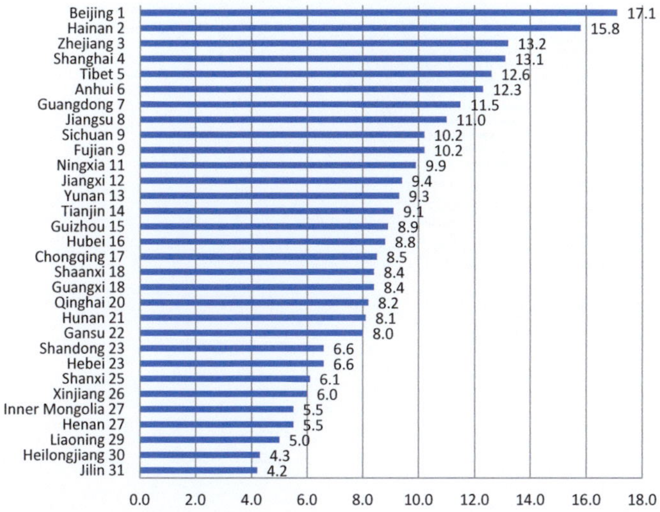

33303 The growth rate of the number of enterprises with e-commerce transactions (%)

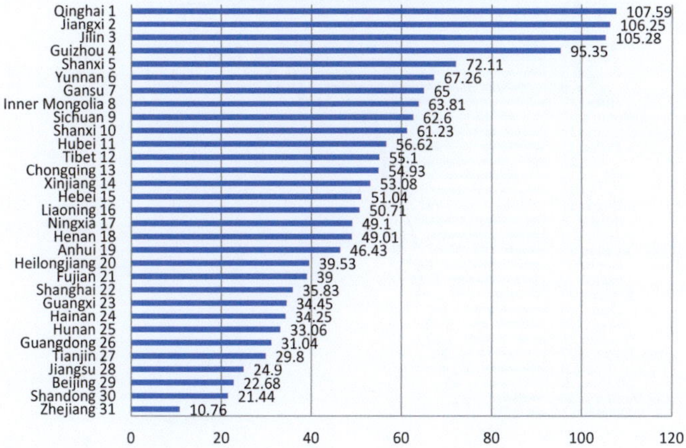

34001 New products sales revenue of large industrial enterprises (100 million yuan)

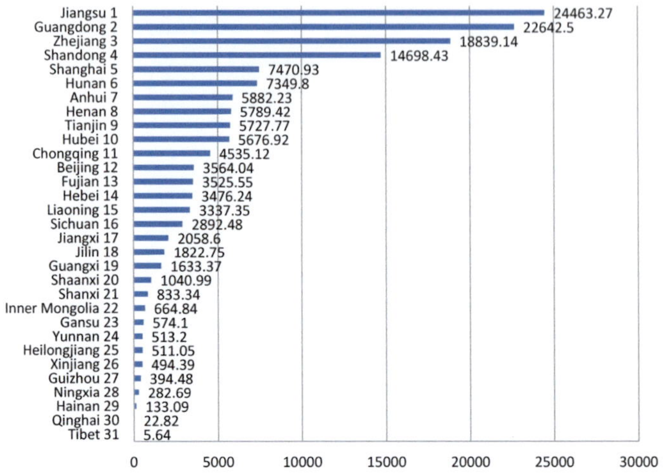

34002 The proprtion of new products sales revenue of large industrial enterprises in total sales revenue (%)

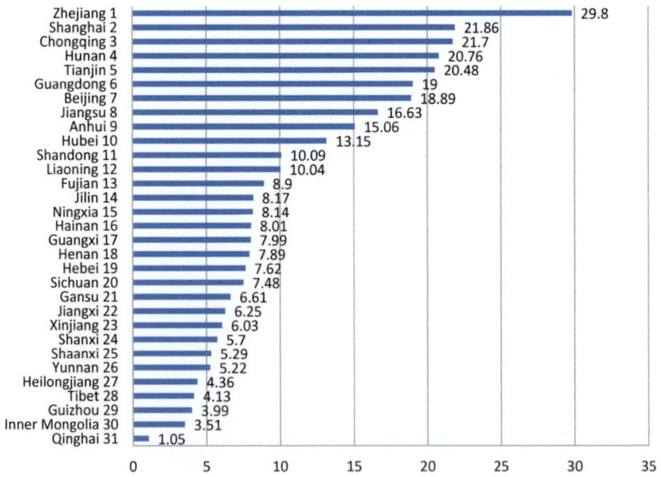

34003 The growth rate of new product sales revenue of large industrial enterprises (%)

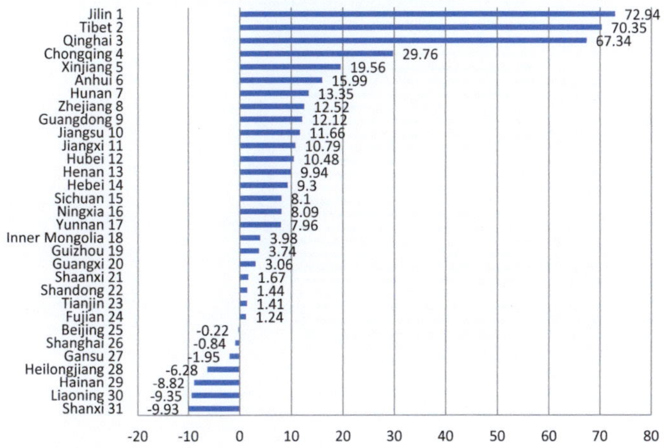

41111 The number of mobile phone users (10,000 users)

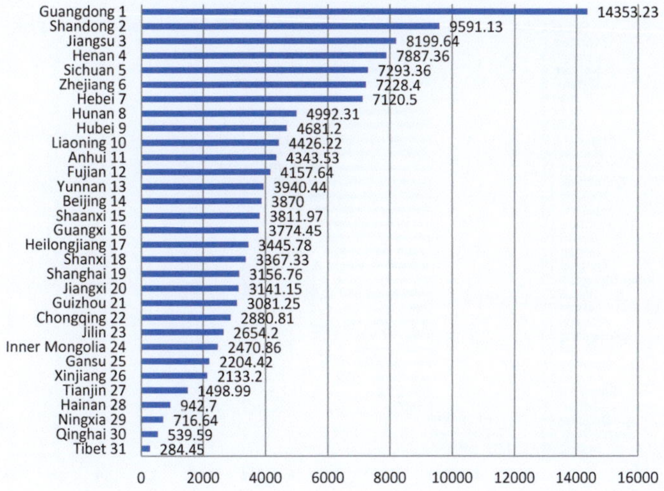

41112 The penetration rate of mobile phone (Nr/100 people)

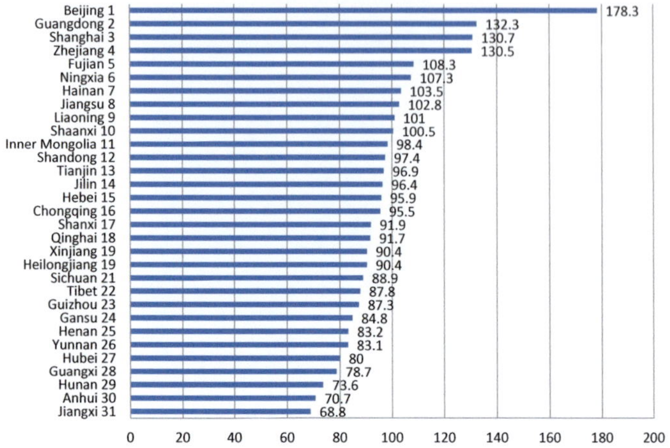

41113 The growth rate of mobile phone users (%)

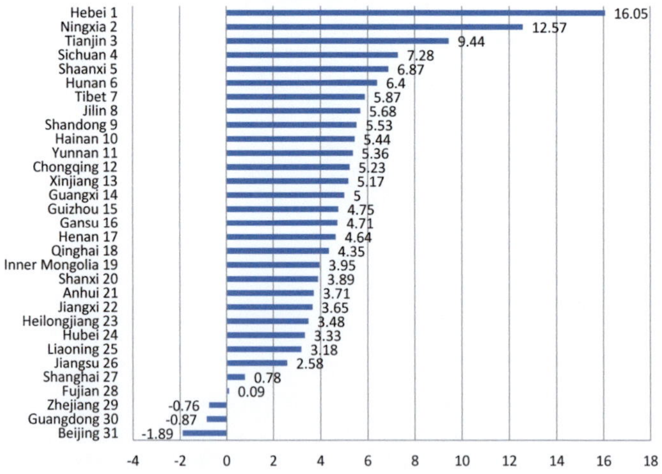

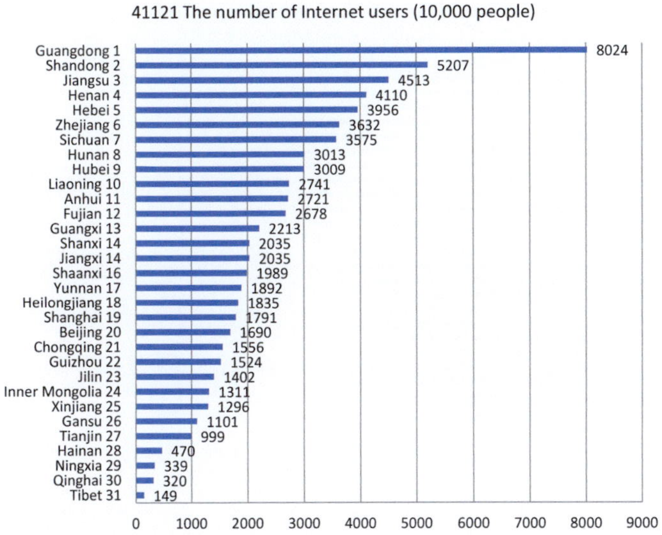
41121 The number of Internet users (10,000 people)
Guangdong 1 8024
Shandong 2 5207
Jiangsu 3 4513
Henan 4 4110
Hebei 5 3956
Zhejiang 6 3632
Sichuan 7 3575
Hunan 8 3013
Hubei 9 3009
Liaoning 10 2741
Anhui 11 2721
Fujian 12 2678
Guangxi 13 2213
Shanxi 14 2035
Jiangxi 14 2035
Shaanxi 16 1989
Yunnan 17 1892
Heilongjiang 18 1835
Shanghai 19 1791
Beijing 20 1690
Chongqing 21 1556
Guizhou 22 1524
Jilin 23 1402
Inner Mongolia 24 1311
Xinjiang 25 1296
Gansu 26 1101
Tianjin 27 999
Hainan 28 470
Ningxia 29 339
Qinghai 30 320
Tibet 31 149
0 1000 2000 3000 4000 5000 6000 7000 8000 9000

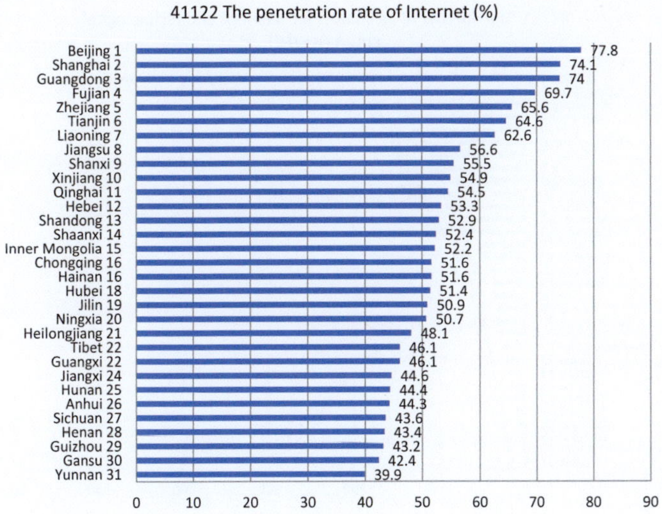
41122 The penetration rate of Internet (%)
Beijing 1 77.8
Shanghai 2 74.1
Guangdong 3 74
Fujian 4 69.7
Zhejiang 5 65.6
Tianjin 6 64.6
Liaoning 7 62.6
Jiangsu 8 56.6
Shanxi 9 55.5
Xinjiang 10 54.9
Qinghai 11 54.5
Hebei 12 53.3
Shandong 13 52.9
Shaanxi 14 52.4
Inner Mongolia 15 52.2
Chongqing 16 51.6
Hainan 16 51.6
Hubei 18 51.4
Jilin 19 50.9
Ningxia 20 50.7
Heilongjiang 21 48.1
Tibet 22 46.1
Guangxi 22 46.1
Jiangxi 24 44.6
Hunan 25 44.4
Anhui 26 44.3
Sichuan 27 43.6
Henan 28 43.4
Guizhou 29 43.2
Gansu 30 42.4
Yunnan 31 39.9
0 10 20 30 40 50 60 70 80 90

41123 The growth rate of Internet users(%)

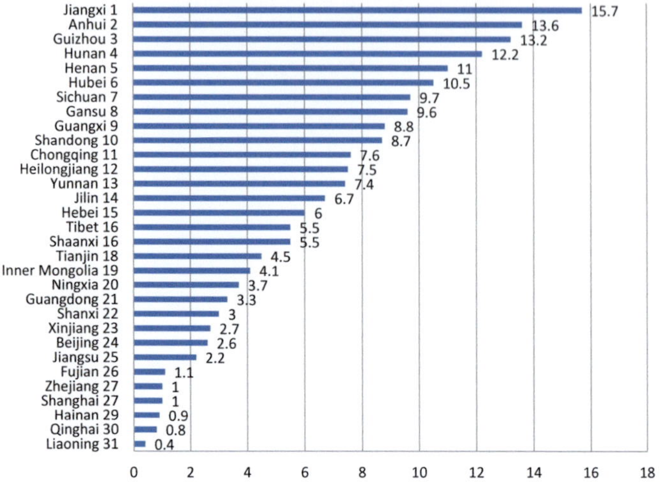

41211 The number of incubators for scientific and technological enterprise (Nr)

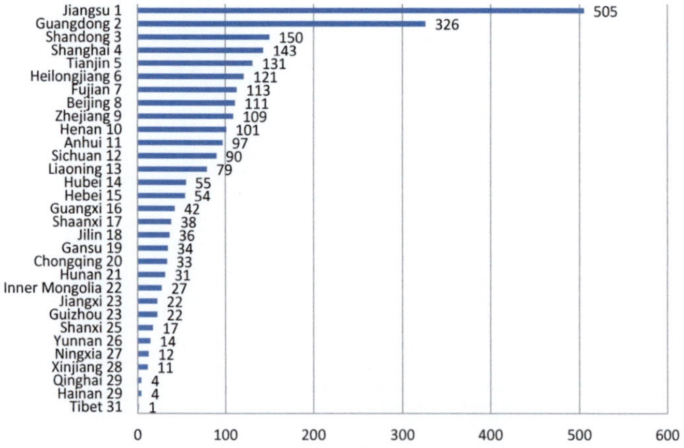

41212 The average number of entrepreneurship mentors in incubator

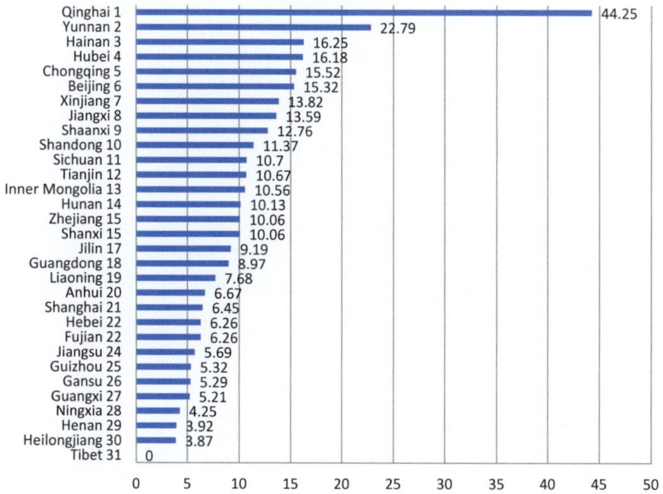

41213 The growth rate of incubators （%）

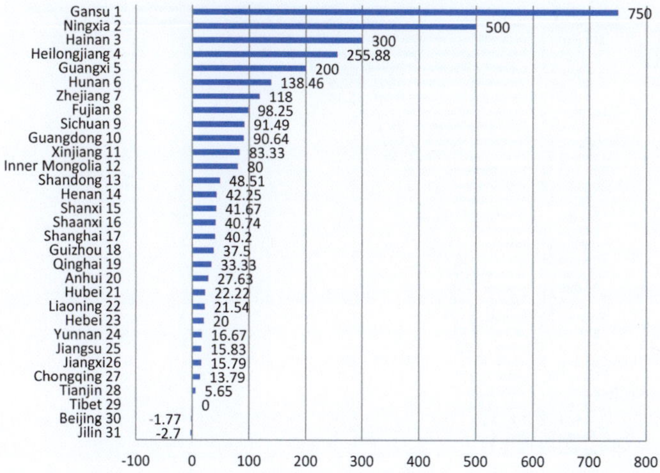

42101 Total export-import volume ($100 million) by destination and source

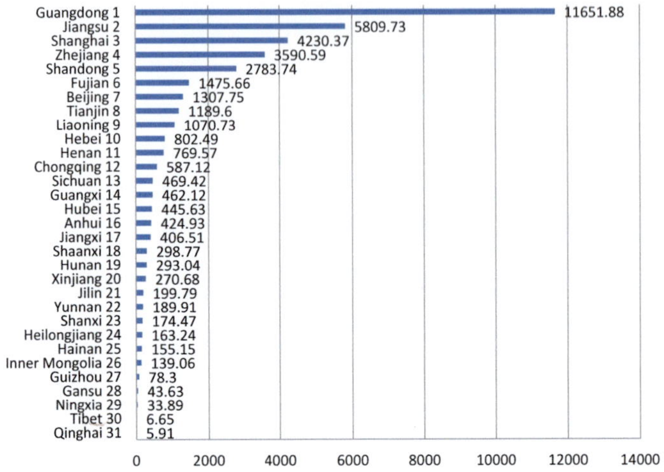

42102 The proportion of total export-import volume / GDP (%).

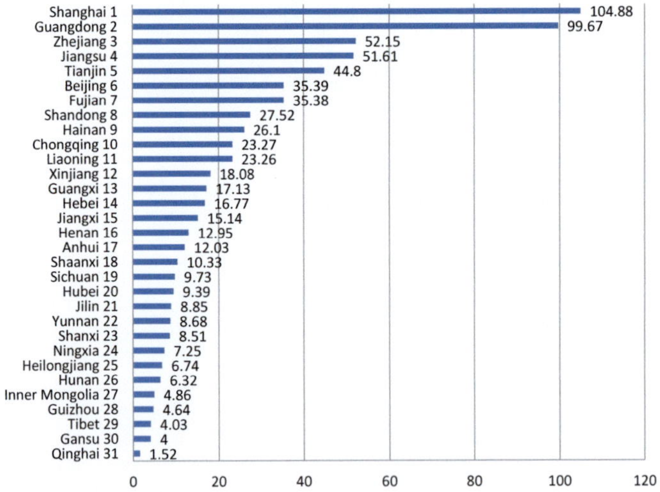

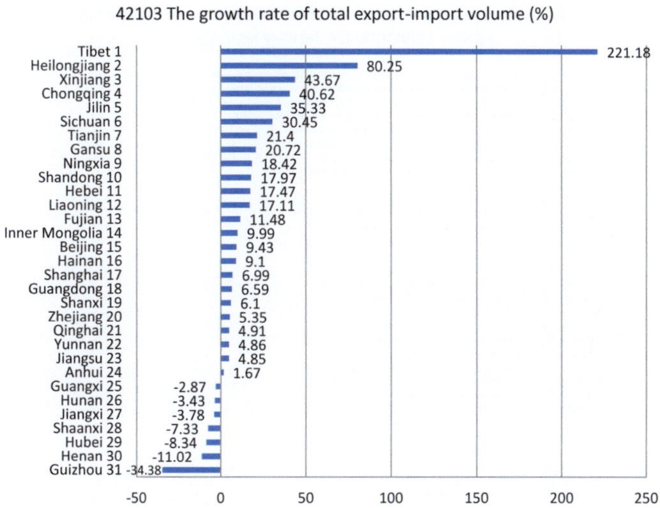
42103 The growth rate of total export-import volume (%)
Tibet 1 221.18
Heilongjiang 2 80.25
Xinjiang 3 43.67
Chongqing 4 40.62
Jilin 5 35.33
Sichuan 6 30.45
Tianjin 7 21.4
Gansu 8 20.72
Ningxia 9 18.42
Shandong 10 17.97
Hebei 11 17.47
Liaoning 12 17.11
Fujian 13 11.48
Inner Mongolia 14 9.99
Beijing 15 9.43
Hainan 16 9.1
Shanghai 17 6.99
Guangdong 18 6.59
Shanxi 19 6.1
Zhejiang 20 5.35
Qinghai 21 4.91
Yunnan 22 4.86
Jiangsu 23 4.85
Anhui 24 1.67
Guangxi 25 -2.87
Hunan 26 -3.43
Jiangxi 27 -3.78
Shaanxi 28 -7.33
Hubei 29 -8.34
Henan 30 -11.02
Guizhou 31 -34.38
-50 0 50 100 150 200 250

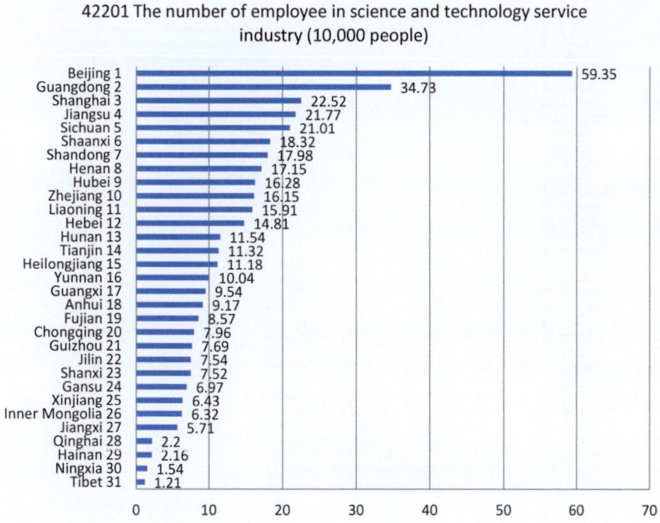
42201 The number of employee in science and technology service industry (10,000 people)
Beijing 1 59.35
Guangdong 2 34.73
Shanghai 3 22.52
Jiangsu 4 21.77
Sichuan 5 21.01
Shaanxi 6 18.32
Shandong 7 17.98
Henan 8 17.15
Hubei 9 16.28
Zhejiang 10 16.15
Liaoning 11 15.91
Hebei 12 14.81
Hunan 13 11.54
Tianjin 14 11.32
Heilongjiang 15 11.18
Yunnan 16 10.04
Guangxi 17 9.54
Anhui 18 9.17
Fujian 19 8.57
Chongqing 20 7.96
Guizhou 21 7.69
Jilin 22 7.54
Shanxi 23 7.52
Gansu 24 6.97
Xinjiang 25 6.43
Inner Mongolia 26 6.32
Jiangxi 27 5.71
Qinghai 28 2.2
Hainan 29 2.16
Ningxia 30 1.54
Tibet 31 1.21
0 10 20 30 40 50 60 70

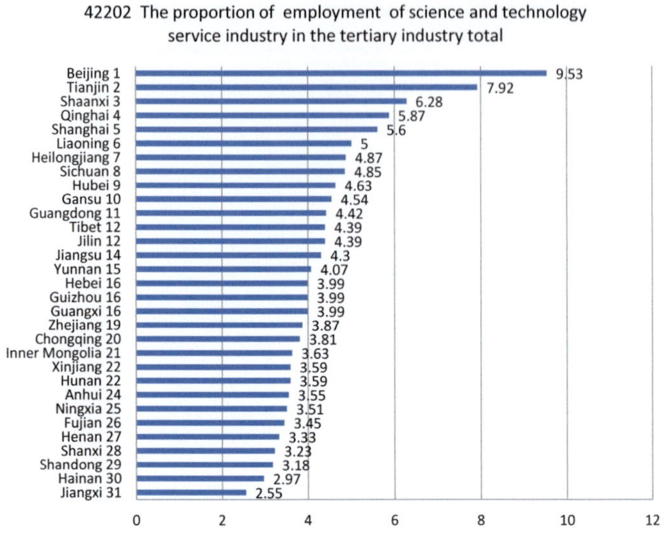
42202 The proportion of employment of science and technology service industry in the tertiary industry total
Beijing 1
Tianjin 2
Shaanxi 3
Qinghai 4
Shanghai 5
Liaoning 6
Heilongjiang 7
Sichuan 8
Hubei 9
Gansu 10
Guangdong 11
Tibet 12
Jilin 12
Jiangsu 14
Yunnan 15
Hebei 16
Guizhou 16
Guangxi 16
Zhejiang 19
Chongqing 20
Inner Mongolia 21
Xinjiang 22
Hunan 22
Anhui 24
Ningxia 25
Fujian 26
Henan 27
Shanxi 28
Shandong 29
Hainan 30
Jiangxi 31
9.53
7.92
6.28
5.87
5.6
5
4.87
4.85
4.63
4.54
4.42
4.39
4.39
4.3
4.07
3.99
3.99
3.99
3.87
3.81
3.63
3.59
3.59
3.55
3.51
3.45
3.33
3.23
3.18
2.97
2.55
0
2
4
6
8
10
12

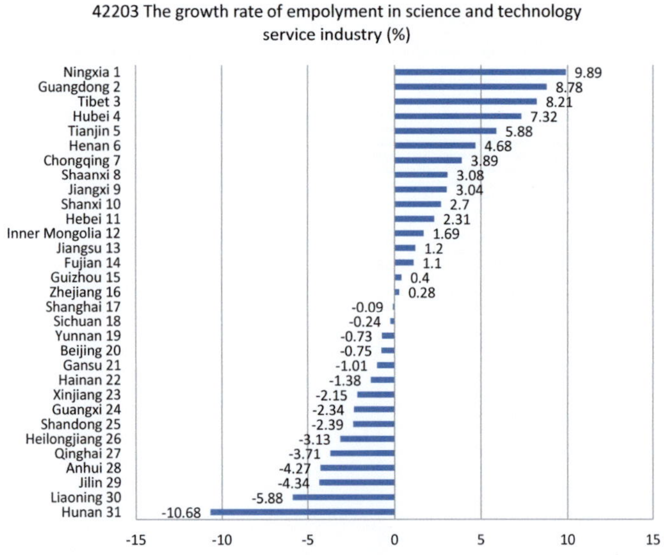
42203 The growth rate of empolyment in science and technology service industry (%)
Ningxia 1
Guangdong 2
Tibet 3
Hubei 4
Tianjin 5
Henan 6
Chongqing 7
Shaanxi 8
Jiangxi 9
Shanxi 10
Hebei 11
Inner Mongolia 12
Jiangsu 13
Fujian 14
Guizhou 15
Zhejiang 16
Shanghai 17
Sichuan 18
Yunnan 19
Beijing 20
Gansu 21
Hainan 22
Xinjiang 23
Guangxi 24
Shandong 25
Heilongjiang 26
Qinghai 27
Anhui 28
Jilin 29
Liaoning 30
Hunan 31
9.89
8.78
8.21
7.32
5.88
4.68
3.89
3.08
3.04
2.7
2.31
1.69
1.2
1.1
0.4
0.28
-0.09
-0.24
-0.73
-0.75
-1.01
-1.38
-2.15
-2.34
-2.39
-3.13
-3.71
-4.27
-4.34
-5.88
-10.68
-15
-10
-5
0
5
10
15

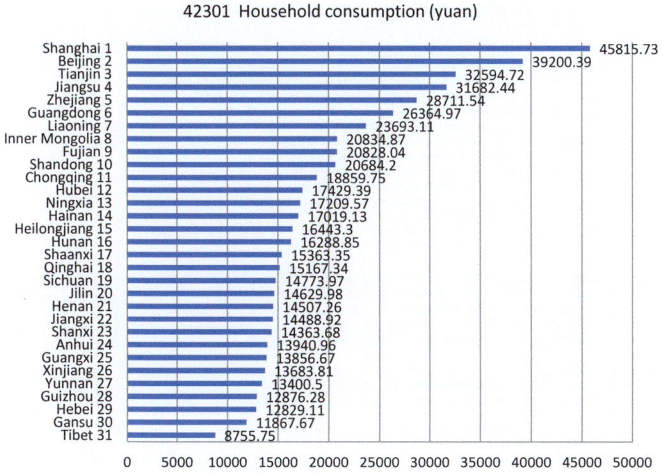
42301 Household consumption (yuan)
Shanghai 1
Beijing 2
Tianjin 3
Jiangsu 4
Zhejiang 5
Guangdong 6
Liaoning 7
Inner Mongolia 8
Fujian 9
Shandong 10
Chongqing 11
Hubei 12
Ningxia 13
Hainan 14
Heilongjiang 15
Hunan 16
Shaanxi 17
Qinghai 18
Sichuan 19
Jilin 20
Henan 21
Jiangxi 22
Shanxi 23
Anhui 24
Guangxi 25
Xinjiang 26
Yunnan 27
Guizhou 28
Hebei 29
Gansu 30
Tibet 31
45815.73
39200.39
32594.72
31682.44
28711.54
26364.97
23693.11
20834.87
20828.04
20684.2
18859.75
17429.39
17209.57
17019.13
16443.3
16288.85
15363.35
15167.34
14773.97
14629.98
14507.26
14488.92
14363.68
13940.96
13856.67
13683.81
13400.5
12876.28
12829.11
11867.67
8755.75
0 5000 10000 15000 20000 25000 30000 35000 40000 45000 50000

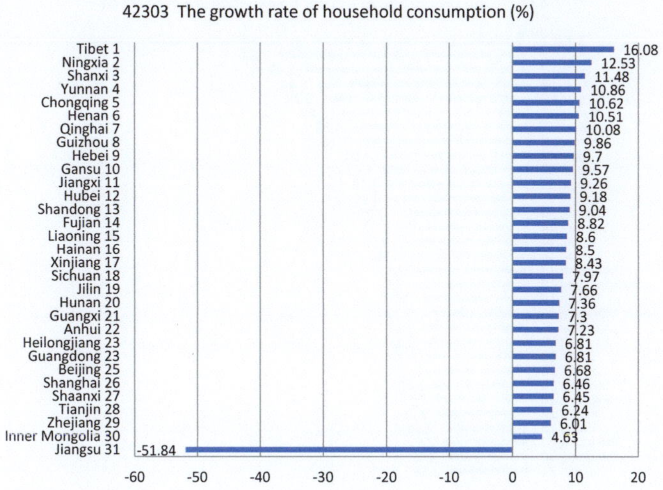
42303 The growth rate of household consumption (%)
Tibet 1
Ningxia 2
Shanxi 3
Yunnan 4
Chongqing 5
Henan 6
Qinghai 7
Guizhou 8
Hebei 9
Gansu 10
Jiangxi 11
Hubei 12
Shandong 13
Fujian 14
Liaoning 15
Hainan 16
Xinjiang 17
Sichuan 18
Jilin 19
Hunan 20
Guangxi 21
Anhui 22
Heilongjiang 23
Guangdong 23
Beijing 25
Shanghai 26
Shaanxi 27
Tianjin 28
Zhejiang 29
Inner Mongolia 30
Jiangsu 31
16.08
12.53
11.48
10.86
10.62
10.51
10.08
9.86
9.7
9.57
9.26
9.18
9.04
8.82
8.6
8.5
8.43
7.97
7.66
7.36
7.3
7.23
6.81
6.81
6.68
6.46
6.45
6.24
6.01
4.63
-51.84
-60 -50 -40 -30 -20 -10 0 10 20

43101 Education expenditure (100 million yuan)

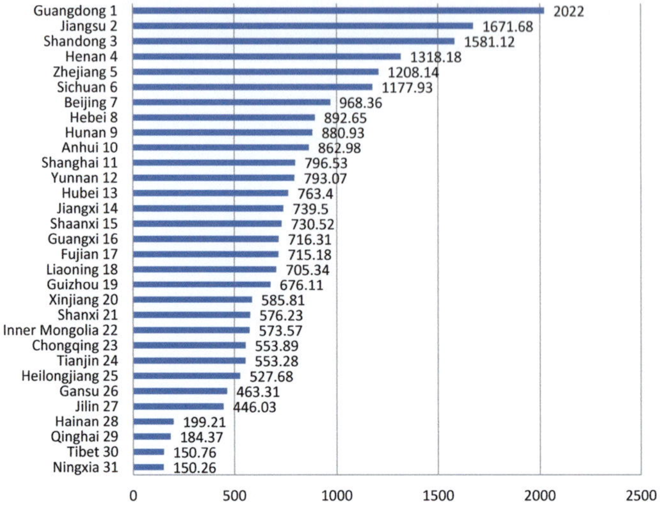

43102 The proportion of education expenditure in GDP (%)

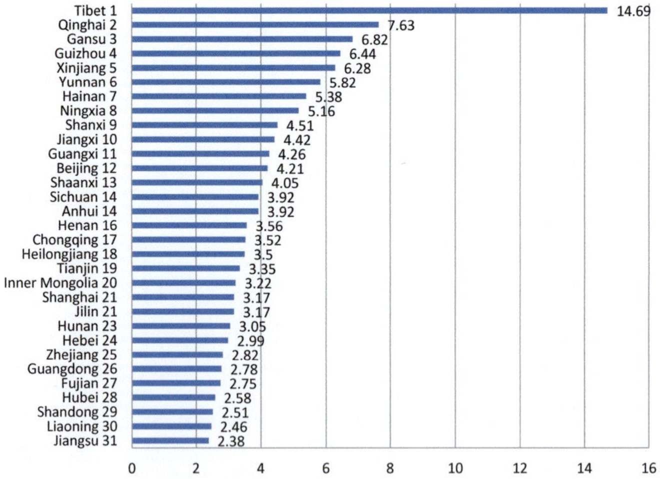

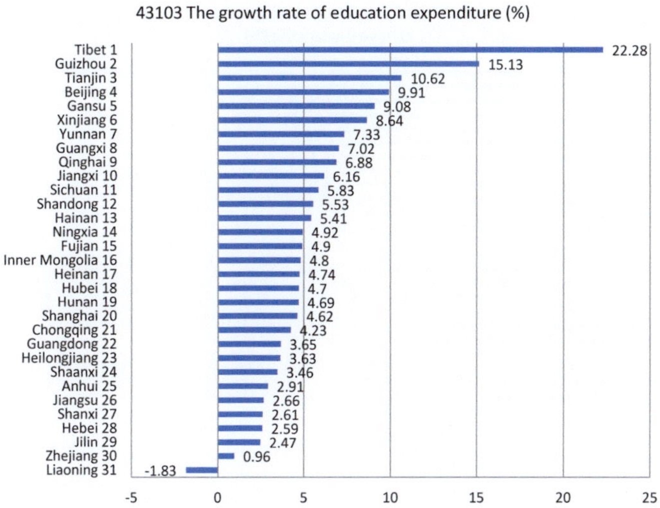
43103 The growth rate of education expenditure (%)
Tibet 1
Guizhou 2
Tianjin 3
Beijing 4
Gansu 5
Xinjiang 6
Yunnan 7
Guangxi 8
Qinghai 9
Jiangxi 10
Sichuan 11
Shandong 12
Hainan 13
Ningxia 14
Fujian 15
Inner Mongolia 16
Heinan 17
Hubei 18
Hunan 19
Shanghai 20
Chongqing 21
Guangdong 22
Heilongjiang 23
Shaanxi 24
Anhui 25
Jiangsu 26
Shanxi 27
Hebei 28
Jilin 29
Zhejiang 30
Liaoning 31
22.28
15.13
10.62
9.91
9.08
8.64
7.33
7.02
6.88
6.16
5.83
5.53
5.41
4.92
4.9
4.8
4.74
4.7
4.69
4.62
4.23
3.65
3.63
3.46
2.91
2.66
2.61
2.59
2.47
0.96
-1.83
-5
0
5
10
15
20
25

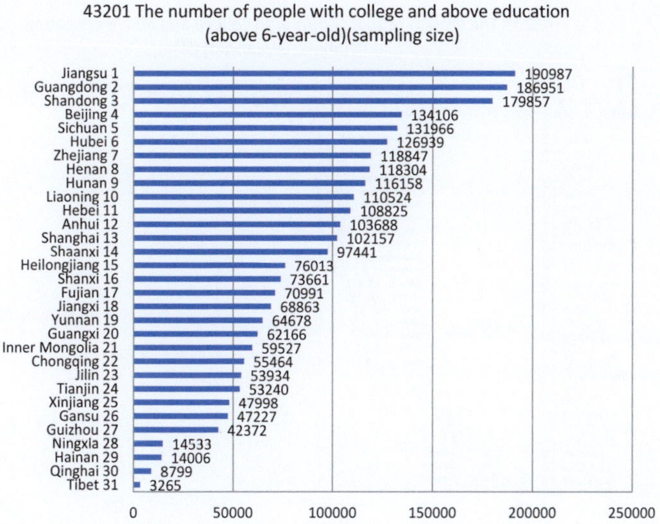
43201 The number of people with college and above education (above 6-year-old)(sampling size)
Jiangsu 1
Guangdong 2
Shandong 3
Beijing 4
Sichuan 5
Hubei 6
Zhejiang 7
Henan 8
Hunan 9
Liaoning 10
Hebei 11
Anhui 12
Shanghai 13
Shaanxi 14
Heilongjiang 15
Shanxi 16
Fujian 17
Jiangxi 18
Yunnan 19
Guangxi 20
Inner Mongolia 21
Chongqing 22
Jilin 23
Tianjin 24
Xinjiang 25
Gansu 26
Guizhou 27
Ningxla 28
Hainan 29
Qinghai 30
Tibet 31
190987
186951
179857
134106
131966
126939
118847
118304
116158
110524
108825
103688
102157
97441
76013
73661
70991
68863
64678
62166
59527
55464
53934
53240
47998
47227
42372
14533
14006
8799
3265
0
50000
100000
150000
200000
250000

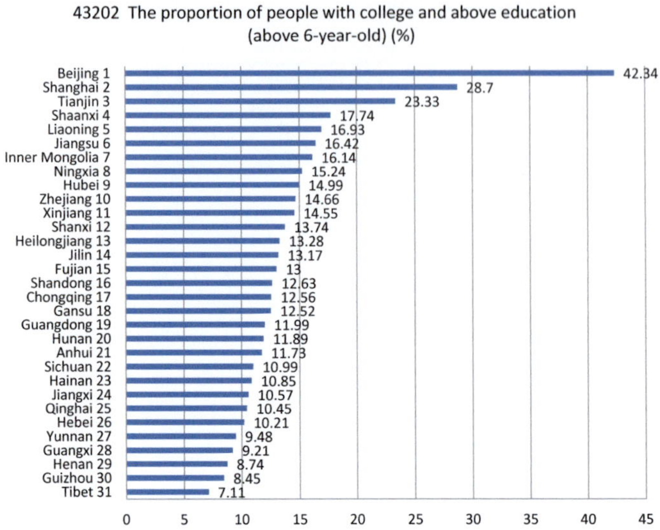
43202 The proportion of people with college and above education (above 6-year-old) (%)
Beijing 1
Shanghai 2
Tianjin 3
Shaanxi 4
Liaoning 5
Jiangsu 6
Inner Mongolia 7
Ningxia 8
Hubei 9
Zhejiang 10
Xinjiang 11
Shanxi 12
Heilongjiang 13
Jilin 14
Fujian 15
Shandong 16
Chongqing 17
Gansu 18
Guangdong 19
Hunan 20
Anhui 21
Sichuan 22
Hainan 23
Jiangxi 24
Qinghai 25
Hebei 26
Yunnan 27
Guangxi 28
Henan 29
Guizhou 30
Tibet 31
42.34
28.7
23.33
17.74
16.93
16.42
16.14
15.24
14.99
14.66
14.55
13.74
13.28
13.17
13
12.63
12.56
12.52
11.99
11.89
11.73
10.99
10.85
10.57
10.45
10.21
9.48
9.21
8.74
8.45
7.11
0
5
10
15
20
25
30
35
40
45

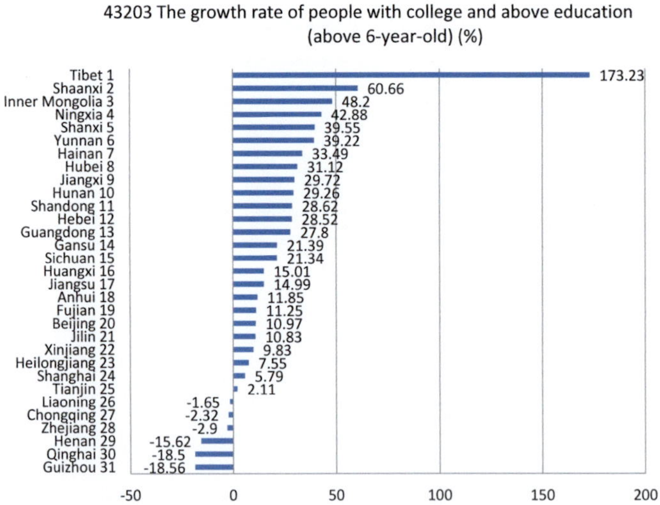
43203 The growth rate of people with college and above education (above 6-year-old) (%)
Tibet 1
Shaanxi 2
Inner Mongolia 3
Ningxia 4
Shanxi 5
Yunnan 6
Hainan 7
Hubei 8
Jiangxi 9
Hunan 10
Shandong 11
Hebei 12
Guangdong 13
Gansu 14
Sichuan 15
Huangxi 16
Jiangsu 17
Anhui 18
Fujian 19
Beijing 20
Jilin 21
Xinjiang 22
Heilongjiang 23
Shanghai 24
Tianjin 25
Liaoning 26
Chongqing 27
Zhejiang 28
Henan 29
Qinghai 30
Guizhou 31
173.23
60.66
48.2
42.88
39.55
39.22
33.49
31.12
29.72
29.26
28.62
28.52
27.8
21.39
21.34
15.01
14.99
11.85
11.25
10.97
10.83
9.83
7.55
5.79
2.11
-1.65
-2.32
-2.9
-15.62
-18.5
-18.56
-50
0
50
100
150
200

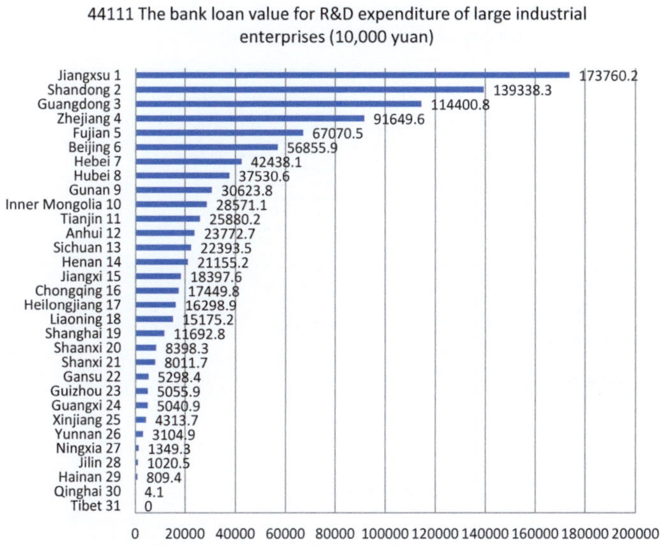
44111 The bank loan value for R&D expenditure of large industrial enterprises (10,000 yuan)
Jiangxsu 1
Shandong 2
Guangdong 3
Zhejiang 4
Fujian 5
Beijing 6
Hebei 7
Hubei 8
Gunan 9
Inner Mongolia 10
Tianjin 11
Anhui 12
Sichuan 13
Henan 14
Jiangxi 15
Chongqing 16
Heilongjiang 17
Liaoning 18
Shanghai 19
Shaanxi 20
Shanxi 21
Gansu 22
Guizhou 23
Guangxi 24
Xinjiang 25
Yunnan 26
Ningxia 27
Jilin 28
Hainan 29
Qinghai 30
Tibet 31
173760.2
139338.3
114400.8
91649.6
67070.5
56855.9
42438.1
37530.6
30623.8
28571.1
25880.2
23772.7
22393.5
21155.2
18397.6
17449.8
16298.9
15175.2
11692.8
8398.3
8011.7
5298.4
5055.9
5040.9
4313.7
3104.9
1349.3
1020.5
809.4
4.1
0
0 20000 40000 60000 80000 100000 120000 140000 160000 180000 200000

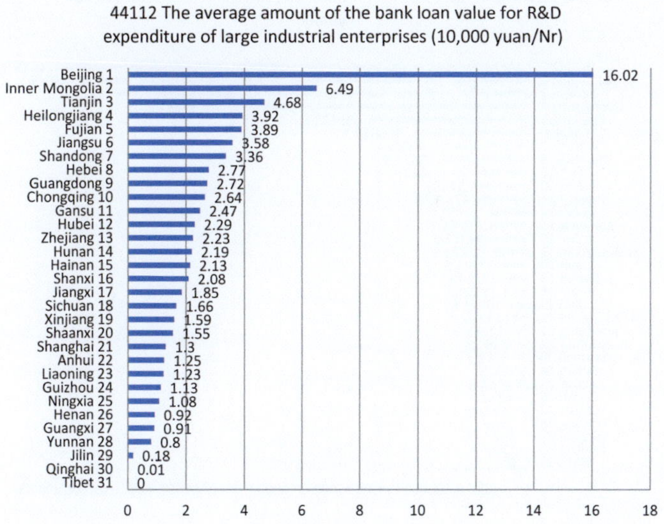
44112 The average amount of the bank loan value for R&D expenditure of large industrial enterprises (10,000 yuan/Nr)
Beijing 1
Inner Mongolia 2
Tianjin 3
Heilongjiang 4
Fujian 5
Jiangsu 6
Shandong 7
Hebei 8
Guangdong 9
Chongqing 10
Gansu 11
Hubei 12
Zhejiang 13
Hunan 14
Hainan 15
Shanxi 16
Jiangxi 17
Sichuan 18
Xinjiang 19
Shaanxi 20
Shanghai 21
Anhui 22
Liaoning 23
Guizhou 24
Ningxia 25
Henan 26
Guangxi 27
Yunnan 28
Jilin 29
Qinghai 30
Tibet 31
16.02
6.49
4.68
3.92
3.89
3.58
3.36
2.77
2.72
2.64
2.47
2.29
2.23
2.19
2.13
2.08
1.85
1.66
1.59
1.55
1.3
1.25
1.23
1.13
1.08
0.92
0.91
0.8
0.18
0.01
0
0 2 4 6 8 10 12 14 16 18

44113 The growth rate of bank loan value for R&D internal expenditure of large industrial enterprises (%)

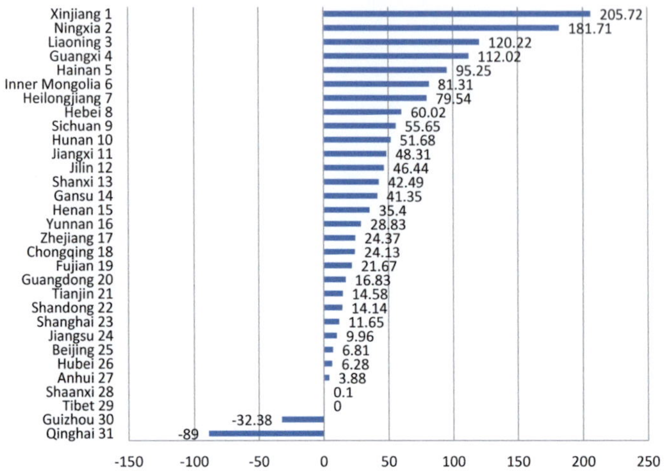

44211 The amount of venture capital investment in incubator (10,000 yuan)

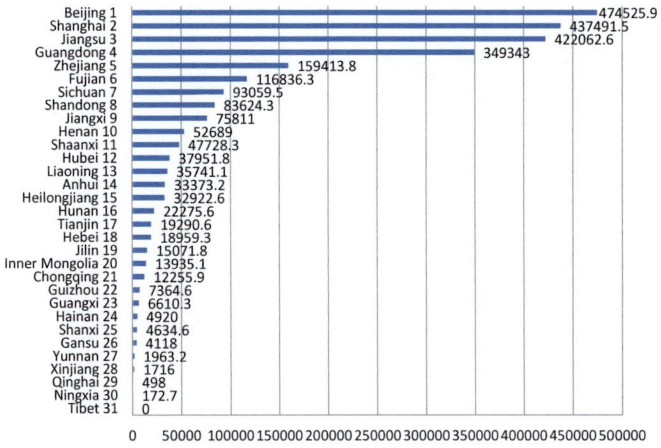

44212 Investment intensity of venture capital in incubator (10,000 yuan/Nr)

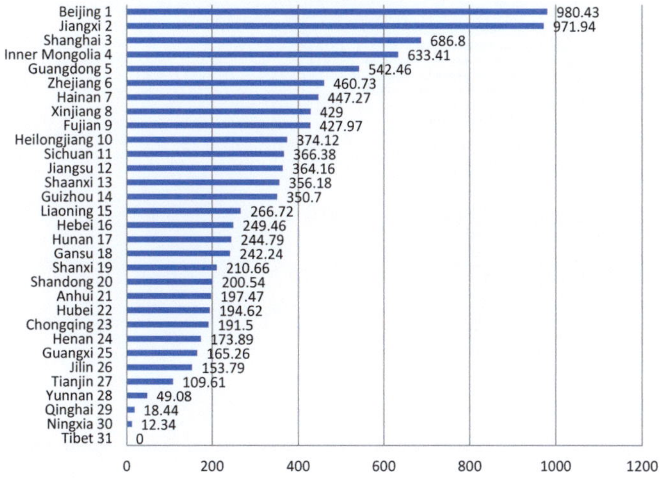

44213 The growth rate of venture capital investment in incubator (%)

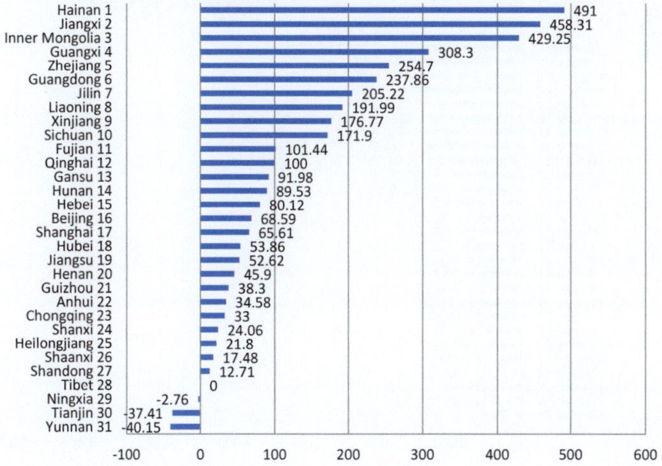

44221 The amount of fund for incubator (10,000 yuan)

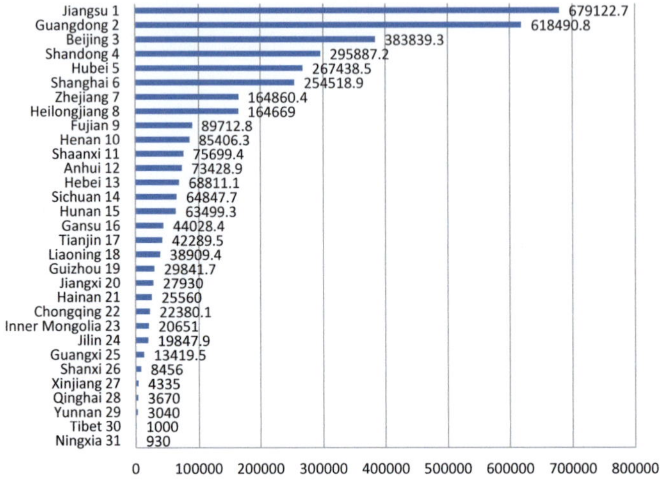

44222 The average fund per incubator (10,000 yuan)

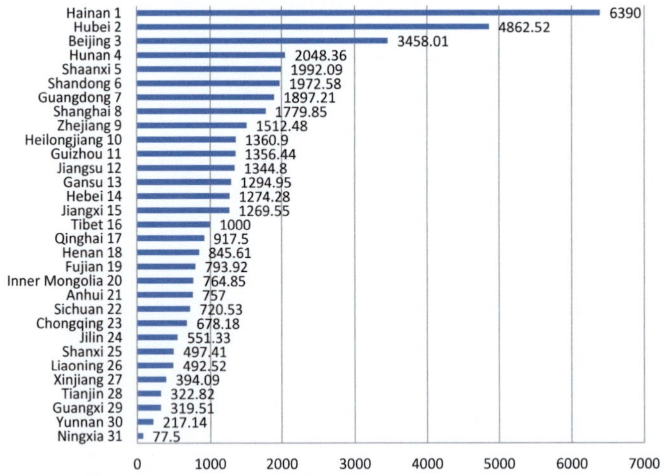

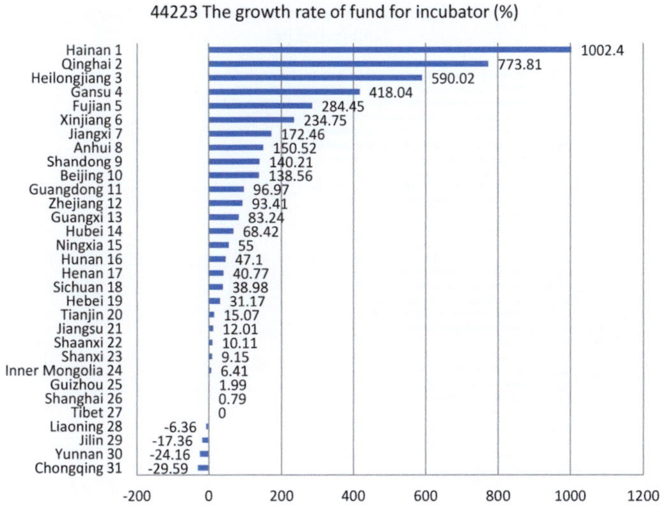
44223 The growth rate of fund for incubator (%)
Hainan 1
Qinghai 2
Heilongjiang 3
Gansu 4
Fujian 5
Xinjiang 6
Jiangxi 7
Anhui 8
Shandong 9
Beijing 10
Guangdong 11
Zhejiang 12
Guangxi 13
Hubei 14
Ningxia 15
Hunan 16
Henan 17
Sichuan 18
Hebei 19
Tianjin 20
Jiangsu 21
Shaanxi 22
Shanxi 23
Inner Mongolia 24
Guizhou 25
Shanghai 26
Tibet 27
Liaoning 28
Jilin 29
Yunnan 30
Chongqing 31
1002.4
773.81
590.02
418.04
284.45
234.75
172.46
150.52
140.21
138.56
96.97
93.41
83.24
68.42
55
47.1
40.77
38.98
31.17
15.07
12.01
10.11
9.15
6.41
1.99
0.79
0
-6.36
-17.36
-24.16
-29.59
-200
0
200
400
600
800
1000
1200

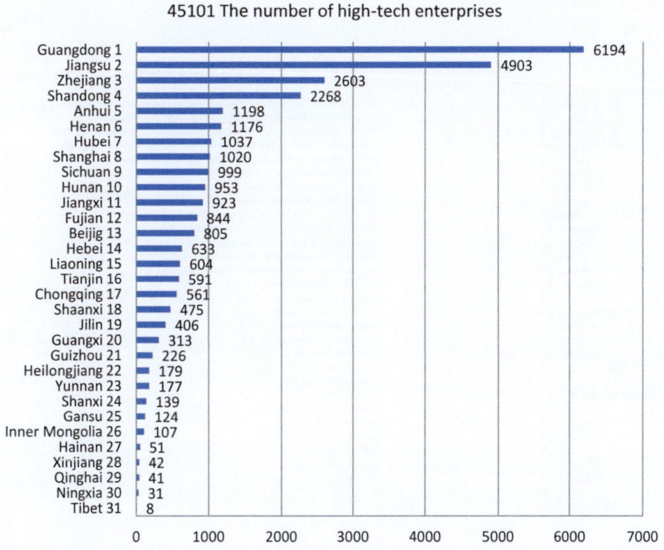
45101 The number of high-tech enterprises
Guangdong 1
Jiangsu 2
Zhejiang 3
Shandong 4
Anhui 5
Henan 6
Hubei 7
Shanghai 8
Sichuan 9
Hunan 10
Jiangxi 11
Fujian 12
Beijig 13
Hebei 14
Liaoning 15
Tianjin 16
Chongqing 17
Shaanxi 18
Jilin 19
Guangxi 20
Guizhou 21
Heilongjiang 22
Yunnan 23
Shanxi 24
Gansu 25
Inner Mongolia 26
Hainan 27
Xinjiang 28
Qinghai 29
Ningxia 30
Tibet 31
6194
4903
2603
2268
1198
1176
1037
1020
999
953
923
844
805
633
604
591
561
475
406
313
226
179
177
139
124
107
51
42
41
31
8
0
1000
2000
3000
4000
5000
6000
7000

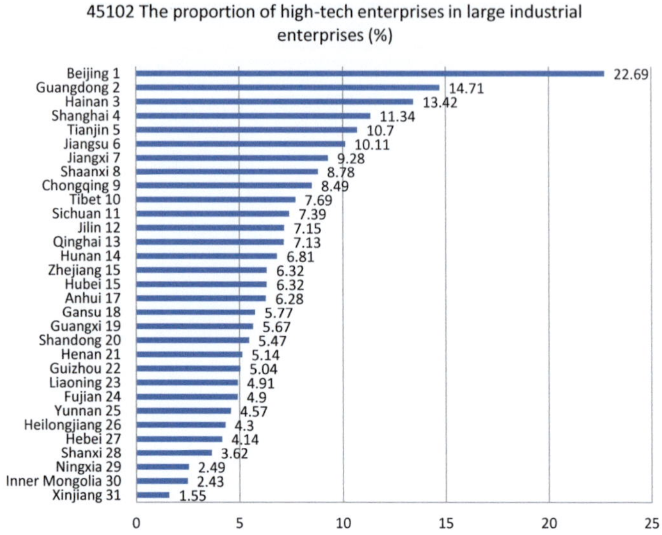
45102 The proportion of high-tech enterprises in large industrial enterprises (%)
Beijing 1 22.69
Guangdong 2 14.71
Hainan 3 13.42
Shanghai 4 11.34
Tianjin 5 10.7
Jiangsu 6 10.11
Jiangxi 7 9.28
Shaanxi 8 8.78
Chongqing 9 8.49
Tibet 10 7.69
Sichuan 11 7.39
Jilin 12 7.15
Qinghai 13 7.13
Hunan 14 6.81
Zhejiang 15 6.32
Hubei 15 6.32
Anhui 17 6.28
Gansu 18 5.77
Guangxi 19 5.67
Shandong 20 5.47
Henan 21 5.14
Guizhou 22 5.04
Liaoning 23 4.91
Fujian 24 4.9
Yunnan 25 4.57
Heilongjiang 26 4.3
Hebei 27 4.14
Shanxi 28 3.62
Ningxia 29 2.49
Inner Mongolia 30 2.43
Xinjiang 31 1.55
0 5 10 15 20 25

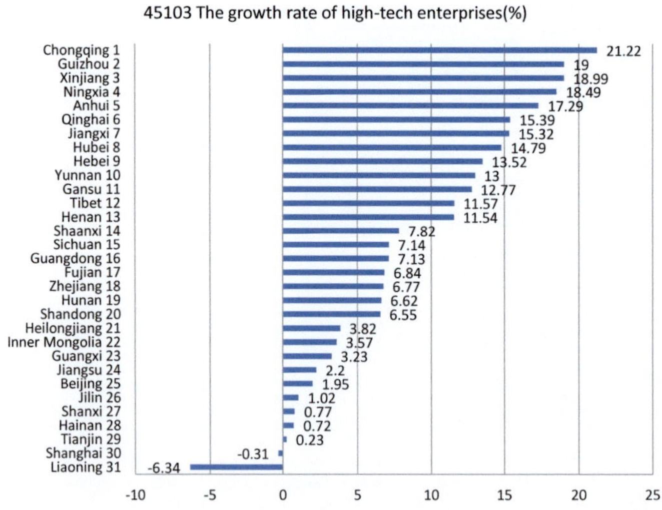
45103 The growth rate of high-tech enterprises(%)
Chongqing 1 21.22
Guizhou 2 19
Xinjiang 3 18.99
Ningxia 4 18.49
Anhui 5 17.29
Qinghai 6 15.39
Jiangxi 7 15.32
Hubei 8 14.79
Hebei 9 13.52
Yunnan 10 13
Gansu 11 12.77
Tibet 12 11.57
Henan 13 11.54
Shaanxi 14 7.82
Sichuan 15 7.14
Guangdong 16 7.13
Fujian 17 6.84
Zhejiang 18 6.77
Hunan 19 6.62
Shandong 20 6.55
Heilongjiang 21 3.82
Inner Mongolia 22 3.57
Guangxi 23 3.23
Jiangsu 24 2.2
Beijing 25 1.95
Jilin 26 1.02
Shanxi 27 0.77
Hainan 28 0.72
Tianjin 29 0.23
Shanghai 30 -0.31
Liaoning 31 -6.34
-10 -5 0 5 10 15 20 25

45201 The number of graduates of enterprises in incubator

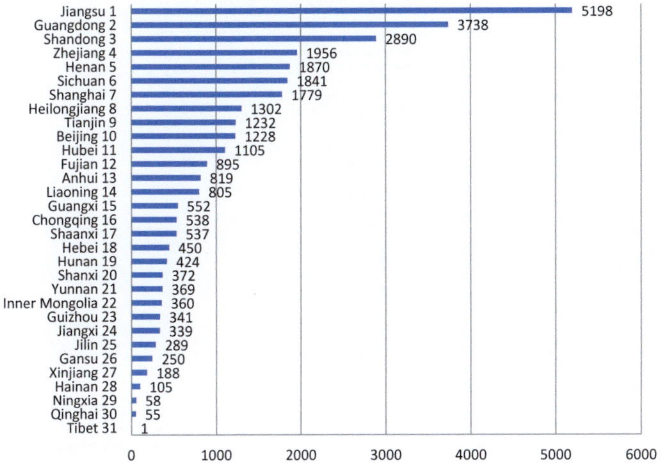

45202 The average number of graduates of enterprises in incubator

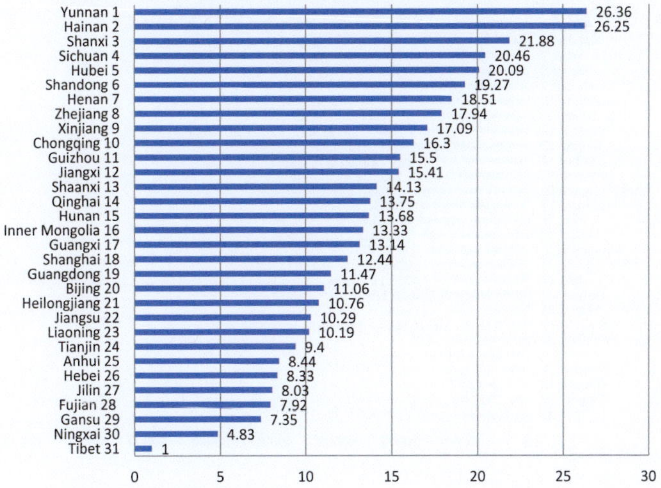

45203 The growth rate of graduates of enterprises in incubator (%)

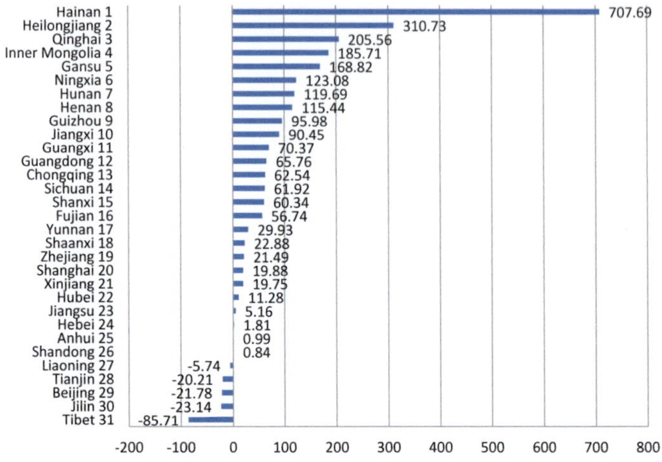

51001 Regional GDP (100 billion yuan)

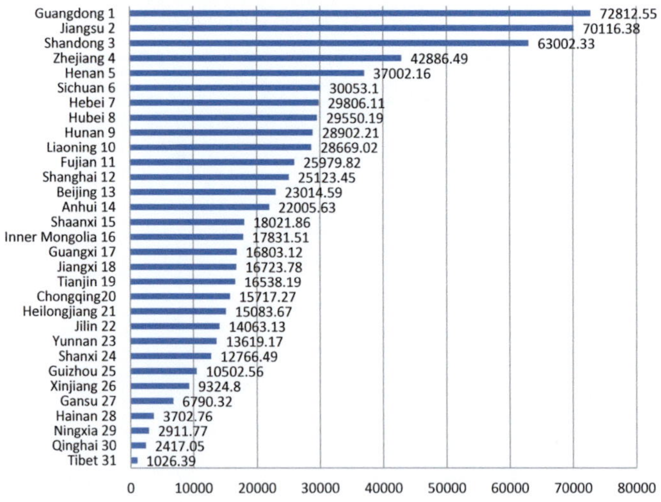

51002 GDP Per capita level (yuan/person)

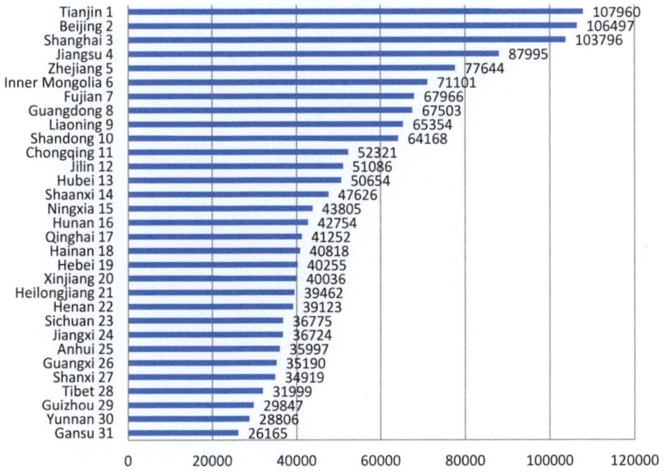

51003 The growth rate of the regional GDP (%)

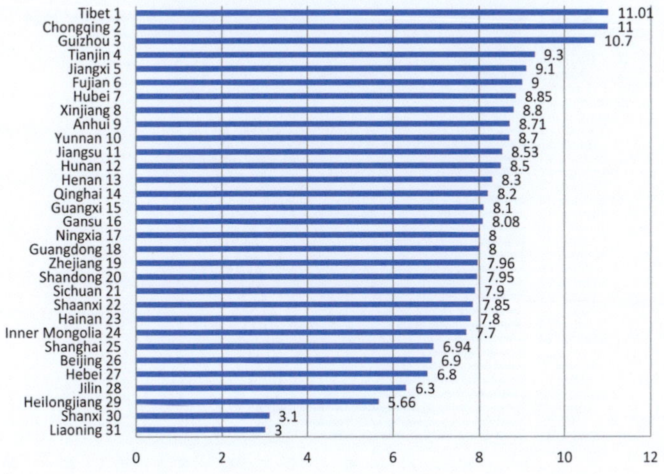

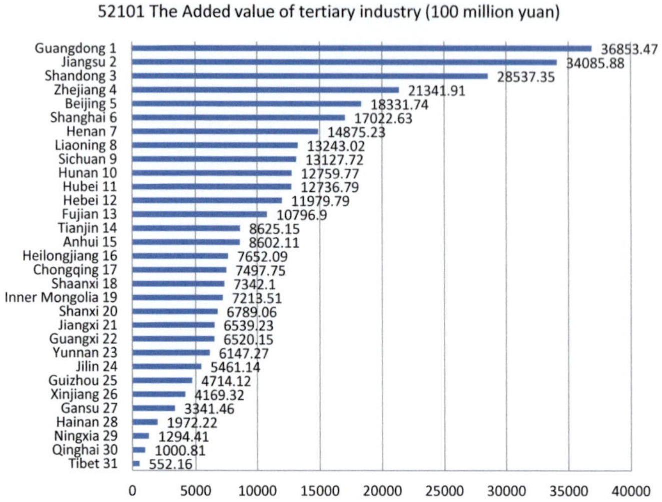
52101 The Added value of tertiary industry (100 million yuan)
Guangdong 1
Jiangsu 2
Shandong 3
Zhejiang 4
Beijing 5
Shanghai 6
Henan 7
Liaoning 8
Sichuan 9
Hunan 10
Hubei 11
Hebei 12
Fujian 13
Tianjin 14
Anhui 15
Heilongjiang 16
Chongqing 17
Shaanxi 18
Inner Mongolia 19
Shanxi 20
Jiangxi 21
Guangxi 22
Yunnan 23
Jilin 24
Guizhou 25
Xinjiang 26
Gansu 27
Hainan 28
Ningxia 29
Qinghai 30
Tibet 31
36853.47
34085.88
28537.35
21341.91
18331.74
17022.63
14875.23
13243.02
13127.72
12759.77
12736.79
11979.79
10796.9
8625.15
8602.11
7652.09
7497.75
7342.1
7213.51
6789.06
6539.23
6520.15
6147.27
5461.14
4714.12
4169.32
3341.46
1972.22
1294.41
1000.81
552.16
0
5000
10000
15000
20000
25000
30000
35000
40000

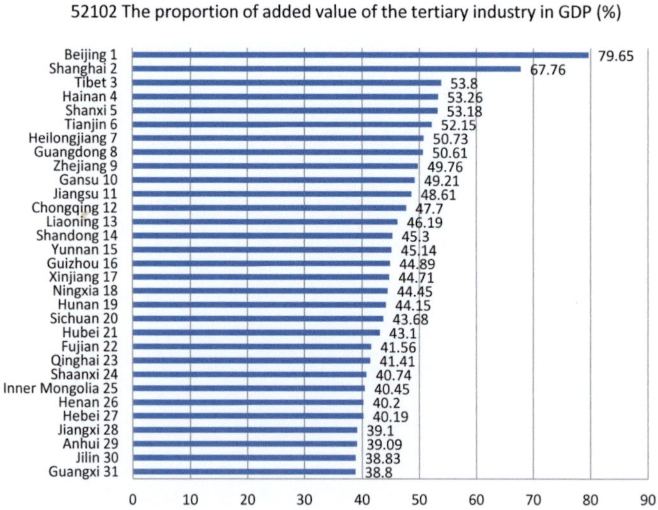
52102 The proportion of added value of the tertiary industry in GDP (%)
Beijing 1
Shanghai 2
Tibet 3
Hainan 4
Shanxi 5
Tianjin 6
Heilongjiang 7
Guangdong 8
Zhejiang 9
Gansu 10
Jiangsu 11
Chongqing 12
Liaoning 13
Shandong 14
Yunnan 15
Guizhou 16
Xinjiang 17
Ningxia 18
Hunan 19
Sichuan 20
Hubei 21
Fujian 22
Qinghai 23
Shaanxi 24
Inner Mongolia 25
Henan 26
Hebei 27
Jiangxi 28
Anhui 29
Jilin 30
Guangxi 31
79.65
67.76
53.8
53.26
53.18
52.15
50.73
50.61
49.76
49.21
48.61
47.7
46.19
45.3
45.14
44.89
44.71
44.45
44.15
43.68
43.1
41.56
41.41
40.74
40.45
40.2
40.19
39.1
39.09
38.83
38.8
0
10
20
30
40
50
60
70
80
90

52103 The growth rate of the added value of the tertiary industry (%)

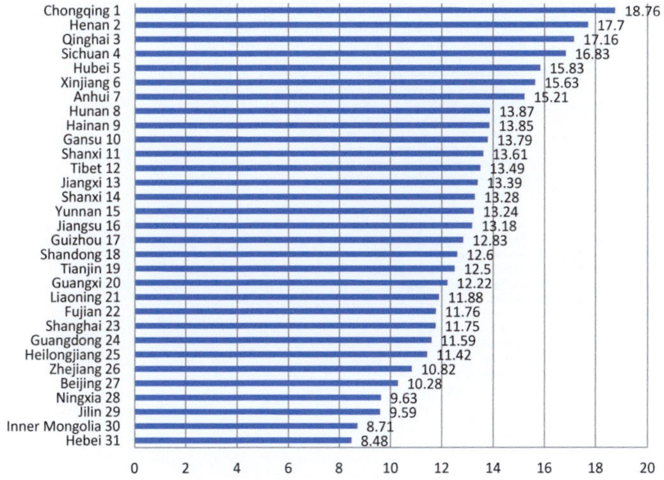

52201 The revenue of high-tech industry (100 million yuan)

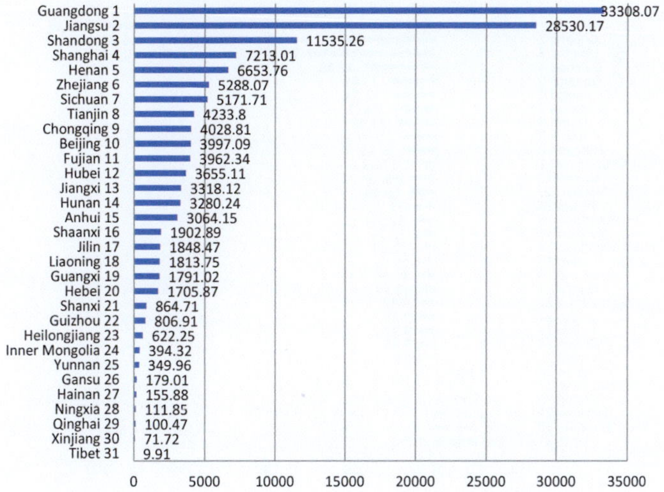

52202 The proportion of the revenue of high-tech industry in GDP (%)

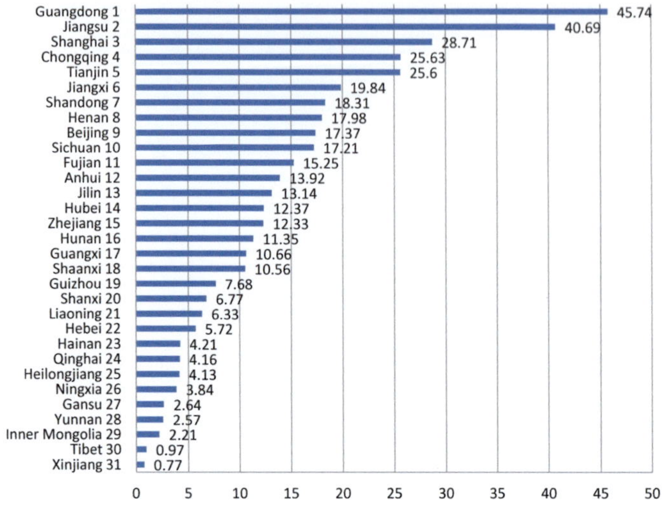

52203 The growth rate of revenue of high-tech industry (%)

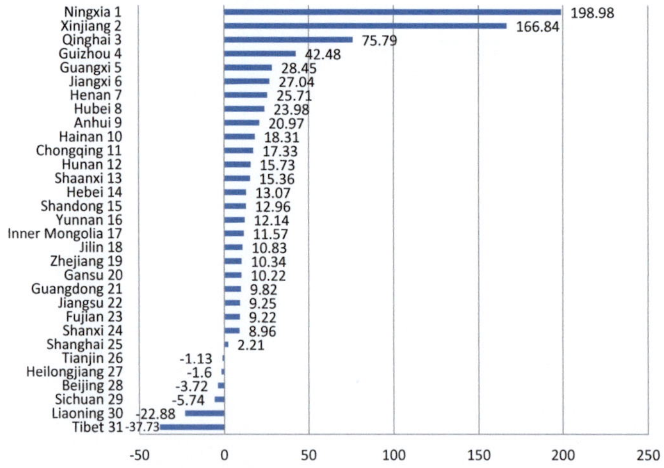

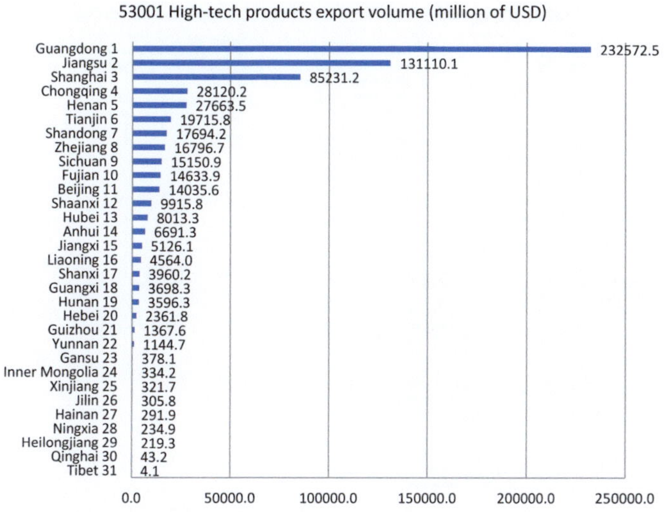
53001 High-tech products export volume (million of USD)
Guangdong 1 232572.5
Jiangsu 2 131110.1
Shanghai 3 85231.2
Chongqing 4 28120.2
Henan 5 27663.5
Tianjin 6 19715.8
Shandong 7 17694.2
Zhejiang 8 16796.7
Sichuan 9 15150.9
Fujian 10 14633.9
Beijing 11 14035.6
Shaanxi 12 9915.8
Hubei 13 8013.3
Anhui 14 6691.3
Jiangxi 15 5126.1
Liaoning 16 4564.0
Shanxi 17 3960.2
Guangxi 18 3698.3
Hunan 19 3596.3
Hebei 20 2361.8
Guizhou 21 1367.6
Yunnan 22 1144.7
Gansu 23 378.1
Inner Mongolia 24 334.2
Xinjiang 25 321.7
Jilin 26 305.8
Hainan 27 291.9
Ningxia 28 234.9
Heilongjiang 29 219.3
Qinghai 30 43.2
Tibet 31 4.1
0.0 50000.0 100000.0 150000.0 200000.0 250000.0

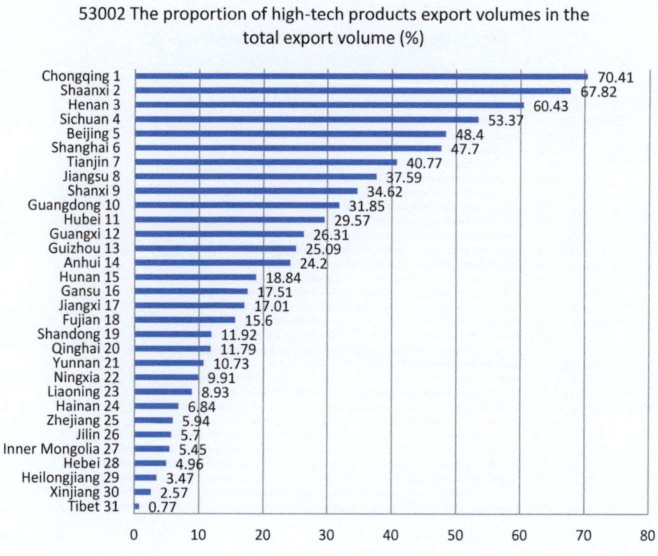
53002 The proportion of high-tech products export volumes in the total export volume (%)
Chongqing 1 70.41
Shaanxi 2 67.82
Henan 3 60.43
Sichuan 4 53.37
Beijing 5 48.4
Shanghai 6 47.7
Tianjin 7 40.77
Jiangsu 8 37.59
Shanxi 9 34.62
Guangdong 10 31.85
Hubei 11 29.57
Guangxi 12 26.31
Guizhou 13 25.09
Anhui 14 24.2
Hunan 15 18.84
Gansu 16 17.51
Jiangxi 17 17.01
Fujian 18 15.6
Shandong 19 11.92
Qinghai 20 11.79
Yunnan 21 10.73
Ningxia 22 9.91
Liaoning 23 8.93
Hainan 24 6.84
Zhejiang 25 5.94
Jilin 26 5.7
Inner Mongolia 27 5.45
Hebei 28 4.96
Heilongjiang 29 3.47
Xinjiang 30 2.57
Tibet 31 0.77
0 10 20 30 40 50 60 70 80

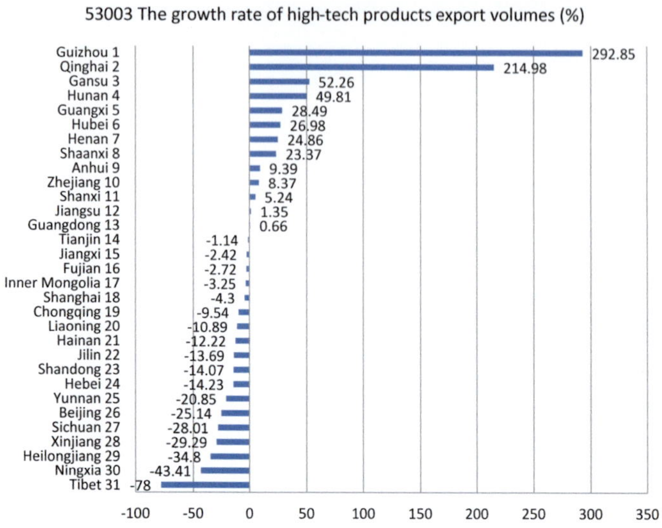
53003 The growth rate of high-tech products export volumes (%)
Guizhou 1
Qinghai 2
Gansu 3
Hunan 4
Guangxi 5
Hubei 6
Henan 7
Shaanxi 8
Anhui 9
Zhejiang 10
Shanxi 11
Jiangsu 12
Guangdong 13
Tianjin 14
Jiangxi 15
Fujian 16
Inner Mongolia 17
Shanghai 18
Chongqing 19
Liaoning 20
Hainan 21
Jilin 22
Shandong 23
Hebei 24
Yunnan 25
Beijing 26
Sichuan 27
Xinjiang 28
Heilongjiang 29
Ningxia 30
Tibet 31
292.85
214.98
52.26
49.81
28.49
26.98
24.86
23.37
9.39
8.37
5.24
1.35
0.66
-1.14
-2.42
-2.72
-3.25
-4.3
-9.54
-10.89
-12.22
-13.69
-14.07
-14.23
-20.85
-25.14
-28.01
-29.29
-34.8
-43.41
-78
-100
-50
0
50
100
150
200
250
300
350

54101 Registered urban unemployment rate (%)
Beijing 1
Gansu 2
Hainan 3
Guangdong 4
Tibet 5
Hubei 6
Xinjiang 7
Guangxi 8
Zhejiang 9
Henan 10
Jiangsu 11
Anhui 12
Qinghai 13
Guizhou 14
Jiangxi 15
Shandong 16
Shaanxi 17
Liaoning 18
Jilin 19
Tianjin 20
Shanxi 21
Chongqing 22
Hebei 23
Inner Mongolia 24
Fujian 25
Yunnan 26
Ningxia 27
Shanghai 28
Hunan 29
Sichuan 30
Heilongjiang 31
1.39
2.14
2.29
2.45
2.48
2.64
2.86
2.92
2.93
2.96
3
3.14
3.17
3.29
3.35
3.35
3.36
3.42
3.5
3.5
3.51
3.58
3.6
3.65
3.66
3.96
4.02
4.04
4.09
4.12
4.47
0
0.5
1
1.5
2
2.5
3
3.5
4
4.5
5

54103 The reduction rate of registered urban unemployment rate (%)

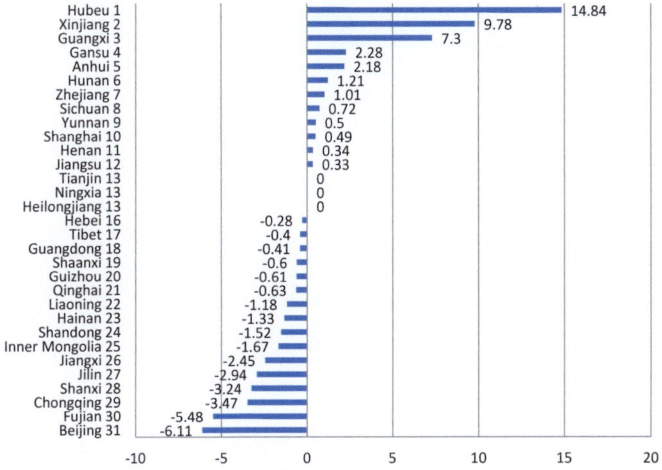

54201 The number of high-tech industrial personnel

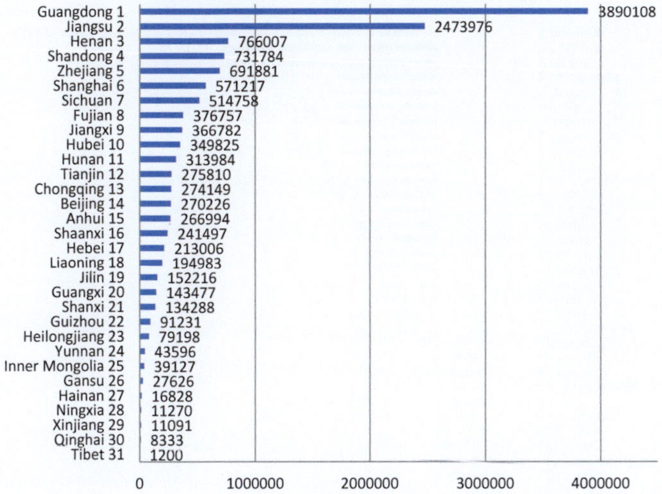

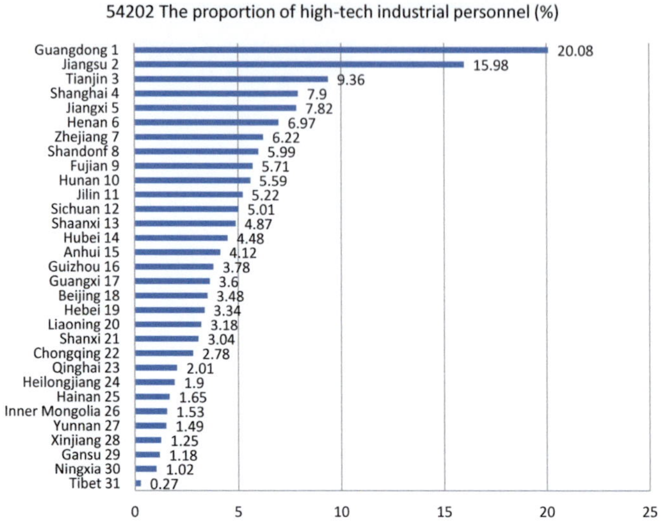
54202 The proportion of high-tech industrial personnel (%)
Guangdong 1
Jiangsu 2
Tianjin 3
Shanghai 4
Jiangxi 5
Henan 6
Zhejiang 7
Shandonf 8
Fujian 9
Hunan 10
Jilin 11
Sichuan 12
Shaanxi 13
Hubei 14
Anhui 15
Guizhou 16
Guangxi 17
Beijing 18
Hebei 19
Liaoning 20
Shanxi 21
Chongqing 22
Qinghai 23
Heilongjiang 24
Hainan 25
Inner Mongolia 26
Yunnan 27
Xinjiang 28
Gansu 29
Ningxia 30
Tibet 31
20.08
15.98
9.36
7.9
7.82
6.97
6.22
5.99
5.71
5.59
5.22
5.01
4.87
4.48
4.12
3.78
3.6
3.48
3.34
3.18
3.04
2.78
2.01
1.9
1.65
1.53
1.49
1.25
1.18
1.02
0.27
0
5
10
15
20
25

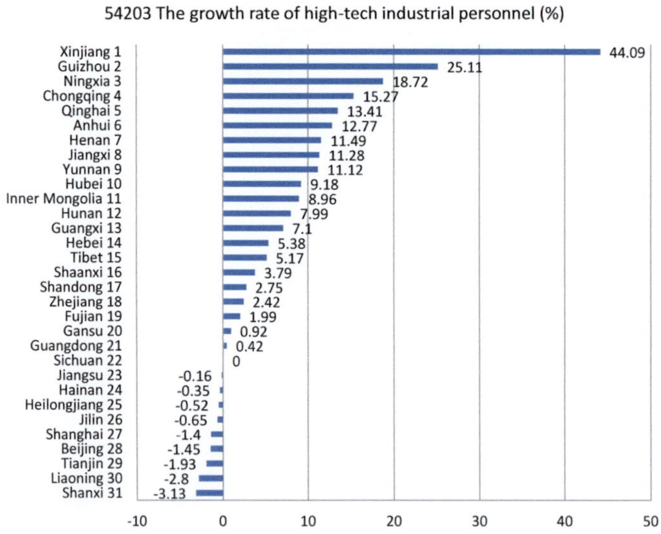
54203 The growth rate of high-tech industrial personnel (%)
Xinjiang 1
Guizhou 2
Ningxia 3
Chongqing 4
Qinghai 5
Anhui 6
Henan 7
Jiangxi 8
Yunnan 9
Hubei 10
Inner Mongolia 11
Hunan 12
Guangxi 13
Hebei 14
Tibet 15
Shaanxi 16
Shandong 17
Zhejiang 18
Fujian 19
Gansu 20
Guangdong 21
Sichuan 22
Jiangsu 23
Hainan 24
Heilongjiang 25
Jilin 26
Shanghai 27
Beijing 28
Tianjin 29
Liaoning 30
Shanxi 31
44.09
25.11
18.72
15.27
13.41
12.77
11.49
11.28
11.12
9.18
8.96
7.99
7.1
5.38
5.17
3.79
2.75
2.42
1.99
0.92
0.42
0
-0.16
-0.35
-0.52
-0.65
-1.4
-1.45
-1.93
-2.8
-3.13
-10
0
10
20
30
40
50

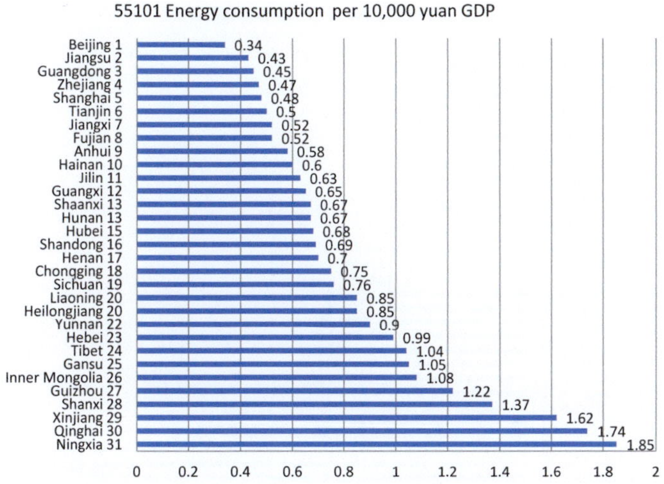
55101 Energy consumption per 10,000 yuan GDP
Beijing 1
Jiangsu 2
Guangdong 3
Zhejiang 4
Shanghai 5
Tianjin 6
Jiangxi 7
Fujian 8
Anhui 9
Hainan 10
Jilin 11
Guangxi 12
Shaanxi 13
Hunan 13
Hubei 15
Shandong 16
Henan 17
Chongqing 18
Sichuan 19
Liaoning 20
Heilongjiang 20
Yunnan 22
Hebei 23
Tibet 24
Gansu 25
Inner Mongolia 26
Guizhou 27
Shanxi 28
Xinjiang 29
Qinghai 30
Ningxia 31
0.34
0.43
0.45
0.47
0.48
0.5
0.52
0.52
0.58
0.6
0.63
0.65
0.67
0.67
0.68
0.69
0.7
0.75
0.76
0.85
0.85
0.9
0.99
1.04
1.05
1.08
1.22
1.37
1.62
1.74
1.85
0 0.2 0.4 0.6 0.8 1 1.2 1.4 1.6 1.8 2

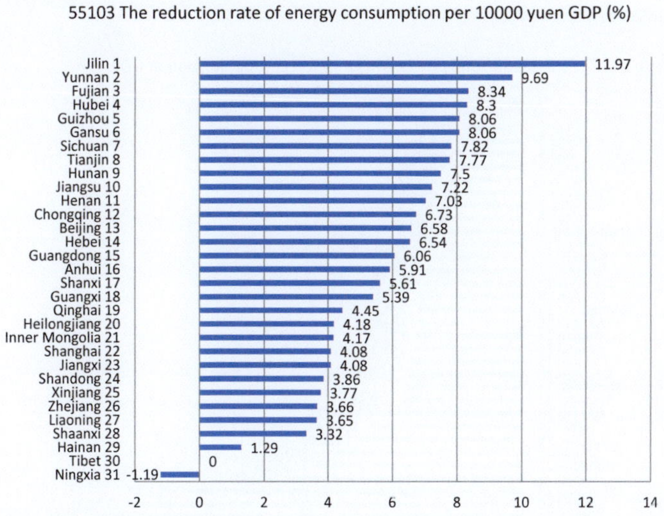
55103 The reduction rate of energy consumption per 10000 yuen GDP (%)
Jilin 1
Yunnan 2
Fujian 3
Hubei 4
Guizhou 5
Gansu 6
Sichuan 7
Tianjin 8
Hunan 9
Jiangsu 10
Henan 11
Chongqing 12
Beijing 13
Hebei 14
Guangdong 15
Anhui 16
Shanxi 17
Guangxi 18
Qinghai 19
Heilongjiang 20
Inner Mongolia 21
Shanghai 22
Jiangxi 23
Shandong 24
Xinjiang 25
Zhejiang 26
Liaoning 27
Shaanxi 28
Hainan 29
Tibet 30
Ningxia 31
11.97
9.69
8.34
8.3
8.06
8.06
7.82
7.77
7.5
7.22
7.03
6.73
6.58
6.54
6.06
5.91
5.61
5.39
4.45
4.18
4.17
4.08
4.08
3.86
3.77
3.66
3.65
3.32
1.29
0
-1.19
-2 0 2 4 6 8 10 12 14

55201 Total power consumption (KWH)

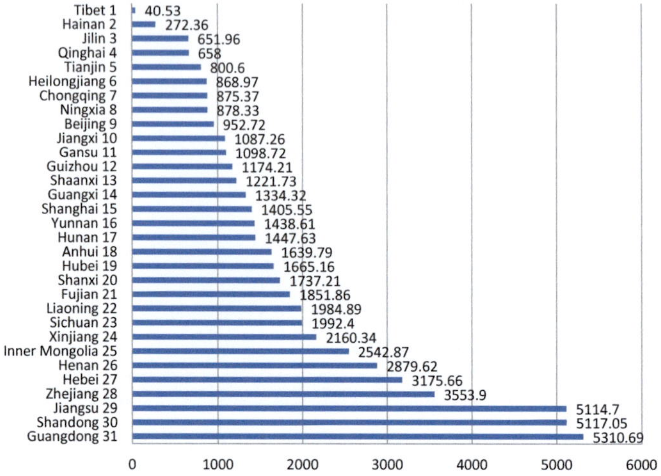

55202 Total power consumption per10,000 yuan of GDP (KWH/10,000 yuan)

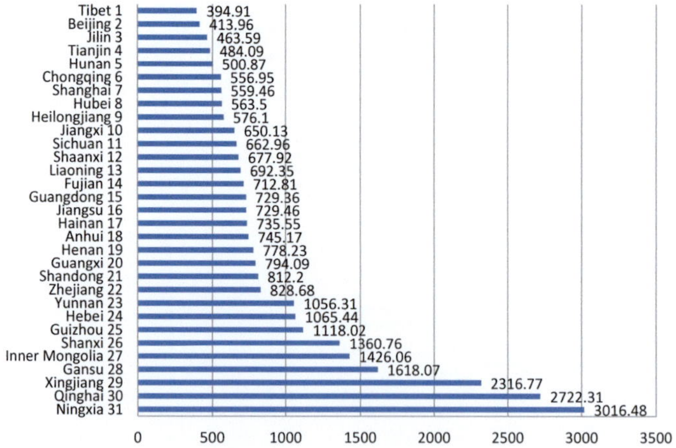

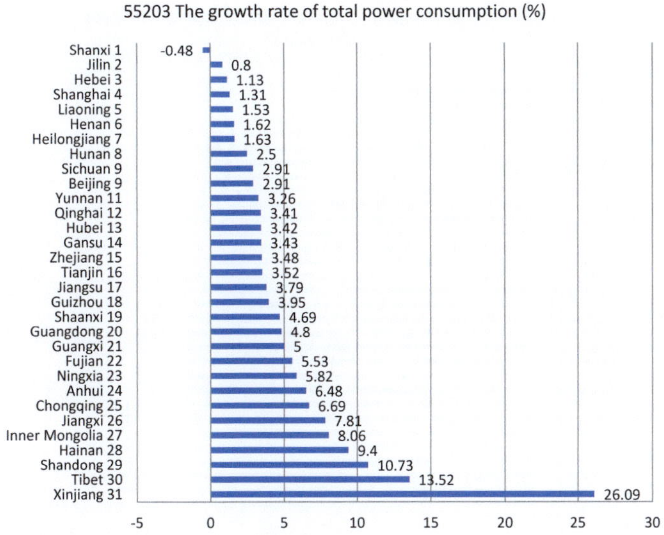
55203 The growth rate of total power consumption (%)
Shanxi 1
Jilin 2
Hebei 3
Shanghai 4
Liaoning 5
Henan 6
Heilongjiang 7
Hunan 8
Sichuan 9
Beijing 9
Yunnan 11
Qinghai 12
Hubei 13
Gansu 14
Zhejiang 15
Tianjin 16
Jiangsu 17
Guizhou 18
Shaanxi 19
Guangdong 20
Guangxi 21
Fujian 22
Ningxia 23
Anhui 24
Chongqing 25
Jiangxi 26
Inner Mongolia 27
Hainan 28
Shandong 29
Tibet 30
Xinjiang 31
-0.48
0.8
1.13
1.31
1.53
1.62
1.63
2.5
2.91
2.91
3.26
3.41
3.42
3.43
3.48
3.52
3.79
3.95
4.69
4.8
5
5.53
5.82
6.48
6.69
7.81
8.06
9.4
10.73
13.52
26.09
-5
0
5
10
15
20
25
30

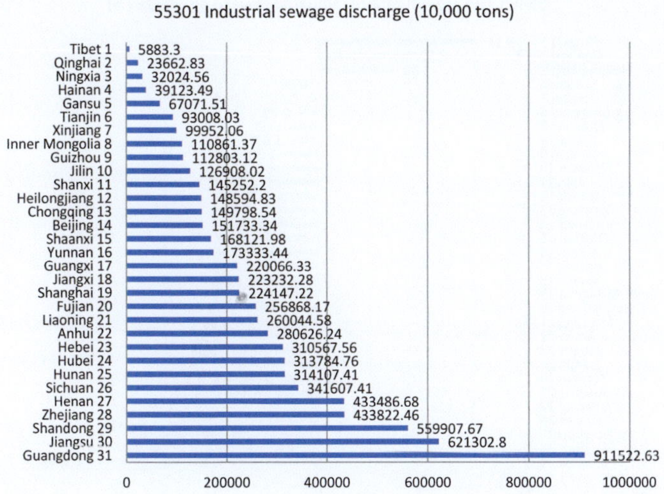
55301 Industrial sewage discharge (10,000 tons)
Tibet 1
Qinghai 2
Ningxia 3
Hainan 4
Gansu 5
Tianjin 6
Xinjiang 7
Inner Mongolia 8
Guizhou 9
Jilin 10
Shanxi 11
Heilongjiang 12
Chongqing 13
Beijing 14
Shaanxi 15
Yunnan 16
Guangxi 17
Jiangxi 18
Shanghai 19
Fujian 20
Liaoning 21
Anhui 22
Hebei 23
Hubei 24
Hunan 25
Sichuan 26
Henan 27
Zhejiang 28
Shandong 29
Jiangsu 30
Guangdong 31
5883.3
23662.83
32024.56
39123.49
67071.51
93008.03
99952.06
110861.37
112803.12
126908.02
145252.2
148594.83
149798.54
151733.34
168121.98
173333.44
220066.33
223232.28
224147.22
256868.17
260044.58
280626.24
310567.56
313784.76
314107.41
341607.41
433486.68
433822.46
559907.67
621302.8
911522.63
0
200000
400000
600000
800000
1000000

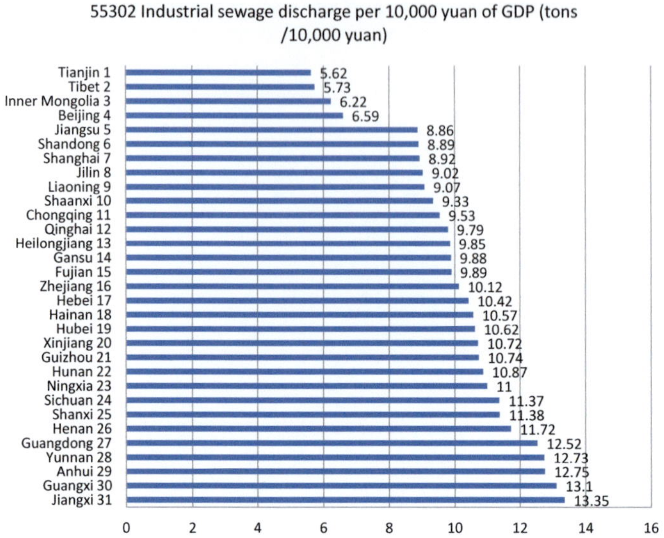
55302 Industrial sewage discharge per 10,000 yuan of GDP (tons /10,000 yuan)
Tianjin 1 5.62
Tibet 2 5.73
Inner Mongolia 3 6.22
Beijing 4 6.59
Jiangsu 5 8.86
Shandong 6 8.89
Shanghai 7 8.92
Jilin 8 9.02
Liaoning 9 9.07
Shaanxi 10 9.33
Chongqing 11 9.53
Qinghai 12 9.79
Heilongjiang 13 9.85
Gansu 14 9.88
Fujian 15 9.89
Zhejiang 16 10.12
Hebei 17 10.42
Hainan 18 10.57
Hubei 19 10.62
Xinjiang 20 10.72
Guizhou 21 10.74
Hunan 22 10.87
Ningxia 23 11
Sichuan 24 11.37
Shanxi 25 11.38
Henan 26 11.72
Guangdong 27 12.52
Yunnan 28 12.73
Anhui 29 12.75
Guangxi 30 13.1
Jiangxi 31 13.35
0 2 4 6 8 10 12 14 16

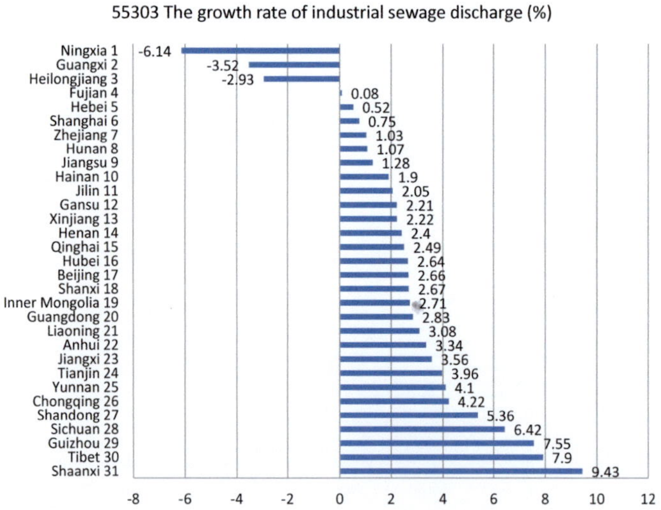
55303 The growth rate of industrial sewage discharge (%)
Ningxia 1 -6.14
Guangxi 2 -3.52
Heilongjiang 3 -2.93
Fujian 4 0.08
Hebei 5 0.52
Shanghai 6 0.75
Zhejiang 7 1.03
Hunan 8 1.07
Jiangsu 9 1.28
Hainan 10 1.9
Jilin 11 2.05
Gansu 12 2.21
Xinjiang 13 2.22
Henan 14 2.4
Qinghai 15 2.49
Hubei 16 2.64
Beijing 17 2.66
Shanxi 18 2.67
Inner Mongolia 19 2.71
Guangdong 20 2.83
Liaoning 21 3.08
Anhui 22 3.34
Jiangxi 23 3.56
Tianjin 24 3.96
Yunnan 25 4.1
Chongqing 26 4.22
Shandong 27 5.36
Sichuan 28 6.42
Guizhou 29 7.55
Tibet 30 7.9
Shaanxi 31 9.43
-8 -6 -4 -2 0 2 4 6 8 10 12

55401 Emissions of major pollutants in exhuast gas (10,000 tons)

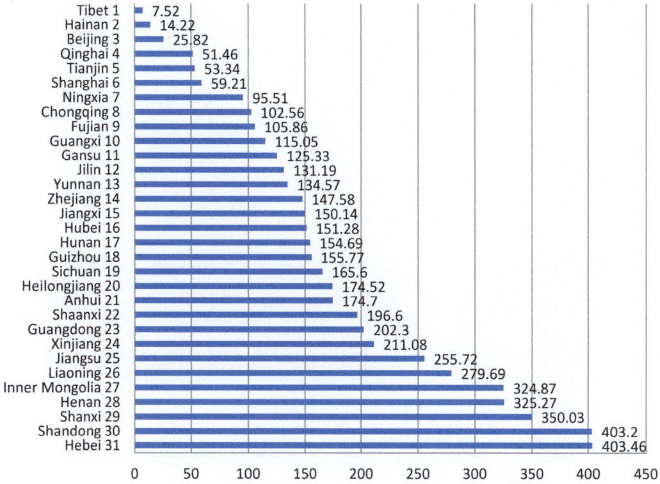

55402 Emissions of major pollutants in exhaust gas per billion of GDP(ton/100 million yuan)

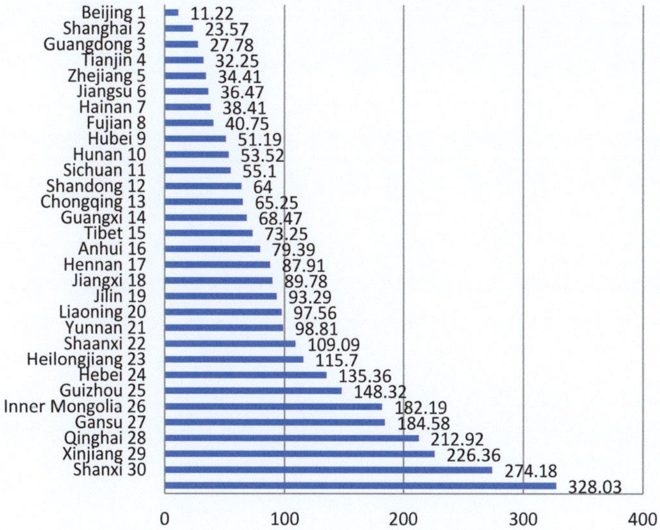

55403 The growth rate of emissions of major pollutants in exhaust gas (%)

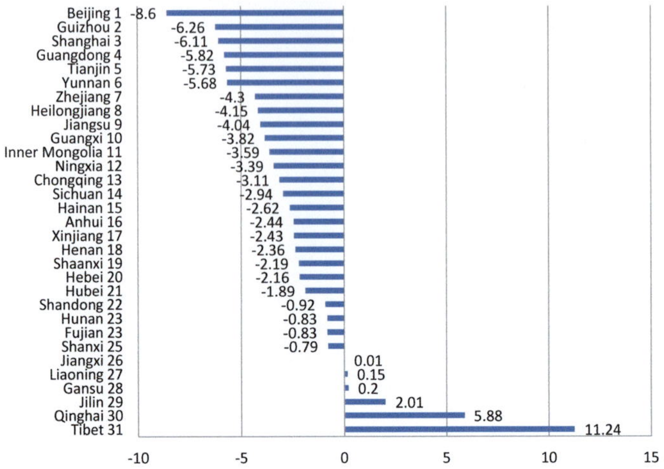

MIX
Papier aus verantwortungsvollen Quellen
Paper from responsible sources
FSC® C105338
FSC
www.fsc.org

If you have any concerns about our products,
you can contact us on
ProductSafety@springernature.com

In case Publisher is established outside the EU,
the EU authorized representative is:
Springer Nature Customer Service Center GmbH
Europaplatz 3, 69115 Heidelberg, Germany

Printed by Libri Plureos GmbH
in Hamburg, Germany